AF540015

Cutting Edge Technology for Agricultural Sustainability

nipa
Browse Subject
Agriculture Sciences (679)
Animal Husbandry and Veterinary Sciences (261)
Climate Change and Environmental Sciences (63)
Community Science (Home Science) (39)
Competitive Examinations and MCQ's (57)
Dairy Technology (91)
Fisheries Sciences (26)
Food Science and Technology (155)
Geological Sciences (30)
Horticultural Sciences (342)
Plant Sciences (191)
Skill and Entrepreneurship Development (47)
eBooks / eChapters / Articles
eBooks
eBooks / eChapters / Articles
eBooks
scan for catalogue
Browse, Search, Read & Buy...
Subject Catalogues
eChapters
Publishers
Print Books
Forthcoming
eBooks
New Titles
NIPA
GENX
ONLINE RESOURCES
Publishing Books and Journals
Ebooks and e articles, Current Affairs
Reasoning, Logic and Aptitude
Competitive Examination Preparation
Language Learning and Development Programme
Document Quality Checker and Improvement Tool
Effective Public Speaking, Presentation and Interpersonal Skills
Online Programmes for Professional Development
Personality Development and Human Values
PAY USING
UPI
PayPal

Cutting Edge Technology for Agricultural Sustainability

A. Zaman

Emeritus Professor
Centurion University of Technology and Management
M S Swaminathan School of Agriculture
Paralakhemundi-761 211, Odisha, India

Sagar Maitra

Department of Agronomy
M S Swaminathan School of Agriculture
Centurion University of Technology and Management
Paralakhemundi-761211
Odisha, India

NEW INDIA PUBLISHING AGENCY

New Delhi – 110 034

NEW INDIA PUBLISHING AGENCY
101, Vikas Surya Plaza, CU Block, LSC Market
Pitam Pura, New Delhi 110 034, India
Phone: + 91 (11) 27 34 17 17 Fax: + 91 (11) 27 34 16 16
Email: info@nipabooks.com
Web: www.nipabooks.com

Feedback at feedbacks@nipabooks.com

ISBN: 978-93-87973-28-2

Composed and Designed by NIPA

Preface

Indian agriculture has reached a momentum which propels economic process. Primarily, the economy of the country remains supported to agriculture and vast majority of population directly or indirectly depends on agricultural activities in maintaining their livelihood. Realizing the importance of revitalization of agricultural growth, the Government of India through execution of Five Year Plans supported agricultural development within the country. The comfortable food production of the country enjoys is that the results of ball-hawking human resource developed, technological capabilities and development of technical skills and their dissemination to the extremely receptive farming community. The country successively passed out in primarily chargeable experiences of white, yellow, and blue revolutions signifying severally increase in food grain, milk, seeds and fish production. This has mostly been attributed to say no in quality of directions imparted to a good extent. There are many factors as a result of this, the chief amongst of them are declining financial resources, lack of ability to access in new and rising frontier areas in depth mating and lack of adequate knowledge, competency, skill and entrepreneurship.

Agricultural Sustainability is a combination of technologies and implementation processes that are used to manage information-related with maintenance of equilibrium condition in the scenario of productivity level in the field of agriculture. Sustainable agriculture is not just about food security, the protection of crops and its cultivation or while it is being processed; it is also about the protection of environment with economic equity and profitability. This is a composite field involving agricultural science, engineering, database and soil-based crop management, user training, and policy issues. A common objective of work in this field is to protect the environment and defend degeneration systems by ensuring resource availability, its integrity, authentication and confidentiality so that the right action can be accessed with the right information and direction at the right time.

The population of our country reached 1330 million and the annual demand for food has reached 260 million tons. To cope with the increasing demand, we need to enhance the agricultural production by 20%, from the present level of 198 million tons in 2020. In a resource poor country like us, if we are dependent on others even for food, apart from other essential commodities, such as oil, natural gas, basic ingredients for fertilizers, heavy engineering materials, nuclear power and computer technologies, there will hardly be anything left in the country to export and settle the import bills. Thus we have to ensure not to import food grains to feed our population with hard earning foreign exchange.

Under normal circumstances, a target to maintain 2% growth per annum can solve the problem. This should not be difficult as we have recorded a higher average annual growth of 2.37% during the last decade. However the strategy adopted in early 70's to boost food production has been well-exploited so far and the productivity has now reached the level of saturation.

Wetherefore, need to look into various factors affecting the crop production and modify our strategy to address these problems, while making optimum use of the available resources and technologies to reach the new target.

We do expect that the book may come as an important contribution to the latest agricultural technologies in bridging up the gap of scientific information to the researchers and policy makers in different relevant field of specialization.

Date: 31 May 2018 **Editors**

Contents

Section 3: Soil Resource Management

Section 4: Sustainability in Agriculture

Section 1
Water Resource Management

1

Watershed Management Technologies in Mitigating Water Crisis and Sustaining Agricultural Productivity

A. Zaman

Emeritus Professor, Centurion University of Technology and Management
M S Swaminathan School of Agriculture
Paralakhemundi-761211, Odisha, India
Email: profazaman@gmail.com

Abstract

Watershed development and management programme appeared to an effective and integrated approach to manage all natural resources like land, water, vegetation, animal and human for sustainable benefit and maintaining balanced eco-system. The main principle of the programme is based on starting from its planning, making design outlay, formulation of technical works, execution and development of land along with creation of water resources developing a hydrological unit at downstream making a common drainage points for the purpose of sustainable development of agriculture. The present case study dealt with the integrated water resources management with specific guidelines to the farmers for clear understanding of installing small-scale irrigation system based on their own resources in accordance with actual land situations as well as judicious use of irrigation with adoption of water-saving technology for sustainable crop production through watershed approach. The work was carried out under National Watershed Development Project for Rainfed Area (NWDPRA), one out of 21 watersheds was identified for development of rainfed agriculture with mobilization of all natural resources like soil-plant-water-animal-human to uplift socio-economic status of the farmers. The effect of watershed management programme was studied to assess the crop water productivity adopting of integrated water management practices with introduction of six rainy season crops *viz*: drilled and transplanted rice, arhar, groundnut, sesame and maize and two winter season crops viz: mustard and wheat at farmers field in Radharamanpur and Genrua in the district of Bankura and Purulia, West Bengal, respectively.The result revealed that selection of suitable crop(s), its appropriate cultivars, introduction of new crops with proper sequences at different land situations the areas through proper utilization of all possible resources increased the land and crop water productivity. The programme was proved both labour intensive that offered high avenues for employment generation in the rural areas and less expensive with high benefit cost ratio of 7:1, which augmented the average income of all the beneficiaries of the adopted villages. This could act as a model of capacity building of the farmers in agricultural education empowering knowledge to increase productivity under rainfed eco-system of the country.

Keywords: Crop productivity; integrated natural resource management; watershed development; rainfed agriculture

Introduction

In India, National Watershed Development Project for Rainfed Areas (NWDPRA) was launched to generate all available and possible natural resources and its effective utilization for sustainable agriculture and the effective management of soil-water resources to uplift the socio-economic status of the rural people wherein emphasis was applied on rain water conservation for development the areas under rainfed conditions. With a view to develop suitable agro-technology for soil water conservation, cropping systems, agro-forestry and livestock production and for its effective transfer under massive scale, the Operational Research Projects for Resources Development and Management on Watershed Basis was formulated to meet the location-specific need along witth need based research programme. At present, out of 141 million hectare net area sown, about 100 million hectare of land was remained rainfed (Tripathi *et. al.*,1990) which contributed about 42% of the total food grain production in the country. Most of the land occupied under rainfed conditions was predominantly based on monocropping where after harvest of rainy season crop(s), nothing could be grown and remained fallow during winter months due to lack of irrigation water or adequate rainfall in these areas (Zaman,1993). Growing second crop(s) with necessary soil and water conservational practices was possible with varying degree of success in such areas (Zaman and Mallick, 1987). Watershed was appeared as integrated management of all possible natural resources like soil-water-plant-animal and human to maximize agricultural production on sustainable basis (Zaman,1993).The performances of various watershed located at different agro-climatic zones in India revealed that the production of food, fodder, fuel, fibre and fruits along with the animal components increased considerably (Bhardwaj and Singh,1990). The achievement of all these watersheds operated in the country was quantified on the basis of results obtained under the Operational Research Projects for Resource Development and Management. With a view to increase the agricultural production on sustainable basis and to bring more area under crop cultivation specially introducing second crops during winter months under appropriate soil and water conservational measures along with increasing or creating irrigation potential collecting run-off, harvesting and recycling of excess rain water, the programme was undertaken at location-specific watershed under red and laterite tract in the districts in Bankura and Purilia in West Bengal. Barron *et. al,* (2003), Zaman (2005) stated that watershed refers to a natural geo-hydrological unit that drains at a common point. In other word it may be described as an area of land from which the runoff water flows through a definite of drain channel, stream or river. This definite point is known as outlet.

The size and shape: the size and shape of any watershed may vary depending upon the morphology of the land surface. Based on the total area of contributing run-off water to be drained at any outlet, watersheds are subdivided into different categories:

a. Catchment	:	Drainage area of a primary river which flows to sea having area more than 1 lakh ha
b. Sub-catchment	:	Drainage area of a secondary river or tribute having area more than 40,000 ha
c. Watershed	:	Drainage area of a tertiary stream having area 4,000-40,000 ha
d. Sub-watershed	:	Drainage area of a quaternary stream having area 2,000-4,000 ha
e. Mini-watershed	:	Drainage area of a streamlet having area 1,000-2,000 ha
f. Micro-watershed	:	Drainage area of a streamlet having area 500-1000 ha

Spectacular improvement in agricultural production through green revolution in several parts of India has been achieved due to high yielding varieties and application of irrigation and fertilizers and the improvement is not sufficient for feeding the increasing population unless a breakthrough in crop production technology takes place particularly in rainfed and dry areas which comprise about 65% of the cultivable land of our country. The increase crop production in those areas is linked with proper soil resource management along with water resource development. The vast rainfed and dry areas are usually associated with soil degradation, aridity, water scarcity, poor soil fertility, uncontrolled grazing, decline of forest, poor socio-economic condition of the human resources. Development of such areas in recent years are undertaken through proper physical and biological management and referred as watershed management. Basically the purpose of integrated watershed management is to follow improved land management for preventing soil loss and enhancing crop productivity, reducing run-off, in-situ and ex-situ water harvesting and conservation, maintain ecological balance between land-water-plant-animal kingdoms (Zaman, 2002).

The main principles of watershed management based on resource conservation, resource generation and resource utilization are as follows:

a) Utilization of land as per capability

b) Production of productive top soil

c) In-situ and Ex-situ water conservation

d) Increasing cropping intensity

e) Safe utilization of marginal lands

f) Enhancing crop productivity

g) Stabilization of farm income

h) Improvement of infrastructural facility

i) Setting up of small scale agro industries

j) Improvement of animal kingdom

k) Upliftment of socio-economic status

Behera and Panda (2006), Bosch (2008) and Zaman (2003) stated that watershed management is a multidimensional developmental approach for steady uplift of masses living in that area. The main objectives of watershed development programme can be described by a word "POWER" which means

P = Production of food-fodder-fuel-fibre-fish-milk

= Pollution control

= Prevention of floods

O = Over exploitation of natural resources to be minimized

W = Water conservation and storage

E = Erosion control

= Economic stability

= Ecosystem safety

= Employment

R = Ground water recharge

= Reduction of draught through management

The above said objectives can be achieved by proper planning and implementation of an integrated watershed development programme. The developmental planning should ensure proper blending of people's requirement, suggestions, technical information about the area to be treated, technically variable solutions and methodology for bringing about desired improvement. The proper planning are usually carried out by the following ways:

A. Collection of soil-climate-geological data base

i. Detailed soil survey of the watershed area has to be carried out to find out soil type, thickness, reaction, horizon differentiation, slope;

ii. Rainfall and its frequency, distribution and duration along with other meteorological parameters;

iii. Existing water resources like pond, dug well, shallow tube well, % of irrigated areas;

iv. Existing drainage pattern, channel system, rate and quality of run-off, geomorphology;

v. Physiography like slope, relief, size and orientation of the area.

Preparation of thematic maps

With the help of 1: 50,000 remote sensing imagery obtainable from NRSA, Hyderabad and based map of the area through GIS techniques following thematic maps to be prepared:

1. Drainage map
2. Relief map
3. Soil map
4. Slope map
5. Geomorphological map
6. Land use map

B. Analysis of secondary data

a) **Natural resources:** secondary data about the existing natural resources like crop productivity, cropping pattern, and varieties used, fertilizer and pesticide use, livestock population, various land uses are to be collected from various sources and Government publications. Certain random field verifications are required to ensure the quality of data. All these data collected to be analyzed to yield much information.

b) **Human resources:** various data on socio-economic aspect like male female ratio, status of education, number of schools, prevailing caste and its percentage to total population are to be analyzed.

c) **Prioritization of watershed for development:** prioritization of watershed is a criterion to select a watershed where developmental work is utmost necessary for economical uplift of that area. Many models are in use to prioritize a watershed. Among these models CSWSI model is commonly followed. In this model various physical and biological resources are used as indicators. A simple formula is used as,

Table 1: SI= Weightage of indicators x Rank

The cumulative SI of different indicators along with its weightage and rank give the prioritization value. More higher is the SI value, the watershed needs development. Some indicators and its weightage and ranks are given below:

S. No	Indicator	Range of parameters			Rank
		Low	Medium	High	
1.	Varieties	Local(0.6)	Improved(0.3)	HYV(0.1)	9
2.	Cropping intensity	100-110(0.6)	110-120(0.3)	>120(0.1)	1
3.	Crop yield (optimum yield)	<50%(0.6)	50-90%(0.3)	>90%(0.1)	10
4.	Irrigation	<30%(0.6)	30-50%(0.3)	>50%(0.1)	4
5.	Marginal farmers	<25%(0.1)	20-50%(0.3)	>50%(0.6)	7

On completion of watershed planning, the developmental work should be initiated in the following ways:

a) Work should be carried out from ridge to valley;

b) Major structure to be constructed;

c) Both mechanical and vegetative measures for effective moisture conservation should go side by side;

d) Long term soil conservation work should be initiated;

e) Awareness programme should be initiated;

f) Co-operative societies to be constructed; and

g) Knowledge based support compared to finance based support should be initiated.

Various state government departments and NGO's are engaged in planning and development of many identified micro-watersheds particularly in drought prone area in a big way. However, sustainability of this development mostly depends on the adoptability of the farmers' friendly technologies co-operative attitude by the beneficiaries and dedication of the implementing agencies.

Water resource conservation

The average annual rainfall in India is about 1170 mm. It is distributed unevenly and erratically over the country's geographical area of 329 million hectares. This account for the total annual precipitation including the snowfall is around 4000 billion cubic-meter. There is large variation in the distribution of rainfall from region to region, season to season and year to year. Rainfall mainly occurs in three to four months during the monsoon season covering only 100 hours in a year. The normal rainfall is as low as 100 mm in the western parts of Rajasthan to over 11000 mm at Mausynram in Meghalaya. The spatial and temporal variation of rainfall is so complex and wide that led to flooding in some areas and drought situation in other places. The increasing disparity between the water availability and growing demands for various sectors needs water conservation on an urgent basis.

It is to be mentioned here that Agriculture is the major user of water and it consumes about 80 per cent of the current level of water utilization. However, with the increasing demand from other competing sectors, the allocation of water is likely to reduce by 10 to 15 per cent in the next two decades. In addition, shrinking water resources due to falling groundwater tables, silting of reservoirs and pollution of water resources from untreated domestic waste, industrial effluents and agro-chemicals are threatening the acute water shortages in the coming years as well as the sustenance of existing levels of agricultural production. Besides, potential agricultural productivity in many areas is not realized due to poor irrigation management. The water use efficiency is as low as 35 to 40 per cent at the field level. Concomitantly, the problems on the quality of groundwater and rapid decline in water table depth due to unscientific exploitation are also noted in many parts of the country which has already been designated as the dark and grey blocks. Thus a proper assessment of the availability of surface and sub-surface water is the cornerstone for proper planning, development and management of this scarce and vulnerable resource. The highest scope of conserving of the precious water resource lies with the agricultural sector followed by other sectors of importance. In addition, it needs to be conserved in all sectors of its uses through various measures from the point it falls as rainfall till it drains into sea for maintaining self-sufficiency in food production and in order to keep India as an economic country with the improved living style.

Aims of water conservation

Water conservation basically aims at matching demand and supply of water. The strategies for water conservation may be demand oriented, supply oriented and management oriented. The demand oriented strategies mostly depend on the field of water use i.e. domestic, irrigation, industrial, navigation and environmental requirements. The supply oriented strategies such as creation of storages, long distance transfer and control of water loss through evaporation are generally common to all the fields of water use. The management oriented strategies include mainly the operational and maintenance aspects for judicious and economic water utilization.

Dimensions of water conservation

Water conservation can be achieved in following three ways:

i) *Water resources conservation*: efficient management of rainwater through storage, allocation and transfer for optimal use and preservation of the quality including its supporting ecosystem.

ii) *Water use conservation*: water supply and distribution with minimum losses and consumption through prevention of wastage.

iii) *Technology and cropping package*: efficient use of water through adoption of water saving technologies and cropping patterns.

Water Conservation Strategies and Technology Options

Some of the conservation measures and technologies available in the agricultural sectors are discussed in the following headlines:

Creation of surface storages

Creation of surface storages is a water conservation measures as it prevents the water flowing waste into the oceans without being put to beneficial use. India with the existing/on-going/proposed major, medium and minor reservoirs in combination has storage capacity of 400 BCM comparing to the ultimate potential of 690 BCM which indicates the expansion of surface storage capacity is critically important to enhance the water availability to cater the needs of various sectors. Greater attention has also been paid to provide better soil condition for retention of monsoon rains in the soil as groundwater storage through adequate soil and water conservation measures on watershed basis.

Rainwater harvesting and recycling

The rainwater harvesting is the process to capture and store rainfall to prevent its runoff, evaporation and seepage for its efficient utilization and conservation in a watershed. The principle of rainwater harvesting is that instead of allowing runoff to cause soil erosion and loss to seas, it should be collected and retained for a longer period for groundwater recharge and subsequent uses. It may be harnessed through proper soil and water conservation measures at places where it is possible and can be utilized at times of need. The technologies available are: i) vegetative barriers, ii) contour bunding, iii) contour stone wall, iv) contour trenching, v) providing check dams, vi) construction of percolation tanks and vii) renovation or excavation of village ponds. The surface runoff if harvested over a large area can yield considerable amount of water for storage and providing lifesaving irrigation to the crop during the dry spells in the monsoon and also for growing a second crop in *rabi* season. By these measures, groundwater potential can also be augmented and problems of deterioration of its quality can be avoided. Further, rooftop harvesting of rainwater from buildings can also be collected and recharged through recharge pits after filtration and should be mandatory in towns and towns. Even small water storage ponds seem to be the most viable strategy in water scarce areas to stabilize crop productivity using the limited quantity of water in most judiciously.

Conjunctive use of surface and groundwater

In canal irrigated areas, because of continuous and intensive application of surface irrigation, the water table gradually comes up at or nearer to the surface resulting in water logging and soil salinity hazards. Similarly, in deep tube well command areas, the excessive pumping of underground water causes decline in the depth of water table as well as deterioration of groundwater quality. The underlying principle of the conjunctive use approach of surface and ground water is establishing a water balance in the area so that total water use from both sources over a period of time is optimized for sustainable crop production and the twin hazards of water logging and salinity are avoided. The planning of exploitation of groundwater is to be executed in such a way that the total withdrawal should not exceed the annual recharge and neither progressive lowering or rising of water table occurs in the command area nor there is deterioration in quality of water. The planning for judicious utilization of the surface and ground water for increasing the crop productivity needs to be strengthened in any irrigation command.

Prevention of water loss through evaporation

About 60 per cent of water is lost through evaporation from the open water surface of tanks, lakes and reservoirs etc. The ultimate spread of storage reservoirs and tanks in the country is estimated to be 2.5 mha and the loss may be around 56 BCM assuming average annual evaporation loss of 2.5 metre. This amount is equivalent to the domestic needs for urban population of nearly

140 crores. It may be possible to check this evaporation loss to a large extent by employing some remedial measures like use of chemical retardants, installation of wind break and artificial recharging the groundwater etc. With the use of known chemical retardation methods about 30% of evaporation loss equivalent to 34 BCM can be checked and conserved for beneficial use. Generally, wax emulsions and free fatty acids are used as anti-evaporants. The use of plastic or rubber floats can reduce the evaporation by 80%. This practice is economically viable in water scarce and drought prone areas.

Prevention of water loss through evapotranspiration

The agronomic measures for water conservation helps to control the evapotranspiration loss from soil and plant surface, reduces the splash effect of raindrops, checks topsoil erosion and runoff losses, permits a better intake of rainwater by improving the better soil physical environment and thereby conserves higher soil moisture for better plant growth and yield.

(a) **Mulching:** It protects the top soil from erosion loss by controlling runoff, facilitates rainwater absorption by soil, increases the consumptive use of water by crops and hence the water use efficiency, regulates the hydrothermal regime of soil, reduces the evaporation component of the evapotranspiration, prevents soil crusting and plant nutritional losses and keep down the weed infestation. Dry soil mulch as created by stirring the soil surface with the interculturing implements helps to minimize evaporation loss from soil surface. Use of cover mulches like black polythene sheets or organic residues or crop stubbles on soil surface will act as barrier to reduce the evaporation losses. Post-rainy season crops (rabi crops) which live primarily on soil stored moisture derive more benefit from mulching.

(b) **Anti-evaporants:** These are used to reduce the evaporation from the soil surface. Straw, sawdust, paddy husk and wood bark are used to arrest the evaporation from soil surface. Hexadecanol, a long-chain alcohol, is also used to retard the evaporation from soil surface. This chemical is resistant to microbial activity and degradation and remains active in the soil for more than a year. The main limitation in soil surface mulching is high cost involvement.

(c) **Control of transpiration loss:** Transpiration loss accounting for 99% of the water taken up by the plant root is transpired to the atmosphere through stomata. This loss from plants can be reduced by the following measures –

(i) Windbreaks or shelterbelts: Plantation of windbreaks or shelterbelts is an important component of improving land quality and soil moisture status in dry land areas. It acts as barriers to reduce the hot air circulation over a crop. The rows of trees are raised on field boundaries across the wind direction to cut down the surface wind velocity. The important characteristics of shelterbelts are longer windbreaks wherein the tall plants were in the middle and shorter plants on both sides. The wind break helps to moderate wind velocity by 20-46%, soil loss by 76% and retain 14% higher soil moisture.

(ii) Weed control: The removal of obnoxious plants can save a significant amount of water which otherwise is absorbed by the undesirable crops leading to lower water efficiency.

(iii) Anti-transpirants: Use of chemical substances like phenyl mercuric acetate (PMA), atrazine, kaolin, lime, *cycocel, mobileaf* can reduce the transpiration either by closing stomata or forming a film over stomata or cooling the leaf by reflecting the solar energy, thereby saves water considerably for crop use.

(iv) Selection of crops and varieties: The crops and varieties with low transpiring rate should be selected according to the adaptability to the limited water and water scarce situation.

(d) **Tillage:** Deep tillage, minimum tillage, zero tillage and conservation tillage are some of the agronomic practices which are intended to conserve rainwater according to the soil, water and climatic conditions. These measures can conserve higher moisture in soil due to increase intake of rainwater and improves the physical and chemical environment of soil. This also prevents the topsoil erosion and runoff loss and holds plant nutrients in available forms, thereby enhancing the yield and water use efficiency of crops. Deep tillage is resorted to conserve the monsoon rainfall whereas shallow tillage is useful for moisture retention in rabi season. On the other hand, conventional tillage may cause excessive harm to the soil health, even pose to land degradation by water and wind erosion. Conservation tillage is found to be useful in conserving soil moisture, maintaining soil physico-chemical environment, reducing soil erosion loss and improving crop productivity on a sustainable manner.

(e) **Land leveling:** Precision land leveling is essential for high and efficient application efficiency, uniform distribution of irrigation water, conservation of rainwater and drainage of excess water in monsoon season. It ensures high water use efficiency and crop yield.

(f) **Selection of cropping pattern:** Maximum effort should be given to select the suitable cropping system in conformity with soil, water and climatic conditions. According to the availability of stored water, crops can be categorized under rainfed crops, limited irrigated crops and fully irrigated crops to maximize production and profit per unit quantity of water. Farmers should be encouraged for growing low water consumption crops and crop diversification in the scheduled cropping programme with inclusion of high value crop should be given top priority.

System of rice intensification (SRI)

Out of 80 per cent water allotted for agriculture, the water-loving crop rice alone consumes more than 45% of it, although the productivity of the crop is as low as 4-5 ton/ha in India as compared to other developing countries such as Korea, Egypt where the average yield is about 10-12 ton/ha. The evapotranspiration requirement for growing paddy is about is 800-1000 mm, whereas in canal/tube well command areas, farmers use as much as 2000-2500 mm which is not only wasteful but also affects the yield due to drainage problem. Considerable water saving and maximum crop productivity can be obtained to irrigate the paddy field to a depth of 3-5 cm as soon as the standing water disappear instead of flooding the crop to a depth of 15-20 cm as usually adopted by the farmers. The system of rice intensification has preference over conventional practices because it advocates keeping the soil moist by regular watering but never continuously flooded for raising the crop. Although it is labour-intensive, but it could save about 25-50 per cent of irrigation water and crop yield manifold as compared to conventional practice.

Concept of tillage in modern agriculture

Tillage is an age old techniques by which farmers influence soil and micro-climate with the objective of improving production. It is an essential component of farming and its usefulness has never been questioned.

The basic concept of soil conservation has turned to mean proper land use, protecting the land from all forms of deterioration, rebuilding eroded soil, conserving moisture, proper drainage and irrigation where needed, building of soil fertility and increasing yield and farm income at the same time. Today the term "conservation" has much broader meaning- embracing conservation of all natural resources such as soil, air, water, woodland, pasture, wildlife, in short the conservation of our environment. Thus the best principle of conservation is to use each unit of resource according to its potential and to treat each unit of resource according to its need All these definitions of conservations means one thing, that is, soil and water conservation means better farming for obtaining yield on sustained basis. Farming accelerates soil erosion. Therefore conservation farming has gained ground with the objective of conservation of soil and water.

Conservation of soil and water in any rainfed area is a strategy towards development of our natural resources. The soil loss by rain water is dominant in agro-climatic zone of humid Assam Bengal plains and Islands of Coastal and Bay of Bengal. The forms of soil erosion by water are sheet erosion, rill erosion and gully erosion. The sheet erosion varies from 4-14 ton/ ha/ year in alluvial soil and 4-10 ton/ ha/ year in red soil. Besides, stream bank erosion, land slide and accelerated erosion due to human activities increase the magnitude more than the permissible rate i.e. 16 ton/ ha/ year. The loss of soil in rainfed area varies with the variation in runoff volume. Runoff is available when rainfall intensity exceeds the infiltration rate of soil. High intensity rainfall events increase with the increase in annual rainfall in rainfed area. Traditional thinking on tillage has been to go in for maximum ploughing and clean cultivation. This has resulted in adverse effects such as accelerated erosion, loss of soil structure, soil compaction, oxidation of organic matter and deterioration of soil biological health, energy crisis, increasing costs and labour shortage, environmental pollution etc. These adverse effects are often aggravated in hot and humid climate of tropics and sub-tropics and in intensive farming. Accordingly, conservation tillage and modified crop establishment practices are being suggested with the availability of new machinery and herbicides for weed control in intensive cropping systems. Modern concept of conservation tillage (any tillage and farming system in which at least 30% of the surface is covered by residues-Soil Conservation Service, USA, 1984) aims at improving resource-use-efficiency with minimum soil disturbance, soil cover through residues and adoption of spatial and temporal crop sequences for achieving higher productivity while protecting natural resources and environment (Sharma and Behera, 2006 and Sangar *et al.*, 2005).

Conservation Tillage

Conservation tillage in a modern sense began in the mid-20th century, with the introduction of effective herbicides. No-till, direct-sowing of crops was first successfully demonstrated in the US in the 1950s (Harrington, 2008). By the end of 1970s the concept of conservation tillage and its effectiveness in overcoming the land degradation crisis has been aptly demonstrated in many countries of Latin America, Asia and Africa. Quality and productivity of agricultural soils and their ability to sequester carbon have been studied by many scientists (Kay and Vanden Bygaart, 2002; Yangand Wander, 1999; Lal, 1997). At the turn of the millennium, conservation agriculture was being practised on about 45 million hectares, mostly in North and South America. In South America in particular, both small-holders and big farmers are rapidly adopting the new technology. In some states in Brazil, it is the official policy. In Central America, Costa Rica has a Department for Conservation Agriculture in its Ministry of Agriculture.

Figures available show that no-tillage - the basic technology for conservation agriculture - is used to cultivate 52 percent of the arable land in Paraguay, 32 percent in Argentina and 21 percent in Brazil. Although in absolute terms the biggest area under no-tillage is in the United States of America, even though it is only slightly over 16 percent of the country's cultivated land.

South American conservation agriculturists are highly organized in respect of regional, national and local farmers' organizations and are supported by institutions from North and South America. Such support is essential in enabling farmers to adopt a new technology that involves a radical change of approach to their life's work.

Outside the Americas, uptake of the new technology has been slower. At present, some large farms in South Africa and Zimbabwe are using conservation agriculture, and no-till farmers' clubs similar to those in South America have been set up in South Africa. Initiatives to promote conservation agriculture for small farmers are under way in Cameroon, Ghana, Kenya, Malawi, Namibia, Tanzania, Uganda, Zambia and Zimbabwe, and other countries.

Efforts to adopt and promote conservation tillage and modified crop establishment practices have been underway since 1990s in India when issues relating to sustainability cropped up and emphasis was given on development of resource conservation technologies. The focus on developing and promoting conservation technologies has been on zero-till seed-cum fertilizer drill for sowing of wheat in rice-wheat system in the Indo-Gangetic plains known as the "bread basket" and the "rice bowl" of the country which occupies the first place in terms of area (11 million hectare) and production (more than 150 million tonne) and accounts for 25% of the total rice area of the country. The twin problems of late sowing and high tillage costs of wheat were overcome and reported to have several advantages over conventional tillage. This technology allowed timely sowing of wheat, enabled uniform drilling of seed, improved fertilizer use efficiency, saved water and increased yield up to 20%. Energy/fuel requirement came down to 1/10th of the conventional tillage and proved effective in controlling the weeds under manageable control. Though, in zero-tillage consistent and satisfactory performance of herbicide becomes a prerequisite as cultivation cannot be taken up to destroy vegetation that escapes herbicide. It has been reported that area planted with wheat adopting Zero-till drill has been rapidly increasing. Zero-tillage in rice-wheat system is now practiced well over two million hectare (Gupta *et al.*, 2006) and crop diversification has been possible by way of growing vegetables, oilseeds etc. after wheat. Raised bed planting and laser land leveling are also being adopted by the farmers of the north-western region of the country (Sharma and Behera, 2006).

Drum seeding (direct seeding of sprouted seeds in puddled soil) of rice has been producing encouraging results for quite some time in India and may be a viable alternative in rice-wheat crop sequence where labour is costly and in short supply (Yadav *et al.*,2006).

Beneficial effects of conservation tillage systems with crop residues or mulching or green manure mulching for improving moisture availability, controlling weeds and regulating soil temperature have been widely investigated in rice-wheat and maize-wheat cropping systems (Acharya and Sharma,1994; Sharma *et al.*, 2001, 2005; Walia *et al.*,1995.). Zero-tillage was also found feasible in *Rabi* crops, namely wheat, mustard, linseed and chickpea following *Kharif* crops of maize, green gram or soybean (IARI, 2004). Conservation tillage with mulching proved beneficial for enhancing water and nutrient use efficiency (Gajri *et al.*, 1992 and Acarya *et al.,* 1998). In West Bengal the practice of zero-tillage or surface seeding of wheat after transplanted *kharif* rice produces encouraging results not only in enhancing farm income and productivity but also proved highly efficient in controlling weeds, soil moistures conservation, water productivity and nutrient use efficiency (Majumder and Das,1986). Rice residue management proved highly beneficial on the succeeding wheat crop (Annual Report of AICW & BIP, 2006-07). Drum seeding (direct seeding of sprouted seeds in puddle soil) of rice in red and laterite zone proved to be highly economic (Researchable issues- *Rabi* maize and some oilseed crops, namely safflower, niger, mustard and linseed grown after rice with zero tillage proved beneficial for increasing productivity or sometimes as good as conventional tillage with respect to productivity (Mandal *et al.*, 1994; Moitra *et al.*, 1996).

Research gaps

Tillage research in India lacks systemic investigation. Research on conservation tillage should always be supported by detailed analyses of soil properties, environmental factors and crop characteristics. Tillage research hitherto conducted in our country lacks a holistic approach. A system approach is essential for wide adaptation of conservation tillage. For conservation tillage system to be adopted successfully in a wide range of soils and environments, they must fit into the overall scheme of the present and the future trends in the farming systems of a region and must meet the rising social and economic aspirations of the farming community. Conservation tillage has become controversial because of its effects on the transport of herbicides and nutrients in surface and subsurface waters. Long-term and large-scale ecosystem studies are necessary to assess the effectiveness of various conservation tillage systems in reducing pollution of natural waters by sediments, nutrients and pesticides (Lal, 1995).

Research priorities

The soil organic carbon (SOC) content in most cultivated Indian soils is less than 5 g/kg compared with 15 to 20 g/kg in uncultivated soils. Total potential of soil C sequestration is 39 to 49 Tg C/y (Lal, 2004).

West Bengal falls under tropical climate. Tropical agricultural ecosystem results in reduction in organic matter content due to decline in the amount of carbon inputs from biomass (Stewart *et al.,* 1990). In many areas of the State, two to three crops of rice are raised per year on the same land; wheat, maize, potato, oilseeds and vegetables after rice are also predominant cropping sequences. As a result of growing exhaustive crop like cereals, soil fertility is declining, thereby adversely affecting the crop productivity. The lower organic matter content, higher cropping intensity, improper cropping sequence, imbalance in the use of fertilizer and faulty management practices are the major causes of depletion of soil fertility. Hence, adoption of conservation tillage is very much essential in West Bengal in order to reduce tillage cost, soil erosion, and environmental pollution as well as to increase the crop productivity and organic matter content of soil. So far suitability of tillage methods was assessed in terms of the economics of crop production on a seasonal or annual basis. However, this is not enough. Appropriate system performance indicators, to assess the sustainability as influenced by tillage methods and soil management, are to be established.

Accurate assessment of tillage effects on soil organic matter (SOC) sequestration is needed. A small difference in SOC can substantially change soil C estimates because the terrestrial carbon pool is so large. Changes in SOC contents have environmental ramifications, including contributions to amelioration of greenhouse gasses (CO_2 and CH_4) and ultimately global warming (Bouwman, 1989). Changes in SOC storage are also closely related to soil quality and the long-term sustainability of agriculture (Doran and Parkin, 1994).

Perhaps the most useful approach to SOM quality is to recognize different pools of organic carbon that vary in their susceptibility to microbial metabolism. The total organic matter content of a soil is the sum of different pools, namely, active, slow and passive fractions. The active fraction which includes living biomass, particulate organic matter, polysaccharides and labile fulvic acids is very readily lost when tillage is intensified. The passive fraction consists of physically protected materials in clay-humus complexes, humin and much of the humic acids remaining in the soil for hundreds or thousands of years. Assessment of these pools has proved useful in explaining and predicting the real changes in SOM level and the attendant soil properties.The impact of the SOC pool on soil fertility is mainly related to the capacity of the SOC to hold and slowly release plant nutrients upon decomposition and/or mineralization. The SOC also plays a critical role in cycling of principal nutrients including N, P, K, and Zn. Soil chemical quality depends on the ability to maintain a favourable balance among macro- and microelements, and on processes governing elemental cycling moderated by the SOC pool. Soil biological quality involves population and species diversity of macro- and microfauna, microbial biomass C, microbial processes leading to transformation of biomass, and denaturing and detoxification of applied chemicals and other pollutants. Soil is habitat for macro- and micro fauna, within and on soil, which transform living and dead organic matter. Biological processes depend on the SOC pool and the characteristics of humic substances in soils. Besides, SOC plays the pivotal role in improving and maintaining the physical tilth of soil suited to supporting high crop productivity.

SOC is thus a win-win strategy, with numerous economic, environmental, and aesthetic benefits. In addition, it is most cost-effective option. SOC sequestration is a bridge to the future, buying time until alternative energy options, based on noncarbon fuels, can take effect (Lal, 2002).

Raised-sunken bed/Raised bed-pond system of cultivation

In high rainfall areas, the cropland is remained fallow in monsoon season due to water logging or prolonged stagnation problem and only one crop is raised in post-monsoon season. These soils are under-utilized, although these are inherently very productive. Their productivity could be enhanced by incorporating raised-sunken bed or raised bed-pond system of cultivation. In the former system, two crops could be raised simultaneously. The water-loving crop paddy can be grown in the sunken areas and less water-requiring high value crops on the raised areas. Similarly, in the second system, the land has been configured into pond or sunken areas and raised bed areas beyond the reach of runoff accumulation. The pond used to store monsoon runoff flow which is being utilized simultaneously for fish rearing and irrigating the dry season crops as and when required basis. Both the systems are useful in augmenting soil, crop and water productivity, increasing the cropping intensity, changing the cropping pattern with inclusion of high value crops and much more remunerative to the local farmers.

Crop diversification

Crop management strategies including the crop diversification and suitable contingent crop planning under condition of limited water supply is prerequisite for obtaining higher water efficiency in command areas leading to considerable water saving. The diversification of crop and cropping pattern should be based on the availability of water resources in the area. Paddy should be grown only where a production of more than 7-8 t/ha can be obtained as it consumes more water. The cultivation of medium and low water requiring crops with high remunerative returns should be given a top priority. The inclusion of legumes in the crop rotation will be must for increasing soil fertility and augmenting crop productivity.

Efficient water management practices

The efficient surface irrigation methods can save about 30 per cent of irrigation water. The advance method of irrigation, such as drip and sprinkler irrigation can save about 30-60 per cent of water with about 30-50 per cent increased yield. These technologies not only use each drop of water most efficiently but also save the soil from getting waterlogged or saline and thus can be adopted in the areas of water shortage particularly for high value crops to save water and increase the area of irrigation. Similarly, by adopting protective/ supplemental irrigation i.e. by providing one/two irrigation at a critical stage of the crop growth, it is possible to increase the production by 100 per cent and also more area can be brought under irrigation.Besides, lining of canals appear to be most effective to prevent losses from seepage and deep percolation. Water can also be efficiently delivered in canal or tube well commands through PVC pipelines to minimize conveyance losses. The scientific findings of irrigating the paddy field after disappearance of ponding water instead of continuous ponding in irrigated command and adoption of aerobic cultivation of rice in water scarcity areas are the options available to improve the management of water leading to increased productivity and conservation of water.

Improving irrigation efficiency

The irrigation sector consumes as much as 80 per cent of available water resources at present. With the competing demand from other sectors rising at a faster pace, the availability of water for irrigation would likely to be reduced by 71 per cent in the coming decades. It is, therefore, necessary to improve the performance of existing system. Higher degree of efficiencies in the management of water use in irrigation sector is required to be achieved to sustain production of crops. Irrigation efficiency should be improved from the present average of about 35-40 per cent to the maximum achievable i.e. around 60 per cent. Wherever water is scarce, but economically advantageous, deficit irrigation may be practiced. Water intensive crops such as sugarcane and paddy should be discouraged in the areas of water scarcity. On-farm irrigation practices mainly by way of flooding or channeling water through parallel furrows causes wastage of water leading to poor irrigation efficiency. Similarly, absence of field channel for adoption of field to field irrigation adds to the problem. Thus introduction of proven water management technologies based on the availability of soil moisture and peak crop water demand, scheduling of irrigation to crops with limited water availability, land leveling, use of PVC pipes and lined channels for water delivery system, rotational water supply system from canal, recycle and reuse of irrigation water or treated sewage and agronomic manipulations for soil and water conservation are some of the options resulting in improved irrigation efficiency. In addition, the adoption of updated efficient water saving technologies such as sprinkler irrigation and drip (micro) irrigation in the fruits and vegetables can achieve irrigation efficiency as high as 90-95 per cent.

Participatory irrigation management

Farmers' participation in management of water resources is one of the prime requirements to improve the irrigation efficiency and enhance food production to meet the future requirement. The objectives participatory irrigation management (PIM) are to achieve equity in water distribution, make best use of rain, surface and ground water conjunctively, develop community responsibility to collect water charges and create a sense of ownership of water resources and the irrigation system among the users so as to promote economy in water use and preservation of system, improve service deliveries through better operation and maintenance and achieve optimum utilization of available resources.

Participation of farmers in the realm or irrigation water management implies an active role of beneficiaries in all the facets of irrigation water management and its attendant linkages with the main system of management in agricultural/ agronomic activities. This role is very different from the traditional passive role of farmers to look to the department for irrigation, for water supply and its distribution. The timing of farmers' involvement is crucial. Farmers' participation is most effective, when it takes place from initial stages of project planning, including the stage of project formulation. Such involvement forms part of ideal conditions for a partnership between farmers and government. The present theme "more crops per drop" is feasible provided the farmers are encouraged and supported to play an active role in the operation and management of the irrigation system. However, to operationalize this concept, there is a need to involve all the farmers within the command area, so that collective and collaborative action becomes possible.

Water pricing

The use of pricing policy is an effective tool for achieving water conservation. A huge amount of national exchequer has been invested for water resources development and distribution. These costs are paid by the state as subsidies using the tax payer's money and are not reflected in the individual's water bill. The water rates should vary according to the availability of water throughout the year. There is an increase during the periods of drought or a decrease during plentiful supply of water. This policy makes the people aware and conscious to use water judiciously and economically. The higher pricing rates especially during the lean/ scarcity period as well as water use beyond a base line induce them to adopt water conservation under compelling situation. However, such policies need to be decided based on the socio-economic and agro-climatic conditions of the area. In India, irrigation water is essentially unpriced. Free electricity for bore wells also has impact on over-exploitation of groundwater. Strict rationing is thus required to allocate the water in the period scarcity. Besides, the imposition of higher irrigation charges to the government operated tube wells, rationalization of electricity charges on privately owned tube wells is likely to be needed so that it could discourage the farmers to grow high water requiring crops like boro paddy in one hand and shifting the cropping pattern towards less water requiring high value crops on the other. Prohibition on the installation of shallow tube wells for lifting water at will is to be implemented in arsenic and fluoride contaminated areas so as to preserve the quantity and quality aspects of groundwater.

People's awareness and participation

The delay and failure of water programme in India is the lack of community participation at every level of its implementation. The success of any water programme depends upon the integration of end users and all stake holders right from programme inception. This is the main principle of integrated water resources development and management. Mass awareness programme and people's participation in the adoption of efficient methods of irrigation by the farmers is the need of hour today. The prevailing low water charges in canal or deep tube well system act as a barrier to adopt these interventions. However, in spite of these limiting factors, farmers can still promote the use of drip and sprinkler systems if proper awareness is created among them about the likely benefits in terms of water saving and more irrigation coverage resulting in increased crop production besides arresting the waterlogging and soil salinity etc. Women are very important stakeholders in water resources management. The workload of women is directly influenced by the availability of water either for irrigation or for domestic purposes. Nearly half of agricultural labour force is provided by women and more than 30 per cent of poor households in rural India are headed by women. With the development of irrigated agriculture, which is labour intensive, increases the workload of women. Similarly, they are required to carry water from far off places if it is not available nearby. It is the women who face lot of problems due to lack of sanitation facilities. Women must be encouraged and allowed to take greater participatory role in all the activities of water management. They should be represented in Water Users' Association, NGO and voluntary organizations. In the local scale, the involvement of women self-help groups in water management needs to be promoted in a major way. The 'Water Resources Day' shall be organized compulsorily starting from school to university level in collaboration with the government and NGO as a means of people's and societal participation in 'conservation of water resources and preservation of its quality'.

Benefits of minimal tillage

1. Reduced labor requirements
2. Time savings more planting index
3. Reduced machinery wear
4. Fuel savings
5. Improved long- term productivity
6. Improved surface water quality
7. Reduced soil erosion
8. Greater soil moisture retention
9. Improved water infiltration
10. Decreased soil compaction
11. Improved soil tilth
12. More milking cows at the cost of bullock
13. Reduced release of carbon gases
14. Reduced air pollution
15. Reduced pest problem
16. More time for social and personal activities

Conclusion

1. No-tillage is an important revolution in several crops including rice cultivation. It can save labor, time, money, increases the farmer's income.
2. No-tillage is a different practice from conventional. Our approach should be simple & affordable to the skeptical farmer to change his mind first.

3. No – tillage is the ideal method for soil and water conservation.
4. Different soil should choose different no-tillage cultivation practices.

It is concluded that no- till and crop rotations constitute a technological revolution for small farmers. No other farming techniques have been shown to have such a high impact on farmers' incomes, reduce their production costs and risks, and at the same time be environmentally sustainable and generate very considerable net social gains to society. To realise these private and social benefits we have a major challenge that will call for considerable effort and dedicated support.

References

Acharya, C.L. and Sharma, P.D. 1994. *Soil and Tillage Research*, 32: 291-302.

Acharya, C.L., Kapur, O.C. and Dixit, S.P. 1998. *Soil and Tillage Research*, 46: 153-163.

Bouwman, A.F.1989. Background Part II. In: Soil and the Greenhouse Effect, Wiley, New York.

Bhardwaj, S.P. and Singh, G.1990. Achievements and performances of twenty operational research projects on integrated watershed management. Proc. National Seminar on Watershed Development and Management, Kanpur Feb 2-4,1990. pp 263-269.

Barron J., Rockström J., Gichuki F., Hatibu N. 2003. Dry spell analysis and maize yields for two semi-arid locations in east Africa. *Agricultural and Forest Meteorology*, 117:23-37.

Behera S., Panda R.K. 2006. Evaluation of management alternatives for an agricultural watershed in a sub-humid subtropical region using a physical process based model. *Agriculture, Ecosystems & Environment* 113:62-72.

Bosch N.S. 2008. The influence of impoundments on riverine nutrient transport: An evaluation using the Soil and Water Assessment Tool. *Journal of Hydrology*, 355:131-147.

Doran, J.W., and Parkin, T.B.1994.*Defining and Assessing soil quality*. 3-21.

Gajri, P.R., Arora, V.K. and Prihar,S.S.1992. *Soil and Tillage Research,* 24: 167-182.

Gupta, R.K.,Jat, M.L. and Gill, M.S. 2006. In: *National Symposium on Conservation Agriculture and Environment*, Oct, 26-28, 2006, BHU, Varanasi

Harrington, L.W. 2008. Conservation Agriculture Newsletter;

Hazra, C.R., 1989. Crop yields increased dramatically in Tejpura watershed. Indian Farming, 39: 37-39. Mittal, S.P. and Dhruvanarayana V.V.,1989. Watershed management increases crop yields in Himachal Pradesh. *Indian Farming*, 39:14-15.

IARI, 2004-2005. Annual Report, Div. of Agronomy. IARI, pp.88.

Kay, B.D. and Vanden Bygaart, A.J. 2002.*Soil and Tillage Research,* 66: 107-118

Lal, R 1995. Tillage Systems in the Tropics.Pp.178.

Lal, R 1997. *Soil Till. Res*. 43: 81-107.

Lal, R. 2002.*Agriculture Practices and Policies for Carbon Sequestration in Soil.* 21-28.

Lal, R 2004.*Climatic Change, Vol. 65, pp. 277-296.*

Mandal, B.K., Saha, A., Dhara, M.C. and Bhunia, S.R.1994.*Soil and Tillage Research*, 29: 49-57.

Moitra, R.' Ghosh, B.C. and Sarkar, S. 1996. *Soil and Tillage Research,* 38: 153-160.

Mazumdar, B. and Das, N. R. 1986. *Proceedings, Indian Science Congress*, 73(3): 2.

Sangar, S., Abrol, I.P. and Gupta, R.K.2005. *Conservation Agriculture: Conserving Resources-Enhancing Productivity*, New Delhi, pp.19.

Sharma, A.R., and Behera, U.K.2006. In: *National Symposium on Conservation Agriculture and Environment*, Oct, 26-28, 2006, BHU, Varanasi.

Sharma, A.R., Singh,R. and Dhyani, S.K. 2001. Annual Report. NATP-CGP Project, CSWCRTI, Dehradun.

Sharma, A.R., Singh,R. and Dhyani, S.K. 2005. *Indian J. Soil Conservation,* 33 (1): 35-43.

Stewart, B.A., Lal, R., El-swaify, S.A. and Eswaran, H. 1990. In: 14th International Congress on Soil Science, 12-18 August 1990, Japan, Vo. VII.

Tripathi, S.K., Dhotre, R.S. and Yoganarsimham G.N., 1990. A model schedule for agricultural watershed management, Proc. National Seminar on Watershed Development and Management, Kanpur Feb 2-4, 1990. pp 314-319.

Walia,S.S.' Brar, S.S. and Kler, D.S.1995. *Environment and Ecology*, 13:3,503-507.

Yang, Xue-Ming, and Wander, M.M.1999. *Soil and Tillage Research,*52: 1-9;

Zaman, A. 2002.Watershed Management for Sustainable Agriculture in Indian Perspectives. *Proc. of the 12th International Soil Conservation Organization Conference (Beijing) on May 26-31*, 2002.

Zaman, A. 2003.Farmer-managed small-scale irrigation systems for sustainable crop production in Uganda, East Africa. *Journal of Food, Agriculture & Environment*, Vo.1 (3-4): 312-15.

Zaman, A. 2005. Integrated water resource management and sustainable crop production: Experience on adoption of watershed approach. *Proceedings of XII World Water Congress held in New Delhi on 22-25 November 2005* (Water for Sustainable Development - Towards Innovative Solution) Vol.5:1-13.

Zaman, A., 1993.Effect of watershed management program on land and crop productivity in drought prone region of West Bengal. *Environment & Ecology*, 11(2): 442-445.

Zaman, A. and Mallick, S. 1987. Feasibility of growing second crop in winter after rainy season crop at rainfed condition in semi-arid region of West Bengal. *Environment & Ecology,* 5(2): 239-243.

2

Rural Livelihoods Improvement through Common Property Water Resource Management

Prabhakar Nanda

ICAR-Indian Institute of Water Management, Bhubaneswar-751023, India
Email: prabhakar.nanda@gmail.com

Abstract

The utilization pattern of common property water resources revealed interesting pictures in different agro climatic regions of the eastern region. The group activities for utilization of water resources in Keonjhar district of Orissa indicate that the 10 different groups of farmers have been benefited differently from the utilization of common property water resources. The group nearer to sources utilized water resources intensely to their maximum benefit. The study reveals the different benefit level for different user groups under WTCER developed water resource. It is revealed that the use of CPR revealed a highest benefit cost ratio of 3.88 against lowest benefit cost of 1.24 for different groups. A case study of use of CPR water resources under irrigation tanks for different economic activities carried out by WTCER in Ganjam, Gajapati of Orissa and South 24 Praganas of West Bengal reveals that, the relatively poor HH (having land holding <1 ha) constituting about 76% of the population under the command of the sources completely depended on the system for livelihoods and for agricultural purpose in comparison to 54% of the households having land holding more than 1 ha. In most of the systems of irrigation tanks, the dependency on common property tanks for livestock rearing was found out to be more than 90% which sometimes goes as high as 100% for poor households. The usufruct sharing (fishing, silt and fuel wood on tank bunds) from the use of common property tanks varied from 5% for Orissa to as high 15% of total agricultural income of households in some of the cases in West Bengal. The contribution of CPWRs towards generation of on farm employment was estimated to be about 452 man days and 180 man days respectively for two different groups of HH in Orissa and West Bengal. There was a stark difference among the two groups of HH with respect to sanitation dependence which revealed that about 84% of poor HH completely depended on the system (tank or elongated water bodies) whereas for relatively better off HH it was only 46% indicating importance of the CPWRs (common property water resources) for the livelihoods of the poor people in the districts.

***Keywords*:** Tenure systems; Community organization; conflict management Pressures; challenges; innovations

Introduction

Common property resource (CPR) management has become focal point in community based management of natural resources under various development schemes for villages. The economic benefit to the community from the access to the common property is substantial. Absence of regulatory mechanism at the village level for use of common properties has led to free access and over exploitation of these resources beyond sustainable limit. The traditional water bodies like village ponds, wetlands and the drainage lines have been issues for conflicts due to clash of interest of users at different points which arises because of competitive exploitation of CPR to maximize private profit. The traditional management mechanism and property right issues got blurred under commercial interest linked to these resources. The emerging scenario of participatory irrigation management in minor irrigation sector along with watershed development programmes put the issues of common property management and community action at the centre stage.

Discussion

CPR refers to those natural resources to which no individual has exclusive property rights as the resource is accessible to whole community with equal access rights. Like village pond, wastelands, pasture lands, wet lands, drainage lines, rivers, rivulets, irrigation canals and village forests etc. Most common property natural resources are land and water bodies within and around local bodies with easy access. There has been a gradual transfer of ownership from kings, local lords, zamindars to private rights/panchyat rights. Reforms measures for land and other community use common resources have come through in last century rapidly for a change in rights of ownership, size and status of holdings. After independence, a host of central and state legislations have come in for better management of resources, revenue earnings for local self-governments. Panchyats exploit their common property land and water resources to earn taxes, duties, tolls and fees.

An estimate by finance commission reveals that about 60% of annual revenue of the Panchayats comes from leasing of the common property. The total market valuation of produce from CPRs in India was estimated to be more than Rs. 9400 crores during the year 1998. Dharia committee (1994) emphasized the quantification of the benefits accruing to different class of people over time and space from CPRs and policy enforcement for their optimal use and management. Little study has been done in India on the actual benefit accruing to the community through use of CPRs especially CPWRs (common property water resources). Chopra *et al.* (1999) attempted one study for forest regeneration under watershed development programme and Selvarajan et al., (2002) for Bhopal lake. Palanisami et al. (2001) attempted for valuation of tank resources under multiple use in South India. These CPRs has economic benefits to the community/individual HH in the form of increase in productivity and production, recreational value, income augmenting values in the private property based farming system and equity impacts (free access irrespective of class and caste if enforced by decree of panchyats). It is estimated that (NSSO 54th round) about half of the rural HH seemed to collect some material or the other from the CPRs. The resource/ proceeds are either used for HH consumption or marketed. The magnitude of accrual of benefits varies in time and space. Some users manipulate its control to their advantage and benefits get skewed though ideally it should be equi-proportional. Indiscriminate/ conspicuous use leads to physical degradation, lessening of benefits to the poor thereby reducing equity impact and ultimately environment degradation.

There are three components for management and governance of CPRs to which a state looks into. These are benevolence (b) revenue maximization (c) policing. Benevolence and policing are having more non tangible benefits where as "revenue maximization" is tangible on which the community/HH is immediately interested for benefit maximisation. The following issues concerning CPRs are of immediate interest to influence state policy domain with respect to natural resource management:

1. Tenure systems and access to CPRs
2. Use, management and production of CPRs
3. Community organization and collective action
4. Conflict and conflict management
5. Pressures, challenges and threats to CPRs
6. Innovations

In the absence of clearly defined property rights, the sustainable use is never in the minds of the present consumers of the CPRs. Absence of regulatory mechanism like commonly agreed price mechanism, regulation on conspicuous consumption, wasteful uses degrade common properties over the years. In a developed watershed, the developed common properties like WHS, community miscellaneous tree plantations, orchards and drainage lines are listed under user's name who might have contributed to their development. But a clear mechanism of usufruct sharing is absent in most of the cases. The watershed development fund is generated for post project maintenance of assets created/ developed during project period. However, a clear cut community driven guideline is not found in the project villages for its use. Rights and collective actions affect decision on investments in most of the village development work where natural resources like water and soil form the basis of interventions. In the case of water resources, the ownership and use rights vary across water sources, usage and states. The water use is guided by multiple property regimes viz. common property, common pool resource, private resource and state owned regime. There is difference in ownership and use of the CPRs. The user of water resources have their own assessment about the resource and accordingly the use pattern is governed which most of the time is based on utility maximisation. The typical use pattern in a drought prone area of the ground water through excessive withdrawals is an example of the short run private profit maximisation using the public investment. In the case of a village pond, while water is considered a common pool resource and there are unwritten customs of sharing it, the water going to recharge the well from the pond becomes a private property. The owner of such a well even in a period of scarcity may not agree to share water with the community. Efforts of a community in conserving water, which help in recharging individual wells, result in building the private poverty which may not be equitably shared with the other members of the community. Recharging of wells under various employment generation and poverty eradication programmes in drought prone areas comes through public investment and augments private profit. Watershed development programmes, is a good example of public investment becoming a private property. Under the watershed development programme and other employment generation programmes where emphasis is given on durable asset creation through public spending in the form of irrigation sources, plantations etc., the benefit sharing mechanism is weak as usufruct rights are ill defined. This leads to non-maintenance of the assets and gradual deterioration of the capital base. The creation of WDF under watershed programme though has potential in taking care of the maintenance of assets after project period, due to absence of operational mechanism; the huge sums deposited in the accounts remain idle.

The analysis of the estimate of NSSO 54th round (Table. 1) reveals that common property land resources constitute about 15% of the total land resources of the country. The land resources are mostly classified under un-cultivable wastelands, village forests wood lots, and pasture lands ctc which are mostly marginal land with no or little irrigation facilities. Per household availability of CPR is estimated to be 0.31 ha and per capita CPR to be 0.06 ha. The households reporting collection of materials from CPRs are estimated to be about 48% (Table .2) and average annual value of collection was Rs. 693 which formed about 3.03 % of total consumption expenditure of the hous holds reporting. The house holds reporting use of CPR water resources for livestock rearing was about 30% and for irrigation purposes was estimated to be 23%.

Table 1: Availability of common property land resources in rural India

Sl No	Items	Estimate
1.	Percentage of common property land resources in total geographical area	15
2.	Common property land resources (ha) per HH	0.31
3.	Average HH size	5.04
4.	Common property land resources per capita	0.06
5.	Distribution of common property land resources (% of total)	Estimate
	Common pasture and grazing land	23 or 3.45% of TGA
	Village forest and woodlots	16 or 2.40% TGA
	others	61 or 9.15% TGA

Source: Estimate from NSSO 54th round

Table 2: Use of Common Property Resources

Sl No	Item	Estimate
1	HH reporting collection of any material from CPRs (%)	48
2	Average value of annual collection per HH (Rs)	693
3	Average value of collection to average value of consumption expenditure (%)	3.02
4	HH reporting to grazing in CPRs (%)	20
5	HH reporting use of common water resources	(%)
	Irrigation	23
	Livestock rearing	30
	HH enterprise	2.8

Source: Estimate from NSSO 54th round

The Case Studies in Eastern Region

A case study of use of CPR water resources like tanks and wells for different economic activities was carried out by WTCER (Presently Indian Institute of Water Management) in Keonjhar, a plateau district of Orissa, reveals that (Table

3 and Chart 3), the majority of relatively poor HH (having land holdings < 0.5 ha) constituting about 75% of the population under the command of the sources completely depended on the system for agricultural purpose. It was estimated that about 33% relatively better off HH (having land holding > 0.5 ha) depended completely on the tank and well system for sustenance of agriculture. It was also estimated that 85% of relatively poor HH used the sources for livestock rearing whereas for relatively better off HH it was estimated to be 78%. There was a stark difference among the two group of HH with respect to sanitation dependence which revealed that about 79% of poor HH completely depended on the system (tank and well) whereas for relatively better off HH it was only 46% indicating importance of the CPWRs (common property water resources) for the livelihoods of the poor people in tribal districts.

Table 3: Use types of CPRs under tank and well system in a project village in Orissa

	Relatively poor HH (%) LH <0.5 ha	Relatively better HH (%) LH >0.5 ha
Complete dependence on tanks and wells for agriculture	75	33
Complete dependence on tanks and wells for sanitation purpose	79	46
Complete dependence on tank and wells for livestock purpose	85	78
Water lifting devices (Mechanical/Manual)	Manual	Manual, Mechanical

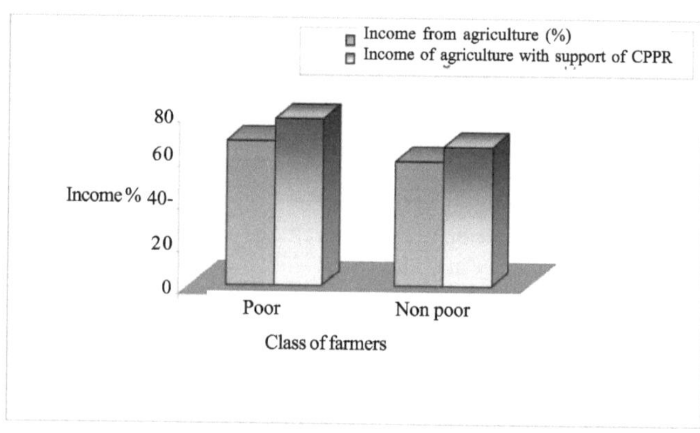

Chart 1: Source wise income distribution in a project village

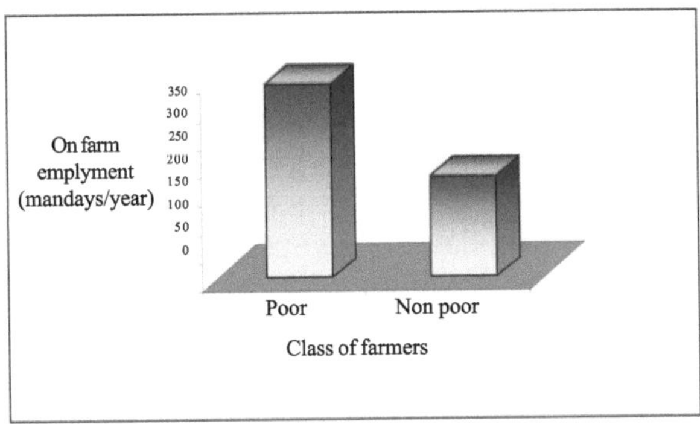

Chart 2: On farm employment generation through use of CPR in a village

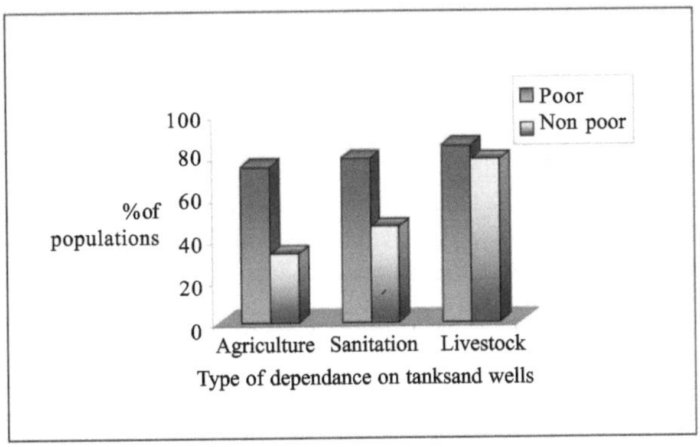

Chart 3: CPR use of tanks and wells

With respect to use of CPRs for economic activities, it is revealed that agricultural income constituted about 67% of total income (Chart 1) of relatively poor HH and the agricultural income obtained from the use of CPWRs was about 78% whereas for relatively better off HH, the agricultural income was estimated to be 58% and the tank cum well based agricultural income was 65%. It is inferred that for economic use also, the poor HH depend more on the CPWRS in comparison to relatively better off households. The contribution of CPWRs towards generation of on farm employment was estimated to by about 342 mandays (Chart 2) and 148 man days respectively for two different groups of HH.

Summary

The common property land resources constitute about 15% of total geographical area of the country accruing an annual income of about Rs 693 per house hold forming 3.03% of total consumer expenditure. About 23% of households used common property water resources for irrigation in India during the year 1998. A case study in Keonjhar reveals that the agricultural income from common property constituted about 78% income of poor households in comparison to 65% of income for non-poor households. The on farm employment generation for poor households through use of CPRs was estimated to be 342 man days per year as against 178 man-days for non-poor households.

References

Chopra, K., Kadekodi, G.K. and Murty, M.N. 1988. Economic Evaluation of People's Participation in the Management of Common Property Resources, Working Paper, No. 3, National Institute of Public Finance and Policy, New Delhi, India.

Chopra, K., Kadekodi, G.K. and Murty, M.N. 1990. Participatory Development: Peopleand common property resources, Sage Publications, New Delhi.

Chopra, K., Kadekodi, G.K. and Murty, M.N. 1989. Peoples participation andcommon property resources', *Economic and Political Weekly* 24(51-52): A189.

Mohan Dharia 1994. Committee Report on Westland Development NSSO.1999. 54th Round Survey.

Palanisami, K, Paramasivam, P, Karthikeyan, C, Rajjopal, A. 2001.Sustainability of Tank Irrigation Systems in South India.Tamilnadu Agricultural University, Coimbatore.

Singh, Subrata. 2003. Conflicts & Disturbance – A reason to change: Lessons from Community based Natural Resource Management Institutions in Orissa, India. "*Forests, Trees and Livelihoods*", Vol. 13. No 3 (2003) pp. 233-246. AB Publishers, UK.

3

Water Scarcity and its Possible Mitigation to Meet Future Challenges for Food Production

Sukla Das

Volunteer, International Institute of Bengal and Himalaya Basin Berkeley California, USA, Kolkata Center, West Bengal, India
Email: das_sukla@yahoo.com

Abstract

Present water scarcity is one of the most important world issues, and according to climate change projections, it will be more severe in future. In several parts of the world, freshwater crisis is largely due to increasing water consumption and pollution by rapidly increasing population and aspirations for economic development, but, attributed by and large to the climate change. Climate change is considered as one of the most serious challenges confronting the present world and is expected to impact hydrological processes such as precipitation, and evapo-transpiration. Increasing human activity and industrialization have led to a broad range of physical, chemical, and biological pollutants entering water bodies and affecting human lives. The groundwater crisis is reaching an alarming stage in many parts of the country. The indiscriminate and every now and then excessive use of groundwater has led to questions concerning its sustainability. The sustainability of groundwater utilization must be assessed from an interdisciplinary point of view, where hydrology, ecology, geomorphology, and climatology play a very vital role. Further, water management strategies ought to be specified keeping aquifer-scales in mind. For the purpose of implementation of this novel approach, reforms are needed in how we assess groundwater resources, map aquifers and monitor water quality. Since water availability and accessibility are the most significant constraining factors for crop production, addressing this crisis is indeed very crucial for areas facing dearth of water. At present, irrigation water use faces harsh opposition from non-agricultural users. Nevertheless, these conditions create important challenges to farmers, managers, engineers and researchers to develop and adopt practices and techniques that favor the sustainable use of water in agriculture. Although a significant relationship exists between the water possessions of a country and the food production ability, assessment of the irrigation requirements is extremely very important for water resource planning in order to meet food requirements and keep away from excessive water consumption. Thus, efforts to develop efficient, economical, and technologically sound methods to produce clean water for developing countries is imperative and 'Hydrogramin Technology' developed by International Institute of Bengal and Himalayan Basins (IIBHB) is expected to serve the purpose.

***Keywords*:** Groundwater; sustainability; geomorphology; hydrology; climatology; ecology; aquifer; irrigation

Introduction

For several decades water has been a vital issue on the international program as the most essential resource for life. Presently, several areas of the world are affected by water scarcity [Alcamo *et al.*, 1997]. Aridity and droughts appear to be the natural causes for scarcity. In addition, as a result of the increased water scarcity and drought due to climate change (Jiménez Cisneros *et al.*, 2014), extensive water use for irrigation is anticipated to take place in the background of increasing competition between agriculture and other sectors of the economy. Rising sea level also affects water availability in coastal areas indirectly by causing water tables in groundwater aquifers to rise, which increases surface runoff at the expense of aquifer recharge.

More recently, man-made desertification and water shortages are aggravating the natural scarcity. Coupled with this, population is gradually increasing and the demand for water faces an increased competition among water users of different sectors and regions. Rainfall is also not sufficient in many regions, thus limiting the quantity of available water resources. Simultaneously, also the quality of water is progressively getting more degraded thereby making water resources to become unavailable for more severe requirements.

Desertification is also aggravating the problems of water resource availability (Fig. 1). Alcamo *et al.* (1997), using the Water Gap model and criticality ratio and their assumptions on how water use will expand with income, have estimated that 4 billion people, or more than half of the world's population, will be living in countries facing high water stress (criticality ratio above 40 percent) by 2025 (Fig. 2).

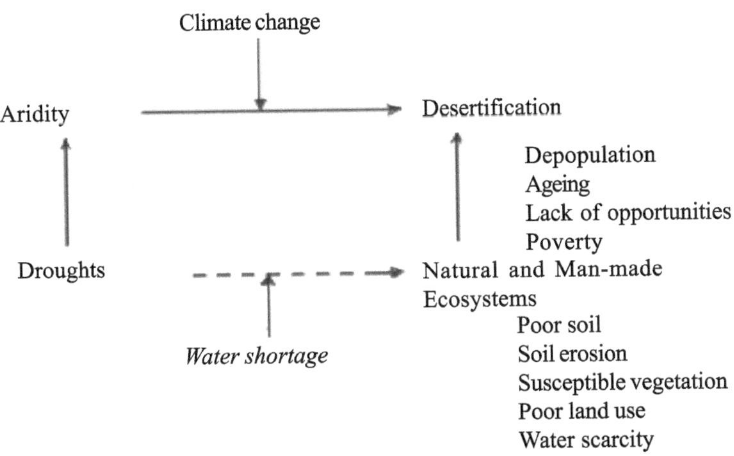

Fig 1: The desertification cycle as influenced by climate, land resources, land and water use, and socio-economic and human resources

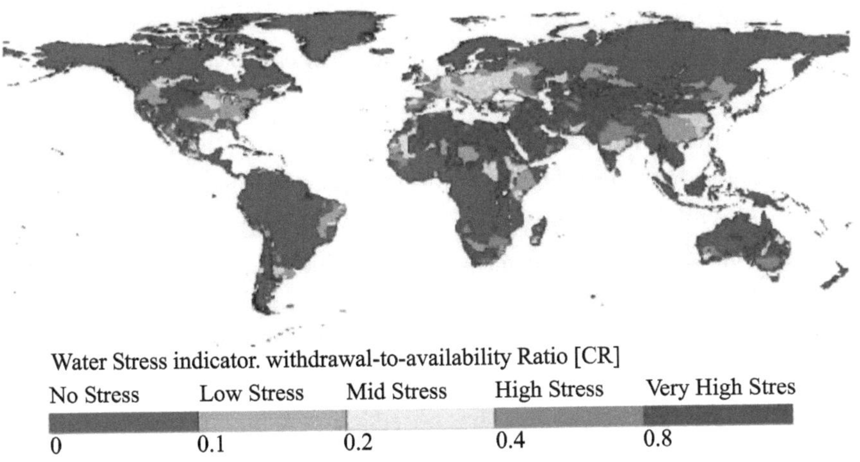

Fig 2: Water stress analysis based on the criticality ratio (Alcamo *et al.* 2000)

Shiklomanov's analysis, based on his forecasts of rising water demands, suggests that water withdrawals will increase by 25%, between 2000 and 2025 from 4,000 to 5,000 cubic kilometre (km^3) (Shiklomanov 1998). The expected increase of the world population growth rate (United Nations Department of Economic and Social Affairs, Population Division, 2013) suggests that higher food demand is inevitable in the future, with a direct effect on agricultural water usage.

Water crisis in India

Water crisis in India is result of three major causes

1. The increasing population is a reason for insufficient water per head. While it has been estimated that the amount of usable water should be between 700-1200 billion cubic meter, India has only 1000 cubic meters of water per head.
2. Water in most rivers is polluted thereby making it unfit for drinking or any other use. Intense urbanization and industrialization have led to a broad array of physical, chemical, and biological pollutants entering water bodies and affecting human lives.
3. Excess extraction by farmers has led to the diminishing groundwater supplies. This is so because access to groundwater is free and everyone has the right to pump water from their own land.
   - Poor monsoon due to climate change has further aggravated the groundwater situation since the latter heavily depends on rains. Poor rainfall compels the farmers to dig further down for groundwater to irrigate the field. This results in pushing the tables deeper down.
   - Unrestrained urbanization has contributed in a big way and despite India being one of the richest nations in water supply; the government and citizens have exploited the water reserves.
   - Quality of groundwater is another issue especially where it is used for human consumption. A number of factors contaminate the groundwater like sewage, run off from landfills, use of pesticides and fertilizers etc.
   - Depleting groundwater has posed a severe threat that cities are now compelled to look for alternate supplies either because of polluted groundwater or that it will cease to exist very soon.

Groundwater depletion in India

Groundwater levels in India are dropping at an alarming rate (Fig. 3) where some states experience water levels dwindling to criticality. Rising sea levels could also reduce water quality and availability in coastal areas. Recent projections of sea-level rise by the end of the 21st century range from 19 to 58 cm.

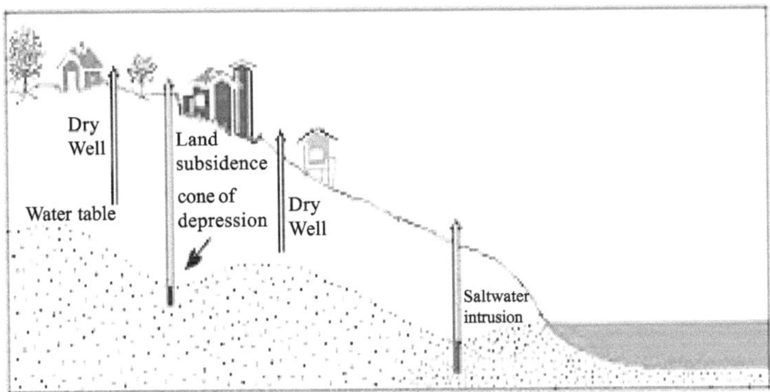

Fig 3: Groundwater levels plunging at an alarming rate

The state-wise status of groundwater resources as on March 2004 is given in Table 1. We can see that in many states, the net draft of groundwater is either in excess of or close to the net available resource, implying that these states are facing a situation of dangerous overexploitation of their available groundwater resources.

Table 1: State-wise Status of Groundwater Resources (2004)

No	States	Billion Cubit Meters (BCM)			Stage of GW Development Net Draft/Net Availabilty*1000
		Annual Replenishable Groundwater Resources	Net Availability	Net Darft	
1.	Andhra Pradesh	36.50	32.95	14.90	45
2.	Assam	27.23	24.89	5.44	22
3.	Bihar	29.19	27.42	10.77	39
4.	Chhatasgarh	14.93	13.68	2.80	20
5.	Cujarat	15.81	15.02	11.49	76
6.	Haryana	9.31	8.63	9.45	109
7.	Jamrmi and Kashmir	2.70	2.43	0.33	14
8.	Jharkhand	5.63	5.25	1.09	21
9.	Karnataka	15.93	15.30	10.71	70
10.	Kerala	6.34	6.23	2.92	47
11.	Madhya Pradesh	37.19	35.33	17.12	48
12.	Maharashtra	32.96	31.21	15.09	48
13.	Orissa	23.09	21.01	3.35	18
14.	Punjab	23.73	21.44	31.16	145
15.	Rajasthan	11J6	10.38	12.99	125
16.	Tamil Nadu	23.07	20.76	17.65	85
17.	Uttar Pradesh	76.35	70.18	48.78	70
18.	Uttarakhand	2.27	2.10	1.39	66
19.	West Bengal	30.36	27.46	11.65	42
20.	Other states	7.67	7.03	0.36	12
Total		432.42	398.70	230.44	58

Source: CGWB61.2006.

Taking quantitative and qualitative aspects together, a total of 347 districts (59% of all districts in India) are vulnerable in terms of safe drinking water in India. This is a matter of serious concern, requiring a new approach.

India is now the biggest user of groundwater for agriculture in the world (Shah 2009). Groundwater irrigation has been expanding at a very rapid pace in India since the 1970s (Fig. 4)

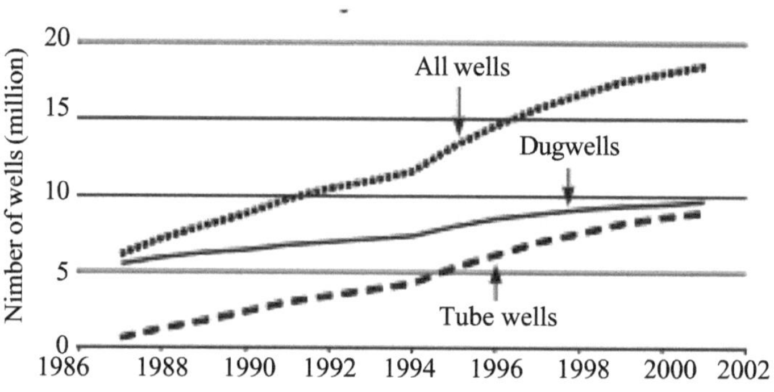

Source: Minor Irrigation Census, 2001

Fig 4: Growth of Groundwater Irrigation Structures (1986-2001)

The share of groundwater in the net irrigated area has also been on the rise. On an average, between 2000/01 and 2006-07, about 61% of the irrigation in the country was sourced from groundwater. The share of surface water has declined from 60% in the 1950s to 30% in the first decade of the 21st century (Fig. 5).

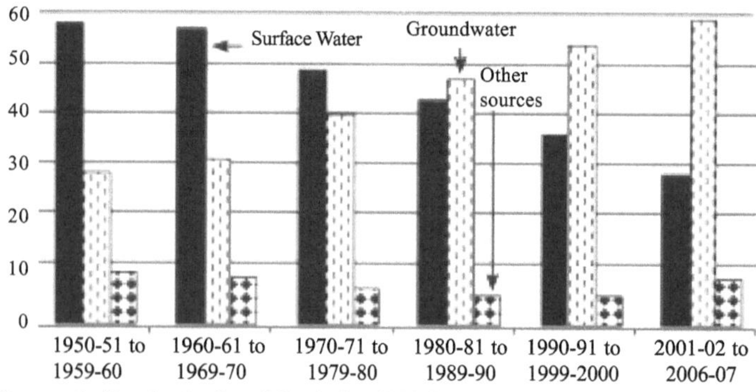

Source: Indian Agricultural Statistics (2006)

Fig. 5: Decadewise Share of Surface Water and Groundwater in Net Irrigated Area (%)

Groundwater depletion is a matter of serious concern, requiring a new approach. It was realized that if indiscriminate use of water go on unabated, adverse effects of degrading hydrogeological regime, will amplify and will engulf the whole state in such a manner that the whole agriculture and industrial development process will suffer a setback.

Growing Demand of Water for Food

Fischer *et al.* [2007] estimated future increases in irrigation water requirements by over 50% in developing regions and by about 16% in developed regions. The largest relative increases of irrigation water requirements are projected to occur in Africa (+300%) and Latin America (+119%) from 2000 to 2080 (Fig. 6). Furthermore, the most critical values of annual renewable freshwater resources are observed in the Middle East and Africa. By 2050, the available water per capita per year will drop below 1000 m^3, not only in the North African belt, but also in eastern and southern Africa and the Middle East (Wallace, 2000). High vulnerability to seasonal water fluctuations can lead to water scarcity for agriculture and domestic purposes. Future population growth is another factor to be taken into consideration. With world population expected to grow by around 2.3 billion people between 2009 and 2050 [United Nations Department of Economic and Social Affairs, Population Division, 2013], up to two-thirds of the world population could experience water scarcity over the next few decades [Alcamo *et al.*, 1997, 2000]. At present, nearly 80% of the world's population is exposed to high levels of threat to water security [Vörösmarty *et al.*, 2010], and the increase of world population will have a significant impact on water usage for food. In a pessimistic low yield scenario, De Fraiture and Wichelns (2010) estimated that 53% more crop water consumption and 38% more land are needed to achieve food production goals in 2050.

Mitigation of Water Crisis through Hydrogramin Technology

Global demand for food production increases progressively at very high rates. Currently, in most countries, continents and regions, agriculture consumes nearly 70% of the total water available. The intensive use of underground water for agriculture in arid and semi-arid countries and regions has aggravated a drastic decrease in the volume of the aquifers and increasing the costs of water extraction. There is an enormous need for reducing this use through the adoption of appropriate technologies.

Since the balance between water demand and water availability has reached critical levels in many regions of the world and increased demand for water and food production is expected in the future, a sustainable approach to water resource management in agriculture is necessary. Our mission is to develop and implement models for proper watershed management, cause remediation of toxic ground and surface waters in Bengal, and to deploy this knowledge throughout the world

Dr. Rash B. Ghosh (Founder of IIBHB) now develops meaningful solutions 'Hydrogramin Model' which injects freshwater to replenish depleted aquifers uniformly across the planet (Fig. 7) which prevents further land subsidence.

'Gramin' is a Bengali word for village and the word implies establishing a water supply system for a community that can be a village or part of a larger urban community. The Himalayan and Bengal Basins and their vast catchment areas are home to one of our planet's most important fresh water resources. This region includes vastly different hydrologies, geologies, and ecosystems, and it spans elevations ranging from the world's highest peaks to one of the lowest lands on the planet. The extreme differences in elevation, geography, and ecology make this region exceptionally valuable for developing water management and remediation technologies for the rest of the planet

The Hydrogramin Model is a watershed management cooperative of a community built around a reservoir built to recharge confined shallow and deep groundwater aquifers. The source of water to supply the reservoir and recharge the aquifer would come from rainfall, melted snows and nearby water courses, if available, and recycled water from the community. Recharge by treated recycled water to the shallow aquifer would be done by percolation through biologically active soils. The design and management of the recharge system would include features to ensure that only unpolluted and toxic free water would reach either shallow or deep water aquifers. The community would then be sustained indefinitely by a reliable supply of water for human consumption, agriculture and industry.

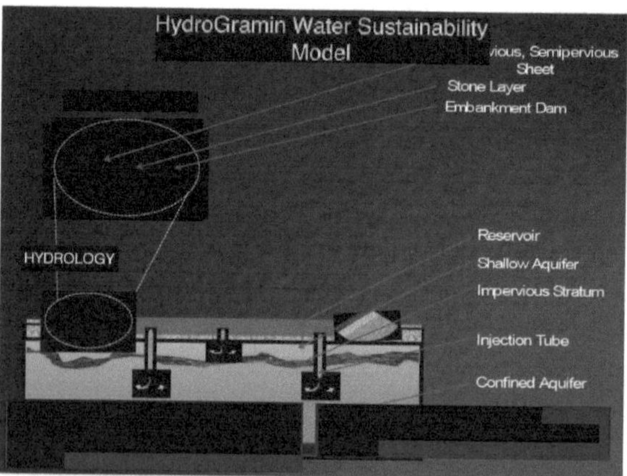

Fig 6: Hydrogramin Model for Water Sustainability

IIBB will engage the services of its other member volunteers to assist in carrying out the research and implementation of Hydrogramin project to inject fresh water to the shallow aquifers for replenishing depleted aquifers to prevent further land subsidence and increase availability of drinking water resources before these natural underground reservoirs are lost forever. The safety measures will consider risk factors to prevent toxic runoff to the aquifers and other water sources before any injection is followed upon consultation with the local experts in the field. The volunteers would be required to further research to refine it and make it cost-effective, site specific affordable fresh water injection project.

Conclusion

Water is a key resource for the progress of any human activity. In many countries, the available water supply and the uneven supply of these resources in time and space are pressing issues. It is projected that up to two-thirds of the world's population will be affected by water scarcity over the next several decades. The availability of water for farming is an essential condition for satisfactory and profitable yields achievement, both in terms of unit yields and quality. The application of efficient water management strategies is a key aspect in increasing water productivity. The correlation between the expected increase in irrigation water requirements, critical values of renewable freshwater resources and economic water scarcity indicates the need for careful water management strategies at the national and site levels.

References

Alcamo, J., Döll, P., Kaspar, F., Siebert, S. 1997. Global Change and Global Scenarios of Water Use and Availability: An Application of WaterGAP 1.0; University of Kassel: Kassel, Germany.

Alcamo, J., Henrich, T., Rosch, T. 2000. World Water in 2025 - Global modelling and scenario analysis for the World Commission on Water for the 21st Century. Report A0002, Centre for Environmental System Research, University of Kassel, Germany.

De Fraiture, C., Wichelns, D. 2010. Satisfying future water demands for agriculture. *Agric. Water Manag.*, 97: 502–511.

Fischer, G., Tubiello, F.N., van Velthuizen, H., Wiberg, D.A. 2007. Climate change impacts on irrigation water requirements: Effects of mitigation, 1990–2080. *Technol. Forecast. Soc.*, 74: 1083–1107.

Genova, R.C., *et al.*, Eds.; Cambridge University Press: Cambridge, UK and New York, NY, USA, pp. 229–269.

International Water Management Institute. 2007. Water for Food, Water for Life: A Comprehensive Assessment of Water Management in Agriculture; Earthscan: London, UK.

Intergovernmental Panel on Climate Change; Field, C.B., Barros, V.R., Dokken, D.J., Mach, K.J., Mastrandrea, M.D., Bilir, T.E., Chatterjee, M., Ebi, K.L., Estrada, Y.O.

Jiménez Cisneros, B.E., Oki, T., Arnell, N.W., Benito, G, Cogley, J.G, Döll, P., Jiang, T., Mwakalila, S.S. 2014. Freshwater resources. In Climate Change 2014: Impacts, Adaptation, and Vulnerability. Part A: Global and Sectoral Aspects. Contribution of Working Group II to the Fifth Assessment Report of the

Shiklomanov, IA.1998. World Water Resources: An Appraisal for the 21st Century. IHP Report. (UNESCO, Paris).

United Nations Department of Economic and Social Affairs, Population Division. 2013. World Population Prospects: The 2012 Revision, Volume II, Demographic Profiles (ST/ESA/SER.A/345); United Nations: New York, NY, USA.

Vörösmarty, C.J., McIntyre, P.B., Gessner, M.O., Dudgeon, D., Prusevich, A., Green, P., Glidden, S., Bunn, S.E., Sullivan, C.A., Reidy Liermann, C. 2010. Global threats to human water security and river biodiversity. *Science*, 467: 555–561.

Wallace, J.S. 2000. Increasing agricultural water use efficiency to meet future food production. *Agr. Ecosystem Environ.* 82:105–119.

4

Integrated Water Resources Management for Agricultural Sustainability and Food Security

A. Zaman

Emeritus Professor, Centurion University of Technology and Management
M S Swaminathan School of Agriculture
Paralakhemundi-761211, Odisha, India
Email: profazaman@gmail.com

Abstract

Water is an indispensable natural resource for sustaining the environment, supporting life and responsible for sustainable agricultural growth and development wherein agricultural sector is the major user of the water resource. The integrated management of water resources could only be possible through adoption of efficient and optimum use of irrigation water, which could only be ensured by judicious and economic use of irrigation potential whatsoever created to increase crop production. The integrated water resources management also includes the concept of rainwater management that has got an immense important on the way to develop the rainfed farming system. Previously, rainfed agriculture was associated primarily with conservational land treatments to check soil erosion and land degradation. But in the areas of medium to heavy rainfall, there is ample scope of tapping excess rain water through suitable water harvesting structures constructed for this purpose for its subsequent uses as irrigation or to apply lifesaving irrigation to the crops which should also come under the purview of water management for crop production. It has got relevancy particularly in installing small-scale irrigation system based on farmers' participatory approach for sustainable crop production for maintaining sustainable growth and development of agriculture. Although there is a difference in the objectives of water management where there is abundant availability of water from continuous flow connected with perennial sources or lakes and rivers and that of from restricted or limited supply from collection of excess rain through water harvesting structures. The former is associated with to obtain as much as yield of crop per unit area under water application and later is associated with increasing productivity of crop per unit of water application even under rainwater recycling and its effective management for the purpose of irrigation.

Hence, both the rainwater and irrigation water are considered to be equally important to context of agricultural development of the country to get higher water use efficiency and proper utilization of water resources. The present paper dealt with the integrated water resources management that should have some specific guidelines to the farmers, which may provide with clear understanding of installing small-scale irrigation system based on their own resources in accordance with actual situations of their available land along with economic and judicious use of irrigation with adoption of water-saving technology for sustainable crop production.

Keywords: Water resources; agricultural productivity; surface and groundwater

Introduction

Water is an essential requirement for sustaining the environment and supporting life and agricultural sector is the major user of the water resource. Despite fact that more than two-third of the earth's surface is covered with water, currently, near about 450 million people in 29 countries are facing water shortage and at least 20 percent more water would be required to meet the total water requirement for additional 3 billion populations by 2025. Amongst the total available water about 97.5 percent of is not available due to its salinity of ocean-sea and 2.49 percent is locked up in ice and only 0.01 percent is technologically available and economically accessible water either in form of surface or ground water sources for human uses including water in river, lake and water bodies (Table 1). As the water is running out due to indiscriminate uses, water resource stands one of the major crises and a threat of dire consequences resulting from poverty, hunger, eco-system destruction, climatic changes tying together other severe world problems are inevitable. As much as two-thirds of the world population could be water stressed by 2025. In this context, conservation of water for agricultural and industrial sectors has got paramount importance. It is stated that agricultural water management must be integrated with an adequate address to water conservation and transfer of water to uses with higher economic return as well as environmental benefits.

Table 1: World water availability pattern

Forms of water	Per cent availability
Sea and Ocean	97.208
Ice and Iceland	2.150
Atmospheric water	0.006
Ground water (Under the ground)	0.625
Surface water (Lake, River, Pond)	0.001

Water resources

Precipitation is the main source of water that varied differently within agro-ecological zones. The amount of annual rainfall exhibits certain typical characteristics of distribution. India received about 80% of the total annual rainfall within three-month time in 80% of the area and average annual rainfall comes about 1170 mm in comparison to 800 mm world average. The total rainfall hours as estimated about 200 hours wherein more than 50% occurred within 20-30 hours time. The wastage of water mainly through run-off became unavoidable. The infiltrated part of rainfall usually converted in to soil water that used by the crops/vegetations, part of which evaporated and rest gone down as ground water or appeared as springs and streams wherein the major part flown as run-off on the land surface. Only about 20 percent of run-off water could be possible to utilize effectively. As a consequence, more that one-third of total area became drought prone in our country.

Status of groundwater

It is reported that nature's way of storing water is under ground, where about 98% of the entire world's liquid and fresh water occurs. And mostly 2% occurs in streams and lakes, which often are fed by groundwater as a dependable source, less affected by the vagaries of climate. Indiscriminate use of groundwater particularly with introduction of high irrigation responsive photo-insensitive short duration rice varieties through out the year caused a drastic depletion of the ground water table and posed a complicated problem. Groundwater reservoir was primarily replenished by the annual precipitation received in a particular area. The rate of entering water to ground water reservoir was depended on pattern of rainfall, run-off, stream flow, permeability of the soils and earth materials present prior to reach the water up to the zone of saturation. The total water resource of the country comes about 400 million ham and total amount of surface flow through out the year in our country is estimated as about 180 Mham, out of which about 105 Mham comes from annual rainfall (Table 2 and 3). The estimated ground water recharge per year is about 67 Mham, out of which about 50 Mham comes from rainfall. Hence there is enough scope to increase the amount of rechargeable ground water adopting ways and means. Introduction high yielding irrigation-responsive varieties of rice and intensive cropping system three to four crops in a year in the same peace of land required huge amount of water. The water was supplied from ground water resources. The deep and shallow tube wells were dug indiscriminately. As results, the depletion of water table occurred to be severe.

In this context, it was very much important to have the access as much as the excess water could be added to under ground reservoir during the period high to medium rainfall; the possibilities of occurrence of flood would be drastically reduced. And as the water table would come to rise, the re-cycling of water during the period of drought would become easier and safer. There was clear indication that several soil conservation and water-harvesting measures could increase the ground water recharge up to a certain limit. In India, groundwater recharge from open wells and pits are very common practices in Orissa, Gujrat and Rajasthan. That's why greater exploitation of ground water prior to rainy season provided scope of greater quantity to infiltrate during high rainfall period. The use of surface water would become also to be more effective and helpful to mitigate the arsenic like problems. As the possibility of contamination of arsenic are very less in surface water; only the flowing surface water is the safest source that is free from arsenic. In this context, it is essential to add ground water as much as possible taking all possible measures for its purification (avoiding all sorts of contamination following de-germs chambers to be constructed for the purpose) and taking off water at need without hampering the water table depletion.

Table 2: Demand and supply of water resources in India

Demand and supply	Km^3
Amount of surface water resources	1880
Amount of usable surface water	690
Amount of usable ground water	418
Total usable water	1108
Water demand in 2002	780
Water demand in 2025 (Estimated)	1200

Table 3: The water resources scenario in India

Annual Average Precipitation	1140 mm
Total Available water	400 million ha
Net area sown	145 million ha
Gross cropped area	175 million ha
Irrigated area	70 million ha
Water demand for irrigation	46 million ha

India is one of the few countries in the world having abundant water resources as evident from demand and supply of water resources in the country. Water resources demand is lesser than supplies. But water scarcity became evident from frequent occurrence of drought at different locations of the country mainly due to lack of proper conservation and maintenance as well as uneven distribution of annual rainfall. As it was not possible to harvest excess rainwater during heavy to medium peak rainfall period, either that caused the disasters in form of flood or flown to the sea and ocean or followed by drought in the period when there is no possibility of occurring rainfall. The erratic and uneven distribution of annual rainfall coupled with lack of proper water harvesting technologies along with improper management of water resources at different locations of the country is responsible for such situations.

Irrigation Potential

The irrigation potential of the country was reportedly created about 22.6 mha with a food grain production of 50 million metric tons in the year 1951 and that reached 96.9 mha having food grain production of 200 million tons resulting per capita food grain production is about 200 kg/year/person. The gap between irrigation potential whatsoever created at the huge investment of national exchequer and its utilization started from the year 1960, is also becoming wider and that was serious concern of the researchers, policy makers and planners.

Irrigated Agriculture

The effect of supplemental water application as irrigation for crop production as it needs during rainless period is well established through out the World. The plant has life, so it requires water for growth and development to give us proper quality and good yield. Water is required for metabolic activities including solublization, absorption and translocation on the way of bio-chemical processes in photosynthesis, respiration and transpiration. The objective of irrigation management is different where availability of water is abundant that of limited or restricted water supply. The former is associated with yield increase per unit of land, irrespective of amount of water used for the purpose, where, later is to increase crop yield per unit of water use. Water requirement of crops and crop response to water use vary with different physiological growth stages. The certain growth stages are more sensitive to water stress situation. So particularly, where water supply is scarce or limited, it is necessary to take account of application of water to the critical growth stages with respect to moisture requirement to get higher water use efficiency. If the crop suffers from water shortage during this period, either quality or yield of the crop reduces drastically. The optimum moisture conditions in the soil profiles are very much essential to maintain for getting the qualitative and quantitative standard of the crop output.

Crop water requirement

The demand for water to crop usually comes from evaporation and transpiration, which is commonly known as consumptive use, associated with unavoidable conveyance losses during application and some special needs during land preparation and transplanting. The supply of water to meet the crop needs usually come from effective rainfall, irrigation water application and contribution from storage moisture at the soil profile.

Equationally, WR=IR+ER+S where WR= Water Requirement, IR= Irrigation Application, ER= Effective Rainfall and S= Storage moisture in soil profile So, IR=WR-(ER+S)

To calculate the water requirement of the crop, it is essential to know the existing moisture status of the soil at sowing and harvesting on which the crop is growing. It needs the knowledge of root system and rooting habit of the particular crop and procedure to determine the moisture content in the different soil profile up to root zone depth of the crop. The crop usually extracts moisture for essential activities through root system. Water comes up to the tip of the leaves to complete the food production processes and goes down to the root again.

So before going to apply irrigation, it is very important to know:

- The crop which the farmer wants to grow with its rooting habits;
- The physiological critical growth stages; and
- The existing moisture status of the soil profiles up to root zone depth.

Water balance

It is very important to harvest excess rain water as runoff during medium to heavy rainfall period for its utilization in crop production during rainless period and development of rainfed agriculture through scientific management of rain water, particularly in the East African countries. The gains of water would come from total rainfall over period of application of water as irrigation. The losses constitute a total runoff (R), downward draining (D), Evaporation (E) and Transpiration (T). Thus the balance sheet shows as Surplus and Deficit over a particular period and it uses to give a changed situation of stored moisture in the soil profiles as well as in the plants.

Equationally

Gains = P (rainfall) + I (Irrigation); Losses = Runoff (R)+ Down ward Drain (D) + Evaporation (E) + Transpiration (T); Gains-Losses = S + V where, S and V are changed moisture status in soil profiles and plants, respectively.

Water use for crop production

To get proper understanding on water use, it needs experimentation and observation on the basis of scientific information on these aspects mentioned above. Only the research findings on water requirement of crops and irrigation water use for crop production would lead to formulate a guide lines on effective utilization of irrigation potential whatsoever it has it has been created. The economic and judicious water application was also ensured by these guidelines. So the entire principles of irrigation, its theories and practices, depends on:

1. How much water to apply? i.e. Measurement of irrigation water;
2. When to apply the supplemental water as irrigation? i.e. proper timings and scheduling of irrigation application; and
3. How best to apply irrigation? i.e. methods of water application to the crop for sustainable production.

Measurement of irrigation water

Establishing the water requirement of crops excluding it from rainfall occurred and moisture present in the soil profile, the requirement is determined. The total amount of water to be supplied as irrigation is usually divided with time of intervals through out the growth period, which indicates the amount of irrigation to be applied at each irrigation. It mainly depends on the rooting habits of the crop and moisture holding capacity of the soil, so it is very important to determine moisture content of the soil profile and also it is very pertinent to know moisture-holding capacity of the soils.

Soil sample placed in dry air oven at 105 to 110°C temperature for 16 to 18 hours up to a constant weight, it gives moisture amount, multiply it with soil depth from where the soil sample was collected (cm) and apparent specific gravity (bulk density, g/cm^3), the moisture content in the soil profile can be determined. This value is subjected to change with tillage operation, cropping pattern and manure application. So it is advisable to check on the bulk density values of the soil, time to time, before calculation of the moisture content in the soil profile.

The followings are the preconditions of irrigation water measurement or amount of water to be applied at each irrigation:

- Occurrence of rainfall, if any;
- Calculation of soil moisture and bulk density;
- Rooting habit of the crop to know effective root zone depth; and
- Moisture holding capacity of the soil at each soil depth.

The moisture holding capacity is commonly known as field capacity and subtracting it from moisture content of the soil profiles, the amount of water can be measured directly by water meter. There are several indirect methods of measuring irrigation water like orifice, v notch, parshall flume. These are very simple devices to measure irrigation water to crop field during its application, which needs only calibration with time factor.

Irrigation scheduling

It is equally important to fix up the time intervals of water application to crop field. Several criteria are usually deployed to find out the proper timings of water application. The soil tensiometer, resistant or gypsum block, neutron moisture meter used to give information on moisture status at particular root zone depth of soil profile which indicated moisture availability in any specific zone. These equipment as well as moisture depletion method could be employed to make irrigation schedule of a crop. The cumulative daily evaporation value from any standard open pan evaporimeter also may be used as indices to irrigate the crop at proper timings. But when the water supply is limited, application of water at identified critical growth stages used to make the irrigation schedule to the crop.

Water application methods

Selection of appropriate method of irrigation water application to the crop field is depended on soil type (clay, silt, sandy loam, loam etc), topography (undulating or flat), sources and amount of water available, crop to be grown and management practices adopted. The appropriate method would lead to give good storage and distribution efficiency.

Surface, subsurface, sprinkler and drip (trickle) are commonly classified methods where the surface method includes border strip, check basin and furrow according to field lay out and availability of water through gravity flow. Surface methods are usually employed to comparatively flat or leveled land with gentle slope having advantages of gravity flow by selecting suitable water lifting devices or making diversion from continuous flow of water from perennial sources.

Sub-surface methods are not very common in practices due to its difficulties in lying out and required construction work. Sprinkler (overhead) method is application of water as similar to rainfall with rotary heads or perforated pipes under the pressure created to the nozzles, holds good to any bunch crops like cereals (other than rice) and oilseeds. The drip (trickle) methods involves carrying water through small diameter plastic tubes with drippers which emits water drop by drop near base of the plants/roots irrespective of topography and holds good and much economic to the horticultural crops like mango, guava, citrus even for coffee and banana. Though initial installation expenditure are more to the sprinkler and drip system, it would be more economic in the long run by increasing yield with much water saving, less power required, much saving in labourers and annual maintenance having effectiveness under any topographical situation at different landscape.

Water conveyance and irrigation efficiency

Water from the sources, when it reaches to the crop field as irrigation, it requires any distributory system like watercourses or channels. It involves a loss of water through the distribution system associated with percolation, evaporation and losses occurred by runoff at borders and furrows. So the ratio of actual water that has been utilized as irrigation out of total water delivered from the source is known as irrigation efficiency. The ratio between the water stored in the root zone depth and the total water reached to the crop field is known as water application efficiency. The ratio between water stored in the root zone depth and the actual requirement of water prior to irrigation water application is known as water storage efficiency. These could be transferred and termed as operational or economic efficiency of the system.

On-farm water management

Adoption of proven and efficient water management technologies are essential to maximize crop production and minimizing gap between irrigation potential created and utilized. Application losses due to conveyance, methods and scheduling of irrigation also are effective in this direction. Selection of crop(s) and crop sequence(s) also played an important role for enhancement of water use efficiency through crop cultivation and related agronomic practices. Conservation, distribution and utilization of irrigation water are the basic parameters of on-farm water management. Optimum scheduling of irrigation, suitable method adoption, conjunctive use of rain, surface and ground water for crop cultivation having improved agro-technology adoption and provision of drainage. Application of proper amount of water at proper time increased the water use efficiency and crop yield maximization with given amount of water reducing evaporation and deep percolation. Scheduling of irrigation with limited water availability is a big challenge to the irrigation experts that needs rigorous research. It is considered the irrigation system as a complex socio-technical phenomenon that involved interaction between a physical, environmental and technological in the process of adoption. However, the participatory action research appeared to be more effective and communicative measures in the developmental processes. It is described that the participatory action research as a method of merging both developmental intervention and research activities as an effort to involve the users in systemic process of change to make the socio-technical system more effective and efficient. Majority of the farmers are reluctant to apply irrigation scheduling technologies as they do not expect a net benefit, out of its application. Some workers elaborated their work on-farm participation and encouraged the farmers to adopt irrigation scheduling as cost-benefit approach. They formulated user-friendly format for a wide variety of crops irrigated in South Africa. The on-farm irrigation management and irrigation service is improved considerably by involving farmers in water management practices and introducing an acceptable water-saving irrigation schedule for growing crops and increasing agricultural production. On-farm water management with emphasis on active participation of the farmers at each and every stage is the corner stone of success.

A series of on-farm water management research activities were conducted as a collaborative project of Indian Council of Agricultural Research, New Delhi and Ministry of Water Resources, Government of India in some of the selected command area either by package intervention or conducting component technology field trials in the irrigation command. On-farm participatory field trials brought positive improvements towards better water use efficiency following the recommended methods of water management practices in the farmers' field in DVC command. Several research workers also reported success story of efficient soil and water management for enhancing agricultural productivity in diverse soil groups of India.

Conclusions

Water resources management is an important consideration to the researchers to address the issues like global climate changes, natural disasters, pollution, environmental degradation, soil health and food security to fit the system in mitigating occurrence of such situations. In addition to the procedures to be adopted for rainwater harvesting and it's effective utilization, it is essential to explore the possibilities of inter basin transfer of water from surplus to deficit basin to increase maximum water resources to get full potential of irrigation development.

1. It is also essential to find out the ways and means to use the contaminated as well as reclaimed water for its effective utilization in crop cultivation;
2. Efficient utilization of water saving irrigation technologies for growing crops;
3. Introduction of supplemental and protective irrigation that could save at least 30% of irrigation water minimizing conveyance and other losses; and

4. Adoption of micro-irrigation could save 30-60% of water with 30-50% increased crop yield. Save water and increase the area under crop cultivation as well as more crop yield with less water application is the basic purpose of adapting integrated water resources management. .

References

Annandale, J.G. *et. al.* 1996. Irrigation management for economic optimization of processing tomato production. Report of Water Research Commission.

Agarwal, M.C. and Khanna, S.S. 1983. Efficient soil and water management in Haryana. Haryana Agricultural University, Hissar, p 118.

Belloncle, G. 1988. Is user participation in hydro-agricultural scheme possible? Integrated Agriculture in Africa. Vol.1 Proc. of the seminar. pp 81-94.

Bouwer. H. 2003. Integrated water management for the 21st Century. *Environment*, 1(1): 118-27.

Bouwer. H. 1978. Groundwater Hydrology. McGraw-Hill, New York. p 480.

Dhindwal, A.S. and Poonia, S.R. 1994. On-farm water management – a success sory. *Intensive Agriculture*. 32(1-2)37-38.

Korten, D. 1980. Community Organisation and Rural Development: A Learning Process Approach. Administrative Review, 40 (Sept-Oct.): No.5.

Pramanik,M. and S. Mallick, 1996. Farmers participatory approach for improvement of present status of irrigation water utilization in DVC canal Command. Proc. of ICID/FAO Workshop, Rome, Water Report 8.

Rajput, R.K. 1993. On-farm water management. Directorate of Water Management Research, Government of India. pp 35.

Robert Svadlenka, 2003. The emerging water crisis and its implications for global security. World Hunger Year, 1:1-5.

Trist, E. 1980. The evaluation of Socio-technical system: A conceptual framework and Action Research Programme. Toronto.

Whyte, W.F. 1991. Participatory Action Research. Newbury Park. California; London & New Delhi, Sage Publications.

Westhurizen, A.J. and Annandale, J.G. 1996. Encouraging irrigation scheduling: A Cost-benefit Approach. Proc. of ICID/FAO Workshop, Rome, Water Report 8.

Xianjun, C. 1996. Introduction of water saving irrigation scheduling through improved water delivery: a case study from China. Proc. of ICID/FAO Workshop, Rome, Water Report 8.

5

Multiple Coping Strategies for Water Crisis: A Review

Rupsha Roy

Dean, International Institute of Business Management, Kolkata and Volunteer, International Institute of Bengal and Himalaya Basin Berkeley, California, USA, Kolkata Center, West Bengal, India
Email: rupsharoy@gmail.com

Abstract

Water Crisis or scarcity can be explained as when demand for fresh water goes high in comparison to the supply of fresh water. The condition is the cause when there is a high rate of aggregate demand from all water using sectors like food, agriculture etc. are compared with available supply under prevailing institutional and infrastructural conditions. It is evident from economic competition from water quantity and quality, disputes between users, irreversible depletion of ground water, and negative impacts on the environment. Water crisis is actually a relative and dynamic concept irrespective of any level of supply and demand. It is a social construct and its causes are directly related to human interference with the water cycle. This scarcity varies over time due to natural hydrological variability. But the stronger reasons could be more of functions of prevailing economic policy, planning and management approaches. The water scarcity can be intensified with most forms of economic development. But if correctly identified many of its causes can be predicted, avoided or mitigated. This paper focuses on the various strategies to cope with water crisis by various organisations, communities and countries through a details study on various articles published so far on water crisis. The water crisis can be resolved by two approaches, one being supply enhancement and the other being demand management. Supply management can be considered when access to conventional water resources is increased. Drainage water and waste water is reused, inter –basin transfers, when water is desalinated and pollution is controlled. Demand management can be defined as a set of actions controlling water demand. This can be done either by raising the overall economic efficiency of its use as a natural resources, or by operating intra- and intersect oral re-allocation of water resources.

Keywords: Water crisis; coping methods; water demand; agriculture

Introduction

Water Crisis or scarcity can be explained as when demand for fresh water goes high in comparison to the supply of fresh water. The condition is the cause when there is a high rate of aggregate demand from all water using sectors like food, agriculture etc. are compared with available supply under prevailing institutional and infrastructural conditions. It is evident from economic competition from water quantity and quality, disputes between users, irreversible depletion of ground water, and negative impacts on the environment.

Water crisis is actually a relative and dynamic concept irrespective of any level of supply and demand. It is a social construct and its causes are directly related to human interference with the water cycle. This scarcity varies over time due to natural hydrological variability. But the stronger reasons could be more of functions of prevailing economic policy, planning and management approaches. The water scarcity can be intensified with most forms of economic development. But if correctly identified many of its causes can be predicted, avoided or mitigated.

Water status of India

Water is the need of all living being. It is also most under prioritized and over used commodity for human beings. Though we are rapidly moving into urbanization and globalization still we are not clearly focusing on water in our planning for infrastructure and resources. India is considered to be the world's oldest civilization. Still our cities and towns have grown without planning for the need of water to availability of water.In 1951, the per capita water availability was about 5177 m3. This has now reduced to about 1545 m3 in 2011 (Source: Water Resources Division, TERI). India is suffering from one of its worst water shortages in decades, affecting over a quarter of its population, emptying its reservoirs and killing the country's crops and livestock. More than 300 districts in 13 Indian states are grappling with drinking water shortages, Minister of State for Drinking Water and Sanitation Ram Kripal Yadav said in Parliament Monday. The government has so far stopped short of describing the water shortage as a drought because the annual monsoon rainfall, due to arrive next month, is expected to provide some relief. But the lack of water is rapidly becoming a big political issue for much of the country. In recent years, India is suffering from water crisis which resulted in droughts in different states like Maharashtra, Karnataka, Andhra Pradesh, Rajasthan, Gujarat, Chhattisgarh, Odisha, Uttar Pradesh and Madhya Pradesh. In these areas, drought has become a yearly phenomenon

Causes of water scarcity

There are primarily three causes for India's water crisis. The first cause is insufficient per capita water availability due to overpopulation. The second reason is poor water quality resulting from inadequate water treatment facilities, and the third problem is dwindling groundwater due to unplanned extraction by industries and farmers for irrigation.

Here are few reasons why India is so parched.

1. **Weak Monsoons:** A deficiency in monsoon rainfall for two successive years resulted in serious water shortages in many states. The monsoon season last year ended in September with a 14% rain deficit, while the deficit in 2014 stood at 12%, according to the India Meteorological Department. India gets 75% of its annual rain in the monsoon that runs from June through September. The rains usually spread across the entire country by mid-July after arriving over the southern state of Kerala in early June. Close to 60% of the country's farmlands are rain-fed and more than half of the workforce is employed in the agriculture sector.
2. **Thirsty Crops:** Excessive cultivation of water-intensive crops such as rice, wheat and sugar cane has exacerbated the water scarcity. As much as 70% of water delivered through India's limited irrigation infrastructure goes to water sugar cane fields in some states. It's no accident that the sugar-producing states of Maharashtra and Uttar Pradesh have been among hardest hit by the water shortage. Producing a kilogram of sugar cane requires up to 5,000 liters of water, according to the World Wide Fund for Nature. Promoting cultivation of less water-thirsty crops such as barley could help, said Himanshu, an associate professor of economics at New Delhi's Jawaharlal Nehru University who uses only one name.
3. **Fallen Forests:** Massive deforestation since the 1990s has had a huge bearing on the South Asian monsoon, causing an 18% decline in precipitation over India, according to a study published in the Proceedings of National Academy of Sciences last year.
4. **Rain Water Wasted**: Rain water harvesting is patchy in India so much of the huge amount of water dumped on the country during the monsoon every year just dries up or flows into the ocean. Rain water harvesting

tries to capture and store more of the rain water in lakes and ponds and steer it towards ground water aquifers. In 2009 India's Central Ground Water Authority asked Indian states to adopt rooftop rain water harvesting systems in government institutions. The Indian government gives financial help for such projects but just 18 of 29 states have so far adopted the practice.

5. **Pitiful Policy:** New Delhi's policies are also to blame for the country's chronic water issues. Its subsidies and price supports encourage the over use of water while it continues to fail to build the nationwide network needed to conserve and distribute water. The government should "ramp up infrastructure to swiftly store rain water," Prof. Himanshu said. Prime Minister Narendra Modi's "imperative responsibility is to quickly assist those suffering from the drought."

6. **Some other major reasons are:**

a. Inefficient use of water for agriculture. India is among the top growers of agricultural produce in the world and therefore the consumption of water for irrigation is amongst the highest. Traditional techniques of irrigation cause maximum water loss due to evaporation, drainage, percolation, water conveyance, and excess use of groundwater. As more areas come under traditional irrigation techniques, the stress for water available for other purposes will continue. The solution lies in extensive use of micro-irrigation techniques such as drip and sprinkler irrigation.

b. Reduction in traditional water recharging areas. Rapid construction is ignoring traditional water bodies that have also acted as ground water recharging mechanism. We need to urgently revive traditional aquifers while implementing new ones.

c. Sewage and wastewater drainage into traditional water bodies. Government intervention at the source is urgently required if this problem is to be tackled.

d. Release of chemicals and effluents into rivers, streams and ponds. Strict monitoring and implementation of laws by the government, NGOs and social activists is required.

e. Lack of on-time de-silting operations in large water bodies that can enhance water storage capacity during monsoon. It is surprising that the governments at state levels has not taken this up on priority as an annual practice. This act alone can significantly add to the water storage levels.

f. Lack of efficient water management and distribution of water between urban consumers, the agriculture sector and industry. The government needs to enhance its investment in technology and include all stakeholders at the planning level to ensure optimization of existing resources.

Coping Strategies for Water Scarcity

The water crisis can be resolved by two approaches, one being supply enhancement and the other being demand management.

A. Supply enhancement can be considered when access to conventional water resources is increased.

During the twentieth century, large multipurpose dams have served the needs of agriculture, energy and growing cities, and helped protect populations from flood hazards. While potential for further dam development still exists in some regions, most of the suitable dam sites are already in use, and the development of new dams is increasingly questioned in terms of economic, social and environmental considerations. On-farm water conservation, particularly the adoption of agricultural practices that reduce runoff, to increase the infiltration and storage of water in the soil in rainfed agriculture, is the most relevant local supply enhancement option that farmers have to increase production. On a slightly larger scale, small, decentralized water harvesting and storage systems contribute to increasing water availability and agricultural production at the household and community levels. However, large programmes of small-scale water harvesting, like the watershed management programmes developed in Andhra Pradesh and other parts of India, have shown significant impacts on the catchment's hydrology and downstream water availability.Groundwater exploitation has grown exponentially in scale and intensity over recent decades. Groundwater's capacity to provide flexible, on-demand water in support of irrigation has been seen as a major advantage by farmers. While intensification of groundwater use has contributed to improved livelihoods of millions of rural people, it has also resulted in long-term aquifer depletion, groundwater pollution and saline intrusion into important coastal aquifers. The adoption of re-cycling of drainage water and wastewater use in agriculture tends to be positively correlated with water scarcity. Re-use of drainage water is a reality in most large irrigation schemes, in particular in the large rice-based systems of Asia. Of lesser global significance, but locally important, is the re-use of urban wastewater (it is estimated that world-wide some 20 million hectares of agricultural land is irrigated with wastewater). Efforts are needed to better assess re-use and its potential, and promote safe recycling of wastewater in agriculture, in particular in water-scarce areas.

B. Demand management can be defined as a set of actions controlling water demand.

The first most commonly perceived option is that of increasing the efficiency of water use by reducing water losses in the process of production. Technically, 'water use efficiency' is a dimensionless ratio that can be calculated at any scale, from irrigation system to the point of consumption in the field. It is generally applied to any management that reduces the non-beneficial use of water (i.e. reducing leakage or evaporative losses in water conveyance and application). The second option is increasing crop productivity with respect to water. This involves producing more crop or value per volume of water applied. The third option is to re-allocate water toward higher value uses through intersectoral transfers (transfers to municipal supply, for instance) or intrasectoral transfers by limiting the irrigated harvested area under a particular crop to reduce evapotranspiration or diverting water towards higher value crops.

Clearly there is scope for managing the demand for water in agriculture in time and in space. But excessive emphasis is often placed on the first option, with efforts aimed at reducing water 'losses' within irrigation distribution systems. Understanding the roles, attitudes and strategies of various stakeholders, including relevant institutions, is a key aspect of demand management strategies. Ultimately, it is at the farmer level that most water will be consumed. Their behaviour and their capacity to adapt will be driven by a carefully selected set of incentives that include both structural and institutional changes, improved reliability and increased flexibility of water supply. Farmers' strategies will be driven by water saving only when water availability becomes their main limiting factor. Policies based on systems of water tariffs aiming to reduce agricultural water demand have proved successful in some cases, but require very constraining conditions and are often difficult to enforce. Approaches based on water quotas and water use (or withdrawal) rights have, in most cases, a higher probability of success.

References

2030 Water Resource Group

Food and Agriculture Organization of the United Nations

Indian Society Blogs

IOSR Journal of Agriculture and Veterinary Science (Jul.-Aug. 2013)

Journal of Sustainable Development; 2014

Mediterranean Journal of Social Sciences MCSER Publishing, July 2015

Technical Documents in Hydrology, UNESCO, 2002

Water Policy and Action Plan for India 2020

www.worldwaterday 07.org

6

Groundwater Management for Agricultural Development in Arsenic Infested Areas - A Case Study

A. Gayen[1], A. Zaman[2] and G. Sen[3]

[1]*Senior Hydrogeologist (Scientist-'D'), National Ground Water Training and Research Institute, Central Ground Water Board, Government of India, Raipur Chhattisgarh-492001, India*
Email: anadigayen@rediffmail.com
[2]*Emeritus Professor, Centurion University of Technology and Management Paralakhemundi-761 211, Odisha, India*
Email: aftab@cutm.ac.in
[3]*Central Ground Water Board for Eastern Region, Kolkata, India*

Abstract

Nadia district covers 3927 sq km of West Bengal State and forms lower part of Indo-Gangetic alluvium of Quaternary age. The district is entirely affected by arsenic contamination in ground water. Population explosion, rapid agricultural growth and inadequate surface water resources has put tremendous thrust on ground water development. 80% of the population is dependent on agriculture. Ground water plays a key role to yield substantial agricultural growth. Irrigational potential of the district is 3356 sq km, of which major contribution belongs to ground water only, i.e., 3100 sq km. Summer season (*boro*) (lean period crop) paddy cultivation demands huge ground water. Shallow tube wells within 80 metre depth are major in numbers than deep (beyond 100 metre) tube wells. Dynamic ground water resource calculation indicates that net ground water available in the area is 198948 ham, of that for irrigation purposes ground water draft is 167617 ham. Net ground water available for future irrigation development is 34996 ham. Out of 17 blocks of the district, six blocks have already been listed under semi-critical category. Ground water withdrawal and switch over to surface water irrigation is needed with due care. Lifting irrigation from adjacent river Ganga coupled with rain water harvesting may be the suitable alternatives of ground water uses. Effective management would help to restrict depletion of ground water table and to maintain the ground water resource for future agricultural and domestic uses. Spreading of arsenic contamination in ground water in the adjoining area may be controlled through augmentation of surface water irrigation. Crop planning in irrigated area is to be taken proper attention to implement location specific proven water saving technologies with conjunctive application of ground and surface water wherein the application of life saving irrigation in growing crop in rainfed areas is required to done water stored surface water.

Keywords: Groundwater; summer season; dynamic resource; river lift irrigation; rain water harvesting

Introduction

As reported, presently total 83 blocks of 08 districts are affected by arsenic contamination in ground water in West Bengal State, India. Of the 83 arsenic infested blocks, this Nadia district alone covers 17 blocks. Arsenic toxicity of contaminated drinking water in humans in many parts of the world, including West Bengal, India, has deservedly received increasing concern. On the other hand, arsenic contamination in agro-ecosystem, that acts as a conduit for the passage of the toxicant to human population *via* food web, came under serious consideration only recently (Sanyal and Nasar, 2002). At present, out of 17 blocks, six blocks have already been enlisted under semi-critical category, which requires due attention towards ground water withdrawal. Apart from this, all 17 blocks of Nadia district are arsenic affected. The district is blessed with 1385 mm annual rainfall, of which mostly escapes as surface run-off. Summer season (*boro*) paddy cultivation demands huge ground water. Shallow tube wells within 80 meter depth are major in numbers that play an important role in irrigation. Therefore, the ground water resource in shallow aquifer is being exhausted. Hence, in future for the required agricultural production in the district, people have to depend on the deeper aquifers, which is a costly affair.

This paper aims at to understand the management options of the available ground water resources to augment the agricultural growth of this entirely arsenic affected district.

Study area

Nadia district of West Bengal State occupies the total geographical area of 3907 Sq. Km and forms the lower part of the Ganga basin (Fig.1). The district.is bounded by the latitude 22° 25' 30" N to 42° 05' 40" N and Longitudes 88° 07' 30" E to 88° 47' 45" E (Fig.1).The Ganga or the Padma with tributaries namely the Bhagirathi, the Jalangi, the Churni from the main drainage. The cultivation is done mainly by groundwater along with the river lift Irrigation.

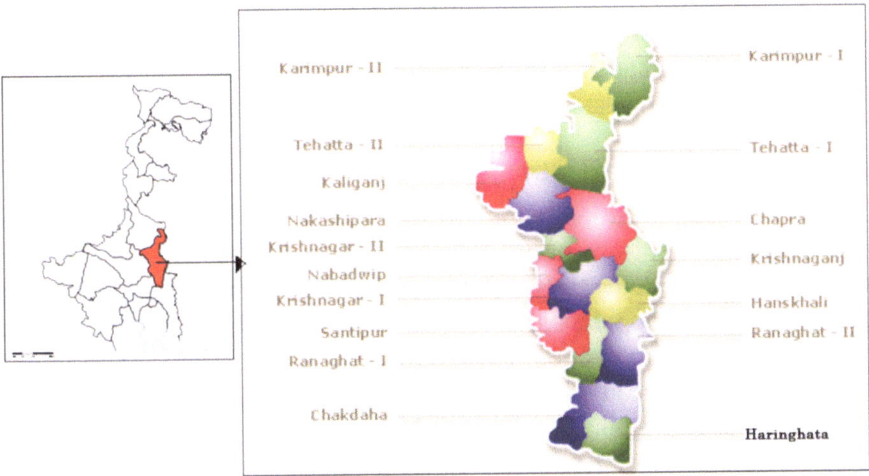

Fig.1: Location Map of Nadia district

Hydrogeology

The district covers an extensive alluvial plain characterized by younger Gangetic flood plain deposits. Elevation ranges from 28.84 meter above mean sea level to 5.40 meter above mean sea level. Numerous small rivers and abandoned river channels are interspersed throughout the district with a number of depressions, meander scars, ox-bow lakes, point bars etc.

Thick Gangetic alluvium of quaternary age is composed of sand of various grades, silt, clay, gravel and kankar and their various admixtures deposited by the Ganga River and its tributaries. It forms the main aquifer system in the district. In general three aquifers have been identified. The shallow aquifer exists down to a depth of 80m. The next aquifer occurs within the depth range of 100 to 162m and the deepest one exists within the depth range of 200m to 335m. But at places, the demarcating horizon in between first and second aquifer could not clearly be detectable owing to pinching and swelling nature of Indo-Gangetic riverine deposits. The flow of ground water is towards south and south-east.

The tube-wells tapping the aquifers of shallow and intermediate depth are yielding maximum ground water and are ranging from 3.21 to 62.11 liter per second (lps). The value of Transmissivity (T) ranges from 1487 to 8607m^2 /day.

The dynamic groundwater resources of the district has been estimated by Ground water Estimation Committee (GEC) 1997 methodology indicates that net ground water availability of the district is 198948ham, wherein existing ground water draft for irrigation' is 167617 ham. Stage of ground water development is 87%. Net ground water availability for future irrigation development has been calculated as 34996 ham. The concentration of arsenic content in ground water ranges from below detection limit (BDL) to 1.18mg/litre.

Groundwater for agricultural use

Summer rice crop (*boro*) is essentially dry season rice crop (spring-summer) under irrigated ecology contributes significantly to food basket of the country as well as making sustained self-sufficient in meeting demand of grain production for ever-increasing population. The extent of summer rice crop indicated farm level groundwater exploitation. The summer rice is being made responsible for ever-growing widespread arsenic contamination in ground water sources. Ground water development by shallow tube wells is more than deep tube wells in the Nadia district. Average depth of shallow tube wells in different blocks of the district ranges from 18 to 30 m bgl. Block wise average depth of shallow tube wells has given in Table 1.

Table 1: Average depth of shallow tube wells in different blocks of Nadia district

Sl.No.	Block Name	Average Depth of Shallow Tube Well (m bgl)
1.	Chakdah	30.2
2.	Chapra	22.36
3.	Hanskhali	20.91
4.	Haringhata	20.63
5.	Kaliganj	20.10
6.	Karimpur-I	19.41
7.	Karimpur-II	20.07
8.	Krishnaganj	21.22
9.	Krishnanagar-I	21'.26
10.	Krishnanagar-II	24.85
11.	Nabadwip	19.51
12.	Nakashipara	19.34
13.	Ranaghat-I	18.7
14.	Ranaghat-II	22.00
15.	Shantipur	21.39
16.	Tehatta-I	19.82
17.	Tehatta-II	20.18

Results

The arsenic contamination in ground water in Bengal delta were bounded by river Ganga-Padma, Hooghly-Bhagirathi (integral parts of Indo-Gangetic Basin)

and as from geogenic origins. Arsenic finds its way into ground water could be due to declining of water table as a result of indiscriminate extraction of underground water especially for extensive summer crop cultivation and lack of recharging of ground water during this period. About 90 per cent of arsenic contaminated water is used for irrigation in agricultural sectors and 10% of such water is used as drinking purpose. The concentration of arsenic was reportedly higher in roots, followed by stems and leaves while edible parts have reportedly been lowest concentration of the toxin.

Application of Farm Yard Manure (FYM) and P_2O_5 could resist the release of native arsenic and could be lowering down to moderate toxic effect in soil-plant system. Application of Zn has reportedly been reduced the concentration of arsenic in summer rice. The application of FYM and phosphate was found to have opposing effect on release of native and applied arsenic in the contaminated soils, with FYM reducing such release, thereby tending to moderate the toxic effect of arsenic in the soil–plant system. Organic manures, namely compost, poultry manure, neem cake, etc., have also been examined as ameliorants, not only for rice, but for other crops as well (e.g., potato, elephant foot yam), with encouraging results. The crop with less water requirement also might be helpful in lowering down the effect of arsenic toxcity. The crop which require less water are considered as substitute for the dominant rice-rice-fallow sequence in the affected areas to minimize the arsenic uptake by food crops. It demonstrates the significance of crop diversification during summer season.

Inclusion of pulses /other legumes /green manure crops in the cropping sequence, coupled with organic manure addition, was found helpful in moderating arsenic build-up in soil and plant parts. Blue Green Algae (*Anabaena* sp. *and Nostoc* sp.) and some bacteria were shown the effect having decontamination ability. The tube wells with activated alumina are being used to get arsenic free drinking water to mitigate the severity of the problem in these areas. Surface water is free from arsenic contamination and safe for use in crop cultivation. The efficient of judicious application of intermittent irrigation water in summer rice crop was found effective as integral part of water saving technologies.

Discussion

Arsenic in soil-plant system

The input of arsenic to soil from various sources may prove detrimental to plant as arsenic uptake would reach to the toxic limit, thereby facilitating its entry into the food chain and human body. There is also the possibility of biomagnifications of the toxin as it travels up in the food web (Sanyal and Nasar, 2002; Ghosh *et al.*, 2004). Several workers including the present authors have reported accumulation and transformation of arsenic by a number of plant species grown in the arsenic affected areas. These crops (such as elephant-foot-yam, green gram, cowpea, sesame, groundnut, etc.) tended to show a build-up of arsenic in substantial quantities in different plant parts (ICAR, 2001; ICAR, 2003; SWID, 2003). Pointed gourd, a vegetable creeper plant, has shown considerable arsenic loading when cultivated in the soils of West Bengal (Panda and Das, 2001). A number of other vegetables, namely cauliflower, tomato, bitter gourd were also noted to accumulate arsenic in their economic produce (ICAR, 2001; ICAR, 2003; SWID, 2003; Ghosh *et al.*, 2004; MoWR, 2004).The distribution of arsenic content in plant parts generally followed the order: root>stem>leaf>economic produce. In a pot-culture experiment, the critical arsenic contents in some paddy soils of China, leading to toxicity to rice plants, had also been worked out (Xie and Huang, 1998). Further, the processing of rice (i.e., parboiling and milling, etc.) was found to increase the arsenic loading in rice for both the traditional and the high yielding cultivars.

Conclusion

Arsenic toxicity of contaminated drinking water in humans in many parts of the world, including West Bengal has received enormous concern for the past many years. On the other hand, arsenic contamination in agro-ecosystem, that acts as a conduit for the passage of the toxicant to human population *via* food web, came under serious consideration only recently (Sanyal and Nasar, 2002). Though only 06 blocks of Nadia district are reported to have depleting ground water resources and falling water table, but all the 17 blocks of the district are arsenic affected. The district is blessed with 1385 mm annual rainfall mostly of which escapes as surface run-off. Summer season (*boro*) paddy cultivation demands huge ground water. Shallow tube wells within 80 meter depth are highly exploited for irrigation resulting in accumulation of arsenic in crop outputs.

References

Anon.2002. Arsenic hazards in ground water in Bengal Basin, Central Ground Water Board, Ministry ofWater Resources, Govt. of India, February, 2002.

Acharya, S.K. 1997. Consultation on Arsenic in Drinking Water and Resulting Arsenic Toxicity in India and Bangladesh, World Health Organisation, New Delhi, India, April 29- May 01, 1997.

CGWB, 1999.High Incidence of Arsenic in Groundwater in West Bengal, Central Groundwater Board, Ministry of Water Resources, Government of India, India.

Das, I. Ghosh, K. and Sanyal, S.K. 2005.*Everyman's Science,* 40 (No. 2): 115-123.

Datta, A., Sanyal, S.K. and Saha, S. 2001. *Plant and Soil,* 225: 115 – 125.

Duxbery, J.M., Mayer, A.B., Lauren, J.G. and Hassan, N. 2003. *J. Environ. Sci. Health Port A Tox. Hazard Subst. Environ. Eng.*, 38: 9-61.

Ghosh, K., Das, I. Das, D.K. and Sanyal, S.K. 2003.*Proc. National Seminar on Development in Soil Science: 2003*, 68th Annual Convention of Indian Soc. Soil Sci. held at C.S. Azad Univ. Agril. Tech., Kanpur, November 4-8, Abstr., 34.

Ghosh, K., Das, I., Saha, S., Banik, G.C., Ghosh, S., Maji, N.C. and Sanyal, S.K. 2004. *J. Indian Chem. Soc.*, 81, 1063-1072.

MoWR.2004. Interim Report: *Ascertaining arsenic mobilization in soil-water-plant system and exploring the possible remedial measures,West Bengal.* [Research Project executed at BCKV, funded by the Ministry of Water Resources, Government of India, Principal Investigator – S.K. Sanyal.

Panda, S. and Das, N.C. 2001. In: *Proceedings of the International Conference onChanging Environmental Scenario of India and Adjacent Countries since Independence*, December 29 to January 01, 2001, Kolkata, *Abstr.*, 107.

Sanyal, S.K. and Nasar, S.K.T. 2002. In: *Analysis and Practice in Water Resources Engineering for Disaster Mitigation,* New Age International (P) Limited, New Delhi, 216- 222.

Section 2

Irrigation Water Management

7

Present Global, National Perspective and Scenario of Irrigation Water Management

Ashwani Kumar

Former Director, ICAR-Indian Institute of Water Management
Bhubaneswar-751024, India
Email:ashwani_wtcer@yahoo.com

Abstract

The per capita availability of fresh water in the World is continuously decreasing. It is reported that in the World about 69 percent of total water resources developed are used for irrigation, 23 percent for industries and the balance is used by the domestic sector. About 60 percent to 70 percent of water used in domestic sector is returned as waste water. Substantial quantities of water used by industries are also returned. To meet the increased food requirements of the global population, there is constant pressure for increase in food production. Since, agricultural productivity depends on availability of water; there is increased demand of water for irrigated agriculture. On the other hand, due to increased demand of water for other sectors, quantum available for irrigation sector is getting reduced. So, for meeting the increased irrigation demand, it is necessary to economise the use of water for irrigation by adopting improved technologies and improving the water use efficiencies. For our country also, similar situation exists. It is estimated that population of the country may stabilize at around 1600 million by the year 2050. The available utilizable water resources would be inadequate to meet the future water needs of all sectors. Therefore, to meet the challenge of water shortage, it is imperative to use the water efficiently and increase the utilisable quantity by all possible means. Vigorous efforts would be needed by all concerned including people from all walks of life in water conservation effort. Water sustains life; it is primarily used for growing crops, household uses, as input to industries, power generation, recreations and for sustaining earth's ecosystems. Presently, this essential natural resource is under threat. Growing national, regional and seasonal water disparities pose serious challenges to various governments, developmental agencies and environmental communities. The challenges of growing water scarcity are heightened due to increased demand of food and fibre for the growing population, wasteful use of developed water resources, depletion of groundwater resources, degradation of land in irrigated areas, increase in pollution level of surface and ground water, etc. Creation of new water resources is highly expensive, limiting the potential for expansion. Thus, high cost of developing new water resources, rapid growth of industrial and domestic water uses and growing recognition of environmental water demands threaten the availability of irrigation water to meet the growing food demand. The scenario has become complex and urge concerted effort for best management options.

Keywords: Water resource; utilizable water; irrigation water availability

Introduction

Water sustains life; it is primarily used for growing crops, household uses, as input to industries, power generation, recreations and for sustaining earth's ecosystems. Presently, this essential natural resource is under threat. Growing national, regional and seasonal water disparities pose serious challenges to various governments, developmental agencies and environmental communities. The challenges of growing water scarcity are heightened due to increased demand of food and fibre for the growing population, wasteful use of developed water resources, depletion of groundwater resources, degradation of land in irrigated areas, increase in pollution level of surface and ground water, etc. Creation of new water resources is highly expensive, limiting the potential for expansion. Thus, high cost of developing new water resources, rapid growth of industrial and domestic water uses and growing recognition of environmental water demands threaten the availability of irrigation water to meet the growing food demand. The scenario has become complex and urge concerted effort for best management options.

Global Water Scenario

About 97.3% of the global water resource is contained in ocean that covers 70% of the globe's surface area. The remaining 2.7% is fresh water. Out of the total fresh water, 68.9% is in the form of glaciers and snow cover, 29.9% is under the earth's surface as groundwater, 0.9% is in the form of soil moisture and 0.3% is stored in lakes, reservoirs and river systems theoretically, though there is plenty of water on earth to meet all foreseeable needs, it is not evenly distributed. The world has "water have" areas like Africa and USA, which constitute one third of the land mass, while remaining two third areas are semi-arid and arid. For adequate living standards as in the western and industrialized countries, a renewable water supply of at least 2000 m^3 per person per year is necessary.

Water resources development is critical for food security in many regions of the world. Today's 250 million hectares of irrigated area, worldwide, is nearly five times the amount that existed at the beginning of the twentieth century. Poor irrigation practices accompanied by inadequate drainage have often damaged soils through over saturation and salt accumulation. It is estimated that on a global scale there are about 20–30 million hectares of irrigated lands severely affected by salinity. An additional 60–80 million hectare is affected to some extent by water logging and salinity.

More than 2,200 major and minor water-related disasters occurred in the world between 1990 and 2001. Asia and Africa were the most affected continents, with floods accounting for half of these disasters. 97 percent of all natural disaster deaths occurred in developing countries. Between 1991 and 2000, the number of people affected by natural disasters rose from 147 million per year to 211 million per year. In the same period, more than 6,65,000 people died in 2,557 natural disasters, of which 90 percent were water-related. Of these water-related disasters, floods represented about 50 percent, water-borne and vector-borne diseases about 28 percent and droughts 11 percent. Floods caused 15 percent of deaths and droughts 42 percent of all deaths from all natural disasters

National Water Scenario

The sub-continent India covers geographical area of 329 M ha, which is 2.42% of the total land area of the world. However, it supports 1027 million population (2001 census) i.e. 16.8% of the world figure. The country's average annual rainfall is 1194 mm which when considered over the geographical area of 329 M ha amounts to 392.8 M ha m (H" 400 M ha m). However, India experiences high degree of spatial variability of annual rainfall. Highest annual rainfall of 11,690 mm is recorded at Mousinram near Cherrapunji, Meghalaya, and lowest of 150 mm at Jaisalmer of Rajastan. Average precipitation of the country during monsoon (June to September) is about 300 M ha m (75% of the annual precipitation). The water resources potential of the country which occurs as natural runoff in the rivers is estimated at about 186.9 M ha m. Considering both uneven distribution of resource over space and time about 112.2 M ha m of the total potential can be put to beneficial use, 69 M ha m through surface water resources and 43.2 M ha m by groundwater.

A storage capacity of 21.28 M ha m has been created in the country taking into consideration the storage due to major and medium projects. The projects under construction will add another 7.6 M ha m and the projects under consideration will add another 10.7 M ha m. The likely storage available will therefore be 39.7 M ha m.

Groundwater played a major role in the success of Green Revolution, and it contributes 60% of the total irrigated area of the country. Annual potential recharge from rainfall in India is about 34.24 M ha m, which is 8.56% of the total annual rainfall of the country. Further, the canal irrigation system recharges the groundwater to the tune of 8.95 M ha m. The net utilizable groundwater resource is 32.5 M ha m. Groundwater potential available for domestic, industrial and other uses is 7.1 M ha m. The present level of groundwater development of the country is estimated at 41.57%. The state Haryana has exploited highest groundwater resources i.e. 112% followed by Punjab (98%) and Rajasthan (86%). The north-eastern states including the union territory of Andaman &Nicober recorded negligible groundwater development.

The agriculture is of fundamental importance in India's economy contributing about 20-21% of gross domestic product (GDP) and generates two third of the employment. Total food grain production, which was only about 51 million tones at the time of commencement of the first five year plan, has gone up to 213.45 million tones in 2003-04. But in spite of all these impressive achievements, net sown area in the country remained at about 141.1 M ha during 2000-02. Total cultivable land of the country is 183.2 M ha. The agricultural land under cultivation, culturable waste, and barren and unculturable land are 155.56, 13.12, and 17.7 M ha respectively.

Per capita water availability is continuously declining from 5176 m^3 in 1951 to 4732 m^3 in 1955, 2209 m^3 in 1991, 1820 m^3 in 2001 and 1703.6 m^3 in 2005. In recent years, India is frequently experiencing natural calamities like floods, droughts, cyclone, etc. Under the National Programme of Flood Management, 34,398 km of new embankments and 51,318 km of drainage channels have been constructed. This has resulted in providing reasonable degree of protection and reducing the intensity of flood to an area of about 14.46 M ha against the total flood prone area of 40 M ha in the country.

Historical development of irrigation in India

Irrigation in India has been practiced since time immemorial (Rao, 1980; Swaminathan, 1981) and in the available history of ancient periods; there are a number of references to the practice of irrigation from tanks, wells, canals and directly from rivers. There are also references to water management practices in Vedas and ancient Indian scriptures. There is evidence that numerous irrigation canals, big and small were dug at various times in Indus system of rivers during high water stage to irrigate the adjacent lands. Utilization of groundwater as well as surface water for irrigation purposes was first started nearly 2000 years ago. During that time, north Indian farmers depended on rivers and canals for irrigation and they did not have much faith on groundwater. On the other hand, southern state farmers used tanks and open wells for irrigating crops. So, in a way, southern state farmers were more advanced in terms of irrigation technology and governance of water resources than their rest of the India counterparts. The Grand Anicut weir across the Cauvery River in south India was built in the second century AD (Sinha *et al.*, 1985).

With establishment of British rule India, rapid development of canal irrigation system in the country started in the earlier part of nineteenth century. Modern large scale water management for irrigation in India started in the third decade of nineteenth century with the construction of Cauvery delta system in southern India and construction of two Yamuna canals in northern India. On the above works, foundation for modern canal irrigation works was laid. Canals were built on natural river channels to carry irrigation water alternatively with flood waters. Large areas became waterlogged due to seepage from poorly designed canal networks which in effect caused malaria epidemics. Costly mistakes were made in the design of river works as river hydrology was little understood. There was frequent siltation of canal leading to heavy expenditure on maintenance. But inspite of all these failures, the works were worthwhile in providing excellent lessons and valuable experience for future developments (Gulati and Smith, 1967).

Encouraged by the above works, many large scale water management works were taken up in India in 19th Century. One of them is the Ganga canal which opened in 1854 with a capacity of 8000 cusec and still is considered as a classic work. The other large water management works in the century are Godavari and Krishna canals in Southern India and upper Bari Doab and Sirhind canals in northern India. Towards the end of the century, Lower Chanab canal (now in Pakistan) was constructed which was considered one of the largest and most successful canals in the world at that time. The canal with about 10,000 cusec capacity was an enormous success which huge benefits for the government as well as new settlers. Between 1869 and 1878, the Khadakvasla Dam which was 1 mile long and 107 feet high was built on uncoursed rubble masonry to conserve water for irrigation lands in western India near Pune.

The Indian irrigation commission appointed in 1901, recommended in1903 a line of policy regarding the selection, financing and maintenance of productive irrigation works. Commission recommended proper maintenance and extension of existing irrigation projects and initiation of large number of new irrigation schemes. The Triple canal project (1905-1915), the first large scale inter-basin project in India (now in Pakistan) comprised two large weirs and one barrage across three different rivers and three large canals with the aim of transferring water from Jhelum River across Chenab and Ravi Rivers to irrigate desert areas on the other side of the latter river. Between 1920 and 1930 many large irrigation projects were constructed in India among them the largest being the Sukkur barrage project with seven canals having capacity of 47,000 cusecs. During the year 1930-45, large scale setting up of deep and shallow tubewells were done in north and central India for irrigation purpose. Well and tank irrigation caught the notice of the planners after seeing their good performance especially in the drought scenarios. A number of large irrigation projects were also completed during this period. They are Damodar canal project of Bengal, Sutlej Valley project in undivided Punjab, Nizamsagar project in Andhra Pradesh, Krishnarajasagar dam in Mysore and Sharda canal project in Uttar Pradesh.

Substantial irrigation infrastructure development had already been done during the pre-independence period. In 1947, there was independence and partition of India. During partition, 80% of population remained in India while one third of irrigated area went to Pakistan's share. Out of the Indus river command area, Pakistan got 8.5 million ha whereas India got only 2.1 million ha of irrigated land. The Sutlej valley project along with the Sukkur barrage went totally in Pakistan's share. After independence, five year plans started where major emphasis was put on strengthening the irrigation scenario. A number of big, medium and small projects were taken up. Prominent among them are Bhakra-Nangal project in Punjab, Nagarjunasagar in Andhra Pradesh, Chambal canal complex for Rajasthan and Madhya Pradesh, Kosi in Bihar, Rajasthan Canal and interstate Gandak project (Bihar and Uttar Pradesh). During the 4th five year plan emphasis was laid on conjunctive use of water as groundwater declined very much in some states and problem of waterlogging and salinity appeared in canal command areas. Due to sustained effort of creation of water resources, a large irrigation potential of 110 million hectares could be achieved by now as against 22.6 million hectares in 1950-51. This is one of the highest growths of irrigation potential in the world.

Future water resources availability

With growing urbanization in the country, there will be more requirement of water from competing sectors like domestic and industrial needs. This is likely to put enormous load on agricultural water requirement of the country. The population of India is expected to stabilize around 1640 million by the year 2050. As a result the per capita water availability will decline from 1820 m^3/year to 1140 m^3/year. For adequate living standards as in western and industrialized countries, a renewable water supply of at least 2000 m^3 per capita per year is necessary (Postel, 1992). Hence with per capita water availability of 1140 m^3/ year India will be water stressed. The total water requirement of the country around the year 2050 will be 1450 km^3/year which is significantly more than the current estimate of utilizable water resources potential of 1122 km^3/ year. Therefore, when compared with present availability of 500 km^3/ year, the water availability around 2050 need to be trebled (Gupta and Deshpande, 2004).

Surface water irrigation systems

Irrigation development is classified into three categories, major, medium and minor irrigation projects. The major and medium irrigation projects are either diversion canal or storage dam based canal irrigation systems. The minor irrigation projects are mostly groundwater based and include to a small extent the tank based irrigation systems. But the irrigation efficiency of canal irrigation system is very low ranging from 40% during peak period to as low as 10% during heavy rainfall months (Pai and Hukkeri, 1979). This means that a large amount of canal water is lost in conveyance and seepage. Availability of irrigation water is another important factor. In a majority of cases plenty of canal water is available in the rainy season, but enough water is not available in the post-monsoon season. So, in a majority of cases canal water is suitable for providing supplemental irrigation during kharif season and taking second crop from the canal water is impossible. The construction of large irrigation projects are also hugely expensive and take years to complete.

When compared with surface water groundwater is readily available at all the time if the energy needs are met. In most parts of the country, cropping intensity in tubewell irrigated areas is 200%. A study of the sources of irrigation in Punjab and Haryana showed that the yields were always higher in well irrigated field than canal irrigated fields.

Groundwater Irrigation System

Groundwater is a preferred source for agricultural use as it is renewable, easily accessible and generally free from pathogens and suspended particles. Groundwater has contributed nearly 80% of additional irrigated area created over the last two decades. However the uncontrolled use of groundwater without any regulation has brought problems of declining groundwater level and deterioration of quality in some parts. Hence, there is concern among planners about sustainability of groundwater resources. Groundwater needs to be developed on a scientific basis with proper understanding of its occurrence in time and space.

The annual replenishable groundwater resource in India is estimated as 432 BCM out of which 398 BCM is available for utilization leaving aside 34 BCM for natural discharge. The total groundwater draft is 231 BCM of which 92% (213 BCM) is used for irrigation and 8% (18 BCM) is used for domestic and industrial use. The overall stage of groundwater development is 58%. But this is not evenly distributed. The sates like Punjab, Haryana, Tamil Nadu, Gujarat, Rajasthan, Delhi have recorded very high level of groundwater development whereas groundwater development in eastern India is quite low. While in some pockets, groundwater level has gone very much down due to overexploitation, in some other regions there is waterlogging and salinity due to heavy seepage from canal water. As per an assessment, about 2.46 Mha of the area under surface water irrigation projects is waterlogged or threatened by waterlogging. Under such scenario, there is need of scientific planning of conjunctive use with optimal use of surface water and groundwater

Almost 90 per cent of the drinking water needs are met from groundwater but only 5 per cent of the total groundwater extracted is needed for domestic use. Irrigation accounts for 85 per cent, and the remaining 10 per cent is used by other sectors, including industries. The rapid increase in groundwater-based irrigation using tubewells that are more than 150 metres deep and high-power pumps in recent times is one of the many reasons for groundwater depletion and the drop in the groundwater table.

The area covered by groundwater-based irrigation has increased from 6.5 million hectares (mha) in 1950-51 to 40 mha. An annual decline in the water table of up to 2 m is considered normal and can cope with even a deficient rainfall the following year. A decline of up to 4 m is a cause for concern and above 4 m is a stress situation. The Central Ground Water Board (CGWB) describes the current situation as a stress situation. However, there are reports that unregulated overdraft has resulted in water tables dropping 10 to 20 m in several areas in Gujarat.

Technology development in deep drilling and pumping methods (even up to 250 m), on the one hand, and populist policies of free or low-cost electricity for irrigation purposes, on the other, have contributed greatly to this massive exploitation of groundwater for irrigation purposes. Coupled with the highly anomalous policy that vests rights to groundwater with the owner of the land, there is no legal limit on the volume of water that a landowner can extract. Unregulated exploitation by rich farmers with large land holdings, who can afford large pumps, leave the poor and the marginal farmers without access to even basic water needs. Landlords become water-lords, resulting in inequity and social conflicts in rural communities.

Several industries have begun to exploit groundwater in an unregulated manner. The industries are located where groundwater is indicated and groundwater gets consumed within the industrial premises for different purposes, including dilution of effluent waste. This rapid groundwater depletion has not only reduced water availability but also affected its quality in terms of excess fluoride and arsenic content and brackishness. This has made State Public Health Engineering Departments (PHEDs) abandon low-cost hand pump-based systems and instead undertake costly and complicated piped water schemes that exploit whatever little surface water that exists in these areas.

There is clearly a need to regulate groundwater extraction so that an adequate share of it is used for drinking water and irrigation purposes. Groundwater is pollution-free, is not brackish and is not lost due to evaporation even in arid regions. But nearly 90 per cent of groundwater exploitation occurs through private hands. A model Bill to regulate and control the development of groundwater has been circulated to all the States since 1970. But, except Maharashtra, no State has enacted this legislation. Gujarat, which faces severe problem of non-availability of fresh water, does have some legislation in place to regulate groundwater use. But recurrent droughts indicate that it is not enough. Implementing and monitoring groundwater use need to be strengthened. Similarly, there have been no moves to notify certain areas where groundwater extraction for private use is forbidden.

The basic problem is, therefore, one of a rapid localized overdraft of groundwater in excess of recharge of subterranean aquifers. Rainwater is the basic source of this recharge potential. This had to be conserved through proper conservation measures and recharging means and techniques such as check dams, percolation ponds and rainwater harvesting. There is a provision in the central funding to States for adopting recharge technologies and measures.

Salinity is a major problem associated with development of groundwater in coastal areas. The groundwater salinity in coastal areas can be attributed to several reasons like inherent salinity, tidal action, irrigation by saline water, up coning and lateral sea water ingress. The saline water intrusion is the result of either overdevelopment of aquifer or unscientific way of exploitation of groundwater. Arsenic and fluoride are two issues that affect the groundwater quality and can lead to health hazards. Hence in areas having poor quality groundwater, a scientific understanding of the area has to be made before development of groundwater. Arsenic problem has been observed quite prominently in West Bengal where in 8 districts, its concentration has gone beyond the permissible limit of 0.05 mg/litre (Romani, S., 2006).

Rainwater harvesting

Rainwater harvesting and groundwater recharge is one of the priority area considering the depletion of groundwater resources. Most of the rainfall in India is received through monsoon, and is obtained from July to September. The high intensity of monsoon rains coupled with poor infiltration rates of the soils result in heavy runoff. As a consequence of heavy runoff, discharge of the rivers also increases. This leads to two options, (a) facilitate groundwater recharge by rainwater harvesting and (b) divert the river water through canals by construction of dams. But as it has been discussed in previous section groundwater is more preferable than canal water as it is readily available for irrigation. Small rainwater harvesting structures have been found cheaper than major irrigation projects in terms of money spent per creation of unit hectare of irrigated land.

Irrigated ecosystem

India is unique in a sense that crops adapted to tropical, subtropical, semiarid and temperate environments can be grown in one or other parts of the countrydue to its varied climate. Crops that are primarily the source of food like wheat, rice, maize, sorghum millets and pulses are grouped as food crops. Sugarcane, coffee, cotton, tobacco, tea and others are classified as commercial crops, while groundnut, mustard, sesamum and rapeseed etc come under oilseeds category. Since 1950's food crops have accounted for 75% irrigated area whereas commercial crops and oilseeds have accounted for approximately 25% irrigated area.

Two distinct irrigated eco-systems exist in India. One is Indo-Gangetic plain comprising the states like Punjab, Haryana, Uttar Pradesh, Bihar, West Bengal and plains of Jammu and Kashmir. The other eco-system comprises coastal areas of Tamil Nadu and Andhra Pradesh. Among the states, Punjab ranks first with 95.2% cropped area under irrigation followed by Haryana (78.2%) and Uttar Pradesh (65.8%) (Yadav *et al.*, 2000). The above seven states share 32 per cent of the total geographical area, but 38% of net cropped area of the country belong to them. The average cropping intensity is 143% in these regions in comparison to national average of 131%. Approximately 50% area under cereals, 31 per cent under pulses, 74% area under sugarcane, 20% area under cotton and 45% area under groundnut in the country belongs to these states (Yadav *et al.*, 2000)

Irrigation scheduling and crop water requirement

Irrigation scheduling is done basically by three approaches. (a) meteorological approach, (b) soil indicator approach (c) plant indicator approach. Out of these three, meteorological approach and soil indicator approach are more common. A considerable work has been done on irrigation scheduling on different crops like rice, wheat, sugarcane, cotton and other crops (Hukkeri and Pandey, 1977). The recommended irrigation requirement and number of irrigations of different crops is given in Table 1. But it is difficult to generalize the recommendations in such a vast country like India. Irrigation schedules have been largely developed in experimental fields, but are rarely used precisely in farmer's field. These scientific approaches require some measurement at the field level and are difficult for farmers to adopt because of their small holdings and uncertainties in availability of water.

Table 1: Recommended irrigation requirement and number of irrigations for different crops

Crop	Irrigation requirement (mm)	Number of irrigations
Cereals and millets		
Wheat	300-400	5-6
Rice	750-1640	10-20
Sorghum	140-300	1-4
Maize: rainy season	100-350	2-5
Summer season	500-900	10-18
Pearl millet	150-250	2-3
Barley	100-300	2-5
Pulses and oilseeds		
Chickpea	80-160	1-4
Pigeonpea	70-140	1-2
Groundnut	150-750	3-10
Mustard	60-180	1-3

Food production

The food grin production of India has increased by four fold, from merely 50.8 million tones in 1950-51 to 260 million tonne by now. This spectacular growth in food production can be attributed to three factors i.e, high yielding varieties, large expansion of irrigated area and fertilizer use. With this increase in production, country is now self-sufficient in its food requirement. This increase in food production has been achieved not only by expansion of cultivated land, but also by increasing the productivity per unit area. However, the average productivity of irrigated land is still low in comparison to international standards. There is still a lot of gap between the crop yield in experimental plots and farmer's field. The total food production can further be increased by attaining the full potential of productivity. Most of the increase in food production has been achieved through production of food grain mainly wheat and to some extent rice. Hence, there is a need for crop diversification and more focus on horticulture.

Irrigated vs. rainfed agriculture

The issue of water and food production has until recently been concentrated mostly on irrigated agriculture, i.e. addition of so called blue water. But most of the countries in the world depend for more than 60% of their cereal production on natural rainfall infiltration i.e so called green water (Rockstorm, 2001). But, now there is a decreasing relevance of clear cut division between irrigated and rainfed agriculture. Irrigated agriculture is partly dependant on infiltrated rain and at the same time rainfed agriculture is also partly dependant on life saving irrigation. In present upgrading of rainfed agriculture, small scale farming is being increasingly supported by supplementary irrigation for the purpose of dry spell mitigation. Therefore, future development solutions tend to be found in between the two extremes of purely irrigated and purely rainfed agriculture.

Water use efficiency

Water losses in the irrigation system comprises of seepage losses and nonproductive evaporation losses from canals and irrigated fields. The seepage losses from canals and irrigated fields can be termed as blue water losses. But in many cases, the seepage and percolation losses from canals and irrigated fields i.e. blue water losses return to the basin and can be reused and hence they are imaginary loses. On the other hand, the green water losses in terms of pure non-productive evaporation losses from canals and irrigated fields are the true losses. On a catchment scale, the green water losses are true losses whereas the return flows are imaginary losses. Hence water use efficiency on a basin scale should only consider the green water losses and not blue water losses.

Surface water deficiency

In the 70 important water reservoirs in India, the storage position is officially described as the lowest in a decade. Ground water levels have fallen considerably in the eight droughts hit states. In a number of districts, says the nodal agriculture ministry, the fall in water levels is at the rate of over 2 metres a year- this includes eight districts in Chattisgarh, 13 in Gujarat, 30 in Madhya Pradesh, 18 in Orissa and 15 in Rajasthan. (*Source:*Catchwater, a CSE newsletter, June 2001). It is interesting to look at the rainfall data for the last couple of years in the regions affected by the current drought - west and east Rajasthan, Gujarat region and Saurashtra, Kutch and Diu, Telengana, Rayalaseema and coastal Andhra Pradesh. In all these regions, the deficiency in the southwest monsoon, which accounts for nearly 90 per cent of the rainfall and hence is the main source of water, was itself not all that severe. However, the monsoon rainfall last year was below normal in all the regions. In fact, the deficiency of monsoon rainfall in regions of Andhra Pradesh was not much at all. Though the immediate post-monsoon (October-December) rainfall was above normal in all the regions except Andhra Pradesh, the winter (January-February) and pre-monsoon (March-May) rainfall during this year has been severely deficient.

Water deficit in the Western States of India

Rainfall deficiency is much less than what it was in the worst drought years of 1987 and 1974. But the ground situation is that the drought is much more widespread in the regions of Gujarat and Rajasthan, which are experiencing a severe "hydrological drought", if not a "meteorological drought". The current drought actually is even more serious than if it was a case of severe meteorological drought leading to a severe hydrological drought. It points to the fact that withdrawal of water from the various resources of the region - which get recharged only by rainfall - has been unsustainably high in recent times leading to a drastic depletion of these resources. The situation is a reflection of a complete failure in water management and a lack of planning in these water-scarce regions. This hydrological drought may result in a severe agricultural drought as well if the monsoon turns out to be deficient this year too for these regions. Preliminary data on some of the relevant meteorological factors up to April do seem to be unfavourable for a good monsoon and even statistically, given the string of 11 "normal" monsoons, the probability for a below-normal monsoon for the country as a whole is high. And, given that these arid and semi-arid regions of western India are drought-prone, deficient rainfall in these regions is indeed likely again this year. Long-range rainfall forecast - in terms of spatial and temporal distribution during the monsoon season - therefore assumes importance.

Green water productivity

The productivity of water used in agriculture increased by at least 100% between 1961 and 2001, mainly due to increases in crop yields. Irrigated rice yields doubled and rain-fed wheat yields rose by 160% in that period, with little variation in water consumption per kilo of output. Globally, FAO estimates that water needs for food per capita halved between 1961 and 2001, a significant saving and an equally significant gain for other water users. By one calculation, a 1% increase in water productivity in food production makes available - in theory, at least - an extra 24 litres a day per head of population, while a 10% increase would equal current domestic water consumption. Investing in agriculture and in agricultural water management, therefore, is an attractive strategy for freeing water for other purposes.

Green water productivity can be increased by rainwater harvesting, improved land management techniques, supplemental and micro irrigation systems, rainfed agriculture

Fishery sector water

With a long 8129-km coastline, 0.5 million km^2 of continental shelf and 2.02 million km^2 of exclusive economic zone (EEZ), India is second in inland fish production and fourth in global fish production. The total fish production increased from 0.75 million tonne in 1950-51 to 6.3 million tonne in 2003-2004. Freshwater aquaculture production increased by 4.5 fold since 1990 (66% of total inland fish production). Over the last four and one-half decades, the annual growth rate in inland fisheries was 11% and that of the marine segment was 9.0%. In India, the total available water body is 6.64million ha. Out of this 2.85 million ha has been intervened while 3.79 millon ha is yet to be exploited.

Industry sector water

Scenario says that agriculture accounts for 80-90% of water (irrigation) (Kumar 2005), while industry accounts for 7-8% and drinking water accounts for 4-5% of available water. There is increasing industrial demand for water while demand in rural areas is expected to increase sharply with the development.

Surface reservoirs

According to the records of the Government of India, reservoirs are generally classified as small (<1000 ha), medium (1000 to 5000 ha) and large (>5000 ha). All man-made impoundments, created by erecting a dam of any description in order to obstruct the surface flow of a river, stream or any watercourse, have been classified as reservoirs. However, water bodies of less than 10 ha in area, being too small to be considered lakes, are excluded.

Table 1: Reservoirs classified as small (<1000 ha), medium (1000 to 5000 ha) and large (>5000 ha). Surface area (ha) of reservoirs by states

State	Small	Medium	Large	Total
Tamil Nadu	315 941	19 577	23 222	358 740
Karnataka	228 657	29 078	179 556	437 291
Madhya Pradesh	172 575	169 502	118 307	460 384
Andhra Pradesh	201 927	66 429	190 151	458 507
Maharashtra	119 515	39 181	115 054	273 750
Gujarat	84 124	57 748	144 358	286 230
Bihar	12 461	12 523	71 711	96 695
Orissa	66 047	12 748	119 403	198 198
Kerala	7 975	15 500	6 160	29 635
Uttar Pradesh	218 651	44 993	71 196	334 840
Rajasthan	54 231	49 827	49 386	153 444
Himachal Pradesh	200	-	41 364	41 564
Northeast	2 239	5 835	-	8 074
Haryana	282	-	-	282
West Bengal	732	4 600	10 400	15 732
Total	1 485 557	527 541	1 140 268	3 153 366

The State of Tamil Nadu has the greatest surface area of small reservoirs in the country (315941ha), followed by Karnataka (228657 ha) and Utlar Pradesh (218 651 ha). Medium reservoirs constitute less than 1% of the total number of reservoirs and 17% of the total surface area. Andhra Pradesh, Rajasthan, and Gujarat have more medium reservoirs than Madhya Pradesh, though the water surface area in these states is much less. Karnataka has a preponderance in number (12) of large reservoirs. Nevertheless, the seven large reservoirs in Andhra Pradesh are much larger and have a total surface area of 190151 ha.

Status of Canal Irrigation Efficiencies

Irrigation efficiency mainly depends on loss of water in conveyance system including distribution system and in the field. Loss of water in conveyance system is mainly due to seepage and evaporation. Evaporation loss takes place from the exposed water surface In hot and dry summer months these losses are maximum but they may seldom exceed about 10 percent of the total conveyance loss. Conveyance and field losses, contributing to overall irrigation efficiency from source to field application, vary widely with the extent of lining, type of strata through which canal system passes, material used and quality of work in the canal lining, preparation of the field, type of soil, stream size, method of water application i.e irrigation practices in vogue etc.

Some studies have been carried out to have a general idea about the project features which influence the irrigation efficiencies. General conclusion of these studies was:

- Area served by one distributary ranged between 400 ha to 3000 ha, conveyance efficiency was higher than in schemes where this area was larger or smaller.
- In schemes where the tertiary units were larger than 200 ha, the average conveyance efficiency was higher than in schemes with tertiary units between 5 and 100 ha.
- In schemes where the size of flow per farm inlet was more than 50 litre/second, the distribution efficiency was higher than in schemes where the size of flow was 50 litre/second or less.

Some data of the existing Indian projects has been collected and efficiencies observed in these projects are tabulated below:

Table 2: Observed irrigation efficiencies of projects

S No	Particulars	Observed irrigation efficiencies of projects (%)					
		1	2	3	4	5	Average
1.	Conveyance efficiency						
	a) Main canal	92	84	85	94	86	88
	b) Distributary/Minor	79	83	75	75	89	80
	c) Field channel	68	64	72	No field Channel	67	68
	Overall conveyance efficiency (EC)	49	44	46	70	51	48
2.	Field Application efficiency (Ea)	77	69	83	58	59(Paddy) 52 (Groundnut)	68
3.	Project Efficiency (Ep)	38	31	38	41	30	36

Some additional information as available in Performance Evaluation reports/modernization schemes reports has been reproduced in the following table:

Table 3: Seepage losses in Canal system

S No	Project	Seepage losses in Canal system (Cusec per Million Sq. ft.)
1	Harsi Canal System(M.P)	5 to 24
2	Sarda Canal System (U.P)	6 to 8
3	Kaldiya Irrigation Scheme (Assam)	17 to 27
4	Pazhassi Irrigation Project (Kerala)	11 to 21
5	Kangsabati Irrigation Project (W.B)	9 to 70

It is observed that such losses are 40 to 50 percent of the discharge at canal head. This includes 15 percent to 20 percent in main canal and branches, 10 to 20 percent in distributaries and 10-20 percent in water courses. In general, it can be said that a considerable amount of water is lost through evaporation and seepage in the canal system from the head of the canal up to the distributaries outlet.

National Commission on Integrated Water Resources Development Plan (NCIWRDP) also tried to compile the data for national level assessment of overall irrigation efficiencies but they could not come to any conclusion. However, the National Commission opined that 35 to 40 percent efficiency in surface water and 65-70 percent efficiency in ground water will be a fair approximation.

Present Efficiencies

The development of irrigation sector, like a few other development activities has its share of problems also. A number of irrigation systems, are operating much below their potential due to poor maintenance, problems of equity, waterlogging, lag in potential created and utilized etc., all leading to actual irrigation efficiencies being much less than the achievable values. This is not to say, that problems of existing irrigation systems can be limited to those related to irrigation efficiency and equity. At times the water availability itself may be inadequate or may have become inadequate due to excessive upstream development. The cropping pattern may have changed from those for which the system was planned. While all these problems need specific consideration, particular problem which is common to all the systems and required to be tackled is the low water use efficiency.The irrigation efficiencies of 35-40 percent in surface water as generally obtained, are considered to be low.

Improving irrigation management

Irrigation system management can be improved to provide more reliable water supply to farmers through storage and improved operation of reservoirs, better distribution of water with improved control structures as well as more responsive management. More reliable water supply allows farmers to invest in better on-farm water management such as better land leveling, zero tillage, or pressure irrigation. Improved management usually requires improved institutions as well as improved technologies. Governments rarely manage to operate and maintain irrigation systems successfully on their own. Farmer participation increases success rates. Schemes have to be financially self-sufficient on operation and maintenance and farmers can, and do, pay for well-performing systems.

Multiple use systems

Single systems for domestic use, agriculture, aquaculture, agroforestry and livestock

Rural people often do not distinguish between water for domestic or livelihood purposes. However, water projects and water experts usually still focus on a single purpose. Increased value can be captured by designing, planning and managing projects in integrated manner.

Enhancing the safe and productive use of wastewater in irrigated agriculture

Growing water demands of rapidly expanding urban areas also create the opportunity to re-use the equally growing wastewater flows. Making an asset out of wastewater for peri-urban small-scale farmers may make sanitation affordable for poor urban dwellers. Development of appropriate treatment systems to make the wastewater biologically safe, while keeping the nutrients that replace fertilizer for farmers, is the challenge. Potential benefits include improved health in urban slums, livelihoods for peri-urban farmers, improved nutrition (vegetables) for the urban poor and reduced pollution.

Water saving through water conservation techniques and making in vogue an efficient system of irrigation would help to avert the negative impact of water deficiency. Environmental concerns must be part of modernization in water use and management. Extraction from rivers and lakes and the construction of irrigation infrastructure invariably displaces natural wetlands, which are highly productive components of agro-ecological systems. Drainage from irrigation often results in loss of water quality, the spread of water-related diseases and soil degradation through waterlogging and salinization. To reduce these impacts, FAO says, modern water management needs to be based on strategic environmental assessments and cost-benefit analysis, constant environmental monitoring and integration of irrigation into the wider environmental context. Hundreds of thousands of fields have become unproductive in the irrigated areas due to widespread problem of water-logging and salinity. Corrective measures are needed if these losses are to be curtailed. These include the lining of canals and water channels.

The sugarcane growing areas along the path of the canal consume enormous amounts of the water. Water conservation measures, demand management, allocation of water for different uses and involvement of communities and local participation in water management have never been given adequate importance.

Region wise Irrigation Water Management Issues

Eastern region

Total geographical area of the eastern part of India is 73.60 M ha, which is 22% of the total geographical area of the country. The region, as per 2001 census, had a population of 353.46 million, which is 34% of the total India's population. About 45% (33.60 M ha) is the net cultivated area as against 46% (141 M ha) cultivated area of the country. Average cropping intensity in eastern India is about 143%.

The average annual rainfall in this region varies from1091 to 2477 mm with a regional average of 1526 mm. Bulk of the rain (about 80%) occurs during the monsoon period. It has erratic temporal and spatial distribution with considerable year-to-year variation.

The eastern region is bestowed with good quality natural resources, but poor management has led to a situation where people of this resource rich region are poor. Ultimate irrigation potential from major and medium irrigation projects in eastern region has been estimated at 19.66 M ha m as against 58.48 M ha m for the whole country. In eastern region, an ultimate minor irrigation potential of 33.07 M ha has been assessed as compared to 54.8 M ha for the whole country. Thus the eastern region possesses total irrigation potential of 53 M ha, which constitutes about 38% of the country's total potential of 140 M ha. Total utilizable groundwater resource for irrigation is estimated at 19.5 M ha m per year. However, average utilization of this resource is only 34.84%. At present net irrigated area in eastern India is 14.4 M ha.

Effective water management in this region is influenced by following characteristics:

- Major irrigation projects of the plain region on Sharda, Gandak, Kosi, Sone and Mahanadi rivers are barrage systems and therefore suffer from the problems of timely supply of water due to fluctuating flow at the source.
- Several medium irrigation projects of the eastern plains are pump canal systems.
- In several canal commands practicing rice based cropping system, seepage from canal network along with the faulty water distribution system and field to field irrigation have resulted in tail end deprivation, low irrigation efficiency, and reduction in net irrigated area. It has also caused water logging in the canal commands and degradation of basic soil and water resources.
- The region has an extensive network of rivers, *nallas* and lakes from which water could be lifted for irrigation by installing appropriate water lifting systems.
- Ample scope is there for further exploitation of groundwater resource as average utilization varies from 6% in Chhattisgarh and 9% in Jharkhand to 46% in Bihar and 47% in eastern UP.
- Under-exploitation of groundwater and rising water table in the canal commands have reduced the underground space for water storage: consequently, there is reduction in groundwater recharge. The reduced amount of natural recharge adds to the rainfall excess, thereby aggravating floods and flood related damages.
- As most of the cultivated areas do not have provision for assured irrigation, even a short spell of drought adversely affects the stability of agricultural production thereby resulting in low productivity.
- In the hills and plateau regions, adoption of appropriate watershed management approach and technology along with the storage of water in small reservoirs is the key for irrigated and sustainable agriculture.

Western region

Western part of India covers 39% of the total geographical area of the country. The region, as per 2001 census, had a population of 265.9 million, which is 26% of the total India's population. Out of the 129.3 M ha geographical area of the western region, about 46% (59.1 M ha) is the net cultivated area. Cropping intensity is 124%.

The average annual rainfall in this region varies from 150 to 3620 mm with a regional average of 1195 mm. Konkan and Goa receives 2831 mm average annual rainfall followed by Gujarat (1057 mm), Maharastra (916 mm), Madhya Pradesh (903 mm) and Rajasthan (883 mm). The Konkan region receives very high precipitation during monsoon.Cyclonic depression induced very high intensity rainfall is a common phenomena in south Gujarat and Coastal parts of Maharastra and Goa leading to excess water situation in flood plains and lowlying areas.

The total net irrigated area in this part of India is 15.86 M ha. Total utilizable groundwater resource for irrigation in the region is estimated at 8.32 M ha m per year.The average ground water development in the region is 38%.

The issues related to effective water management in this region are as follows:

- The region is having considerable cultivable area under poor irrigation management. Thus, even a short drought spell adversely affects agricultural production in the region.
- There is insignificant natural recharge to groundwater leading to depletion of groundwater resource. Water-table in the region is very deep which at places has gone down to 130 m.
- In sharp contrast, the Indira Gandhi NaharPariyojana command area faces the problem of waterlogging and salinity.
- Vertisols with high swell-shrink potential, constitute large area in Maharashtra and Gujarat. Soils are fertile and retain very high amount of water, but pose serious drainage problems if not managed properly. Due to intrinsic potential micro irrigation is very successful on these soils.
- Medium and coarse textured soils of Rajasthan and Kutch of Gujarat are poor in fertility and retain water for a very short period. For effective water management, special efforts to maintain organic carbon pool will be of great importance.
- Difficulties in realization of the ultimate potential since suitable sites for dams are fewer, irrigation development is cost prohibitive, and the environmental concerns are strong.

- There is significant degradation of existing irrigated crop land of high production potential due to waterlogging and soil salinisation. In addition to production loss, the irrigation potential created from this land at a huge cost remains unutilized.
- Rainfed agriculture suffers from low agricultural productivity. Even though annual rainfall is reasonably sufficient the agricultural production suffers due to water stress on one hand and damage due to flooding on the other hand. Better water harvesting, simultaneous shift to improve water management technologies and adoption of suitable cropping system under overall framework of watershed management are important to improve the situation.

Northern region

Total geographical area of the northern region is 34.21 M ha (excluding J & K), which is 10% of the total geographical area of the country. The region, as per 2001 census, had a population of 177.4 million, which is 17% of the total India's population. About 59% (19.6 M ha excluding J & K) of total area is the net cultivated area with cropping intensity 172%.

The average annual rainfall in this region varies from 627 mm to 1660 mm with a regional average of 988 mm. Total utilizable groundwater resource for irrigation in the region is estimated at 10.24 M ha m per year and extent of groundwater development is 44%. Net irrigated area in northern India is 15.98 M ha.

The effective water management in this region is influenced by following facts:

- States like Punjab and Haryana overexploit groundwater resources resulting in lowering of groundwater table. Groundwater quality is also of serious concern.
- Because of exhaustive cropping systems and intensive cultivation, soils of Haryana, Punjab and Western Uttar Pradesh exhibit poor soil quality indicators. Injudicious use of irrigation water and irrigation with poor quality water lead to soil sodification.
- In canal commands, better recharge of ground water through the canal system has prompted farmers to utilize ground water resources that led to overexploitation of ground water and decline in water table in many areas.
- Overdraft of ground water at rates higher than recharge has caused increase in pumping lift and costs due to the lowered water table which implies greater disadvantage to the resource poor farmers.

- Implication of over mining of ground water is intrusion of saline water into aquifer in coastal areas resulting in degradation of water quality or can lead to subsidence in some cases.
- With increased irrigation intensity, the problem of water logging and soil salinity will aggravate and get further accentuated with adoption of unsustainable cropping such as cultivation of rice in northern India.
- Energy shortage will be a dominant aspect of water resource management. Rechargeable ground water use, despite being more efficient, may be less sustainable in the long run because of shortage of energy and increasing cost of pumping.

Southern region

Total geographical area of the southern region is 19% of the total geographical area of the country. The region had a population of 223.4 million, which is 22% of the total India's population (2001 census). Out of the 61.3 M ha geographical area of the southern region, about 44% (26.8 M ha) is the net cultivated area as against 46% (141 M ha) cultivated area of the country.

Average annual rainfall ranges from 891 to 2777 mm regional average of 1593 mm. Total utilizable groundwater resource for irrigation in the region is estimated at 7.28 M ha m per year. Net irrigated area in southern India is 8.57 M ha, out of which the net irrigated area in Andhra Pradesh is 3.63 million ha, followed by Karnataka (2.38 M ha), Tamil Nadu (2.18 M ha), and Kerala (0.38 million ha).

The effective water management in this region is influenced by following issues of concerns:

- The hard rock formations in the region with extreme heterogeneity, lacks primary porosity; and have restricted storage capability. The groundwater is restricted in areas where the structure of the rock formation favours storage and transmission.
- South India is predominantly a hard rock region with low water availability and excessive withdrawal of water is causing fast decline in water table.
- In several canal commands practicing rice based cropping system, seepage from canal network along with the faulty water distribution system and field to field irrigation have resulted in tail end deprivation, low irrigation efficiency, and reduction in net irrigated area. It has also caused water logging in the canal commands and degradation of basic soil and water resources particularly in the states of Karnataka and AP.

- Due to paucity of assured irrigation, even a short spell of drought adversely affects agricultural production, thereby resulting in low productivity.

Issues related to water management

Impediments

- Demand for more water
- Inefficient use and low productivity of surface and rainwater resources
- Overexploitation of groundwater and declining water tables
- Pollution and contamination of surface and groundwater resources
- Waterlogging and secondary salinisation in canal commands.
- Continuing degradation of land, water and biotic resources
- Weak and inadequate institutions for water governance and management
- Inadequate integration of crops, fisheries and other production sectors

Efforts

- Recycling of water
- Water saving through water conservation techniques
- Environmental concerns must be part of modernization in water use and management
- Corrective measures for waterlogging and secondary salinization in irrigation command areas
- Development of holistic integrated water resource management system.
- Scientific planning of conjunctive use with optimal use of surafe water and groundwater
- Multiple use of water.

Recycling of water

Recycling water is used for agriculture; industries with a principle to adopt that, no user should be allowed to put any polluted water into a fresh-water body. The user that pollutes the water should clean up the water before it is put back into the system. This principle has been used across the world for a very long time. It has to be done at every level. We have to "make water" by saving and cleaning water from basic domestic units, right up to managing entire systems.

Demand for more water

The entire world is going on to use less water and grow more crops. Historically, water management for agriculture was equated with the development and operation of water systems and structures, largely for irrigation. However, the rapid growth of urban centers and industry has led to increasing competition for water across sectors. Thus, the key challenge now for agricultural water management is achieving "more crop per drop" % an approach that marked a paradigm shift in IWMI's thinking on how to increase food production for a growing population while simultaneously meeting the water quality and quantity requirements of other economic and environmental sectors. In agriculture, it takes an average of 3000 litres of water to grow one kilogram of rice in the Indian subcontinent. Typically, 20 litres are consumed per person in developing countries.

Water saving through water conservation techniques and making in vogue an efficient system of irrigation would help to avert the negative impact of water deficiency. Environmental concerns must be part of modernization in water use and management. Extraction from rivers and lakes and the construction of irrigation infrastructure invariably displaces natural wetlands, which are highly productive components of agro-ecological systems. Drainage from irrigation often results in loss of water quality, the spread of water-related diseases and soil degradation through waterlogging and salinization. To reduce these impacts, FAO says, modern water management needs to be based on strategic environmental assessments and cost-benefit analysis, constant environmental monitoring and integration of irrigation into the wider environmental context. Hundreds of thousands of fields have become unproductive in the irrigated areas due to widespread problem of water-logging and salinity. Corrective measures are needed if these losses are to be curtailed. These include the lining of canals and water channels.

The sugarcane growing areas along the path of the canal consume enormous amounts of the water. Water conservation measures, demand management, allocation of water for different uses and involvement of communities and local participation in water management have never been given adequate importance.

Water management under climate change scenario

Agriculture, water and poverty are inherently linked. It is highly likely that the rising water insecurity from climate change will mainly affect the poorest and most vulnerable by further limiting their access to the diminishing quantity of water available for domestic and productive purposes. Apart from affecting agriculturalists, water scarcity will also have a deleterious impact on agriculture-related activities like animal husbandry and fisheries. Uncertainty, risks and farmers distress has increased due to higher frequency and intensity of extreme weather events like heat waves, cold waves, unseasonal rains/snow fall, floods, hailstorms, cyclones, tsunamis, etc. Since climate change is now an inevitable prospect, we must also consider how best to adapt it, so as to minimize its harmful consequences. Global warming, melting of glaciers, altered river flows, redistribution in rainfall pattern, rise in sea level and ingress of saline sea water into coastal regions calls upon adaptation and mitigation technologies.

Improved physical infrastructure can afford some protection against phenomenon associated with climate change such as floods, extreme weather events or coastal erosion. A switch to new crops, seeds or agricultural practices can moderate the impacts on agriculture of changes in temperature and water availability. Education, training and rural extension services can facilitate adaptation efforts. Improved weather and flood forecasting and better communications can assist evacuation, relief and rehabilitation.

For a developing country, rapid economic and social development is an essential requirement for successful long-term adaptation to climate change. For example, in the agriculture sector, a wide variety of adaptive actions are being taken to lessen or overcome adverse effects of climate change on agriculture. At the level of farms, adjustments may include the introduction of short duration crop varieties or species, switching cropping sequences, sowing earlier, adjusting timing of field operations, conserving soil moisture through conservation tillage etc., and improving irrigation efficiency. Some options such as switching crop varieties may be inexpensive while others, such as introducing irrigation (especially high-efficiency, water-conserving technologies), involve heavy expenditure

New initiations in Irrigation water management Programs

- Policy on extraction of groundwater
- Policy on prevention of polluting waterbodies.
- Promotion of multi-disciplinary integrated approach to catchment treatment
- Improvement of land capability and moisture regime in the watersheds.
- Promotion of land use to match land capability
- Integration of water use and land use policies.
- Implementation of rotational water distribution system and supply of water on a volumetric basis.

- Policies on involving farmers in management of irrigation systems (water distribution and collection of water rates).

Apart from this major activities to implement the Sustainable Agriculture and Rural Development (SARD) policy are as follows:

1. Development of crops based on regionally differentiated strategy
2. Development of Horticultural crops
3. Adequate and timely delivery of core inputs
4. Integrated Pest Management
5. Greater use of bio-fertilizers and bio-technology
6. National Agricultural Technology Project
7. Rained farming and Watershed Management
8. Soil and Water Conservation
9. Animal Husbandry and dairying
10. Development of fisheries
11. Agricultural research and education
12. Development of Human resources

The major thrust of the agricultural development programs in India is on improving the efficiency in the use of scarce natural resources, namely, land, water and energy. This can be achieved only through improved productivity in a cost-effective manner, which alone could increase the welfare of the farmers and agricultural labor. Balanced and integrated use of fertilizers, agricultural credit, institutional support, accelerated investments in agriculture, enhancing the competitiveness of agro-exports, creation of additional irrigation facilities etc. have been given encouragement through various schemes and activities of the Government of India.

A wide range of approaches have been employed to address problems of land degradation, some of which include:

- Prevention of soil loss from the catchments
- Promotion of multi-disciplinary integrated approach to catchment treatment.
- Improvement of land capability and moisture regime in the watersheds.
- Promotion of land use to match land capability

- Reduction of run-off from the catchments to reduce peak flow into the river system.
- Upgrading of skills in the planning and execution of watershed development programme.
- Increase of productivity of land affected by alkalinity for increasing sustainable agriculture production.
- Identification of critical degraded areas,
- Generation of data on land suitability and capability for regulating land use.
- Preparation of soil resource map and inventory of soil and land resources.
- Development of technical skills in soil and water conservation
- Building up and strengthening of land capability of State Land Use Boards.

Conclusion

It is evident that the water availability for agriculture is declining and to enhance agricultural production more irrigation potential needs to be created. The concerted efforts of Governmental agencies is required in increasing the overall water use efficiency at system level which would be achieved through various measures like timely execution of projects, minimizing the losses, better operational efficiency through stake holders participation, implementation of on-farm water management technologies, conjunctive use of water and changes in irrigation policy. Simultaneously, the efforts of R&D institutions are required in development of water management technologies, suitable database development, economic studies of various irrigation systems, policy guidelines for on-farm water management and adoption of participatory irrigation management. The serious efforts of developmental agencies as well as research institutes are required to develop a suitable water productivity Plan for various regions in the country for its implementation.

References

Gulati, N. D., Smith, W.C. 1967.Integrated Agriculture: An Historical Review, *Irrigation of Agricultural Lands*. Ed. R. M. Hagan, H. R. Haise, T.W. Edminister, No. 11 in the series Agronomy, Publisher- Madison, Wisconsin, USA.

Gupta, S.K. and Deshpande, R.D. 2004. Water for India in 2050: First-order assessment of available options. *Current Science,* 86(9): 1216-1224.

Hukkeri, S.B. and Pandey, S. L. 1977.Water requirement and irrigation of crops. In *Water Requirement and Irrigation Managementof Crops in India*pp 163-273. Mongraph no. 4 (new series). Water Technolgy Centre, Indian Agricultural Research Institute, New Delhi.

Kumar, Rakesh, R.D.Singh and K.D.Sharma.2005."Water Resources of India".CurrentScience,Vol 89,No-5,pp 794-811.

Postel, S. 1992.Last Oasis.Worldwatch Institute, Washington DC.

Rao, G.V.K. 1980.Irrigation development in India-tasks for future.*LalBahadurShastriMem.Lect., 12th,* Indian Agricultural Research Institute, New Delhi.

Rockstorm, J. (2001) Green water security for food makers of tomorrow: Windows of opportunity for drought prone Savannahs. In *Water Security for 21st Century- Innovative approaches.*Stockholm Water Symposium 2000. Water, Science and Technology, 43: 71-78.

Romani, S. 2006. Economic sustainability of groundwater resources- Need for scientific management. Key note address, *Groundwater-2006, All India Seminar on Role of Groundwater in national Economy* held on 18th January' 2006 at Bangalore.

Sinha, S. K., Aggarwal, P.K. and Chopra, R. K. 1985. Irrigation in India: A physiological and phonological approach to water management in grain crops.,*Advances in Irrigation, Volume -3,* Edited by D. Hillel, Academic Press, Orlando, Florida. pp 130-212.

Swaminathan, M. S. 1981. Irrigation and our Agricultural Future. *A. N. Khosla Lect.*, 1st University, Roorkee.

Yadav, R. L., Singh, S. R., Prasad, K., Dwivedi, B. S., Batta, R. K., Singh, A.K., Patil, N. G. and Chaudhary, S. K. 2000. Management of irrigated ecosystem, *Natural Resource Management for Agricultural Production in India,* Ed. J. S. P. Yadav and Singh, G. B., pp 775-870.

8

Gravity Drip Fertigation on Crop, Water Productivity, Nutrient Use Efficiency and Economics Towards Livelihood Security in Bengal Basin

Sanmay Kumar Patra

Professor and Chief Scientist, AICPR Irrigation on Water Management
Bidhan Chandra Krishi Viswavidyalaya, Gayeshpur 741234
West Bengal, India
E-mail: sanmaykumar@rediffmail.com

Abstract

Water is the elixir of life, livelihood, food security and indispensable to all economic activities. It is going to be a constraint in agriculture in near future due to the competitive demands from other sectors. Agriculture is the largest user of limited water resources. The Gangetic Bengal basin once described as the areas of excess water now suffers from acute dearth of water during the drier months. Water management holds the key for maximum agricultural production through optimizing the utilization of available water resources in a systematic manner. Drip-irrigation provides unique agronomic, water and energy conservation benefits that address many of the challenges facing irrigated agriculture. In intensive agriculture, both fertilizer and irrigation management have contributed immensely in increasing the yield and quality of crops. The method of fertilizer and irrigation application affects the efficiency. Drip-fertigation is the most efficient and less expensive agronomically sound method of fertilizer application along with irrigation water. It has the potential to ensure that the right combination of water and nutrient is available at the active root zone, in synchrony of the plants' total and temporal requirements and stabilizes the yield and quality of produce. Fertigation saves fertilizer nutrients as it permits applying fertilizer in small quantities at a time matching with the plants' nutrients' need. To address these problems, AICRP on Irrigation Water Management,

BCKV center has developed many cost-effective gravity drip irrigation and fertigation technologies on different fruits, vegetables and flowers for increasing crop and water productivity and higher nutrient use efficiency. These low-cost technologies could provide higher income generation and accessible to the farmers. A concerted effort is therefore needed for its successful adoption on larger scale for deriving maximum benefits in terms of crop productivity, input use efficiency and corresponding economic returns.

***Keywords*:** Gravity drip fertigation; crop production; water productivity; nutrient use efficiency

Introduction

Water is one of the most important non-renewable natural resources. Among all the natural resources, water is considered the prime importance due its multifaceted requirements for agriculture, domestic, industry, energy and ecology of the nation. It is the elixir of life, livelihood, food security and sustainable development. It is becoming an increasingly scarce and valuable resource. Water fulfils most basic human and ecological needs and is indispensable to almost all economic activities, including agriculture, energy production, industry, and mining. It has greater impacts on health, gender equity, education, livelihoodand the alleviation of poverty. Our planet is apparently rich in water. About 97.5 per cent of its water resource is salt water which is unsuitable for drinking or irrigation purposes. The volume of fresh water available is only 2.5 per cent of the total and that too is not readily accessible. About 68.7 per cent of fresh water is frozen in glaciers and polar ice caps in high latitude and altitude and 29.9 per cent is remaining as ground water and soil moisture. India receives annual precipitation of 4000 billion cubic metres constituting only 4 per cent of world's renewable water resources with 2.4 per cent of world's land area to mitigate the demand of almost 16.3 per cent of the world's population and 18.0 per cent of livestock population. With rapid growing population, improving living standards and social developmental activities, the pressure on limited fresh water resources is increasing and per capita availability of water resources is reducing day by day and the demand for fresh water is exponentially increasing. Water is going to be the scarce commodity in coming years due to the over rising population growth, rapid urbanization and industrialization, expansion of irrigation networks and economic development. Growing water scarcity along with widespread misuse of freshwater also pose serious threats to the sustainable development. Competition among agriculture, industry and domestic sectors for limited water resources is already constraining development activities in India. Rainfall is the maximum contributor enriching both the surface and groundwater resources. Ironically, due to spatial and temporal variability in precipitation, the country faces the problem of flood and drought syndrome almost every year. Over-exploitation of groundwater is leading to reduction of water flows in the rivers, declining of the groundwater resources, and salt water intrusion in aquifers of the coastal areas. The exhausted aquifers are losing their capacity to hold water and lands are subsiding. Waterlogging and salinization are diminishing the productivity of irrigated croplands. The impacts of climate changes are affecting the population and development directly by frequent extreme events such as floods and droughts, rising sea levels, changes in the seasonal distribution and amount and type of precipitation. In this perspective, the judicious and optimal management of water resources with adoption of water-saving technology for sustainable crop production is of much relevant in view of dwindling financial resources and increasing cost of irrigation water. The challenge is to produce more with less water by increasing substantially the use efficiency with which available water resources are allocated.

In water management practices, irrigation scheduling is considered as a vital component to produce higher water use efficiency under any irrigation system, as excessive or inadequate irrigation has deleterious effects on crop productivity (Panigrahi *et al.*, 2011). It can be managed precisely to meet crop demands, holding the promise of increased yield and quality. Moreover, scheduling irrigation is influenced by many complex factors such as soil, crop, environment, water supply and cultivation practices. In the Indo-Gangetic plain of eastern India, the resource poor farmers generally follow the conventional surface method of irrigation and soil fertilization for crop production, which causes heavy losses of water and nutrients in runoff and deep percolation, besides aggravating water logging, poor soil aeration and heavy weed infestation (Bangar and Chaudhary, 2004; Pramanik and Patra, 2014). The meteorological based irrigation scheduling approach such as pan evaporation replenishment, cumulative pan evaporation, ratio between irrigation water and cumulative pan evaporation (IW/CPE) and soil moisture depletion have been used widely due to its simplicity, data availability and higher degree of adaptability at the farmer's level.

Drip irrigation is often preferred as an efficient and economically viable method over surface irrigation method because of its ability to apply small but frequent water directly into the rooting zone (Rajurkar *et al.,* 2012). Adoption of drip irrigation might help in increasing area under irrigation, productivity of crops and increase the water use efficiency (Tarawalie *et al.*, 2012). It provides the daily consumptive use of water of plant and maintains a high soil matric potential in the rhizosphere to reduce plant water stress (Ayas *et al.,* 2011). It also provides unique agronomic, water and energy conservation benefits that address many of the challenges facing irrigated agriculture. Farmers and other drip-irrigation users are continually seeking new options, such as waste-water reuse, that will continue to provide new challenges for designers and irrigation managers. The required quantity of water is provided to each plant daily at the root zone through a network of piping system. Hence, there is no loss of water either in the conveyance or in the application. Evaporation loss from the soil surface is also very little since water is given only to the root zone and crop canopy provides shade to prevent evaporation. The water saving in fruit crops under drip irrigation appeared to be around 30-60 per cent as compared to surface irrigation method and increase in fruit yield was 20-40 per cent over surface method under adequate supply of water. In general, the modern improved irrigation methods give overall water use efficiency as high as 90 per cent as against only 30-40 per cent under the conventional methods. Besides, subsurface drip irrigation produced highest yield and best quality of fruits as compared to surface drip irrigation system (Deshmukh and Sen, 2000; Sharma and Kumar 2007).

The practice of supplying crops in the field with fertilizers via the irrigation water is called fertigation (Fanish and Muthukrishnan, 2013). It is a modern agro-technique which provides an excellent opportunity to maximize yield and minimize environmental pollution (Hagin *et al.*, 2002) by increasing fertilizer use efficiency, minimizing fertilizer application and increasing return on the fertilizer invested. In fertigation, timing, amounts and concentration of fertilizers applied are easily controlled. It enables the application of water soluble fertilizer and other chemicals along with irrigation water uniformly and more efficiently in the active root zone of crop. However, drip-fertigation (application of fertilizer solution with drip irrigation) is the most advanced method of precise application of irrigation water and fertilizer nutrients simultaneously in the root zone, satisfying the plants' total and temporal requirements of these two inputs (Mohammad and Said, 2003; Raina *et al.*, 1999; Singandhupe *et al.*, 2003; Patel and Rajput, 2010; Sandal and Kapoor, 2015). This improves the water and nutrients use efficiency, reduce runoff and surface evaporation and minimize leaching and volatilization losses with least groundwater contamination due to minimal deep percolation (Brahma *et al.*, 2010; Raina *et al.*, 2011). Use of plastic mulch along with drip irrigation/ fertigation further added the advantages of water savings, weed control, minimum nutrient loss and incidence of reduced disease and pest infestation besides moderation of soil hydrothermal regimes within the effective root zone of the plants (Tiwari *et al.*, 1998; Pattanaik *et al.*, 2003).

Materials and Methods

Experimental site, soil and water

The field study was conducted at the Central Research Farm, Bidhan Chandra Krishi Viswavidyalaya, Regional Research Station, Gayeshpur, West Bengal,

India during winter (*rabi*) seasons of respective years. The experiments on gravity drip irrigation and drip fertigationvis-a-vis farmers' traditional surface irrigation and soil fertilization were performed on different crops during the period from 2005-06 to 2015-2016. The experimental site is located between 22°58'31" N latitude and 88°26'20" E longitude with an altitude of 9.75 m above mean sea level. It belongs to the New Alluvial Zone of West Bengal encompassing the lower Indo-Gangetic alluvial plains. The soil of the experimental site was sandy clay loamin textureclassified taxonomically as TypicFluvaquept. The soil had pH 6.83, EC 0.37 dS/m, organic C 5.6 g/kg and available N, P and K content 153.7, 21.5 and 160.2 kg/ha, respectively. The amount of average annual rainfall received was 1500 mm and three-fourth of it is received during the four monsoonic months from June through September. The average pan evaporation varied between 0.91 and 1.93 mm per day. The depth of groundwater table varied from 2.8 to 5.6 m bgl. The irrigation water pumped out from a deep bore well was used in the experiment. The pH, electrical conductivity and SAR value of irrigation water was 7.1, 0.22 dS/m and 0.18, respectively, which could be used safely in irrigation.

Agronomic manipulations

Standard agronomic practices in respect of the selection of land and its preparation through primary and secondary tillage with proper land leveling, choice of suitable variety of crop(s), appropriate plant geometry, recommended dose of fertilizers and manures, plant protection measures and other intercultural operation were uniformly followed. The early date of sowing and corresponding harvesting were top prioritized for availing of the high market demand of crop. The indigenous and advanced soil mulching was adopted for soil water conservation. The mulching with black polythene film of 25 μ thickness having small holes of desired diameter at a right spacing of crop was spread over the prepared field and the seeds (okra) were planted in the holes. Paddy straw mulch @ 5 t/ha on air-dry basis was spread uniformly at 4-5 cm thickness.

Estimation of crop water requirement

The amount of irrigation water given to crop through drip system was calculated by using the equation as below:

$$I = Ep \times Kp \times Kc \times P$$

Where, I is the irrigation water (mm), Ep is the pan evaporation (mm), Kp is the pan coefficient, Kc is the crop coefficient and P is the percentage of wetted area (%). The wetted area was taken as 100% assuming that lateral interval is equal to the spaces between drippers. The evaporation data was obtained from a USDA Weather Bureau Class A Pan located inside the experimental site on a wooden support at a height of 15 cm above the soil surface. The pan coefficient value was taken as 0.7. Crop coefficient values for different growth stages of crop viz., early growth, mid season, and late season etc was adoptedfrom Allen *et al.* (1998).The drip irrigation system was operated at 3 days interval in summer and 5 days interval during winter based on variable evaporation replenishment or, crop evapotranspiration. Irrigation water was scheduled at 22-35 days interval in surface irrigation scheduling at IW/CPE 1.0 with 50 mm depth or, farmers' practice of flood irrigation at 30-45 days interval. Required amount of surface irrigationwater was measured by Parshall flume at the head of each plot.

Seasonal water requirement of crop was computed using the water balance equation as,

WU = I + Re + Cp – Rf ± "S

Where, WU is the total water use (mm), I is the irrigation water (mm), Re is the effective rainfall, Cp is the contribution through capillary rise from groundwater (mm), Rf the surface runoff (mm) and ±S the change in soil water storage in the profile between planting and harvest time (mm). The monthly effective rainfall during the cropping was calculated by soil moisture depletion by concerned crop. The contribution from groundwater, Cp was assumed negligible. There was no Rf due to low depth of water application. Thus, WU = I + Re ± "S

Water use efficiency

The water use by the crop is expressed in terms of water use efficiency (WUE). It was computed as the ratio of crop yield to the amount of water depleted by the crop in the process of evapotranspiration. In other words, water use efficiency is the yield per unit area of land per unit depth of water used by the crop and is calculated as,

Water use efficiency(WUE) = Y/WU or, ETc (kg/ha-mm),

Where, Y = rhizome yield (kg/ha),

CU or ETC = crop water use or crop evapotranspiration (mm)

Fertilization schedules

The required amounts of fertilizers for the respective crop were selected for fertigation study. All N, P and K were applied through urea (46% N), phosphoric acid (31.68% P) and muriate of potash (60% K_2O), respectively through drip irrigation in several splits based on growth stages of plant. The concentration of nutrient solution passing through irrigation water was around 1.1 to 1.7%.In surface method of irrigation, full doses of P and K and half of N were applied manually one week before planting and remaining N was applied in 2-3 splits at weekly intervals. The amounts of required fertilizers were dissolved in 10 litre (L) of water and the stock fertilizer solution was dissolved in a 500 litre capacity over head tank placed at a height of 3.2 m above ground level to facilitate water flow by gravity action. The desired amounts of fertilizers as per schedule along with irrigation water were applied in all the drip fertigation treatments by maintaining different time durations. For the drip system 12 mm diameter LLDPE (Linear low density polyethylene) pipes used as laterals having emitters at 40-50 cm apart with a discharge of 1.8 lph were placed in between the two rows. The lateral line interval was equal to the dripper interval for each plot. The discharge of drippers was taken to adjust time of operating the drippers according to the volume of water exactly required on the basis of varying irrigation requirements throughout the cropping. The discharge from all in-line drippers at selected laterals of drip system operating at pressure of 1.2 kg/cm^2 was measured for a specified period. The emission uniformity of the drip system was computed prior to delivering water. The values ranged between 91.2 and 92.5% implying the excellent functioning of the drip system in supplying irrigation water uniformly and steadily through the lateral lines.There was one overhead micro-microsprinkler for every four plants (pineapple) with a discharge rate of 38 lph. In the subsurface irrigation system, laterals (16 mm diameter) were buried at 25 cm depth in soil along the crop rows with one porous pipe (15 mm diameter) between two plants having a discharge of 9.6 lph per meter of pipe length.

Benefit-cost analysis

The economic feasibility of the drip irrigation system with fertigation vis-à-vis surface irrigation with conventional soil fertilization was worked out through benefit: cost ratio analysis. Seasonal system cost of drip irrigation included depreciation, prevailing bank interest rate and repair and maintenance cost of the system. The interest rate was taken 13 % and repair and maintenance cost of the system was taken 2% per annum of the fixed cost. The life span of drip system was considered to be 10 years. The cost of cultivation includes expenses incurred in land preparation, intercultural operation, fertilizer, crop protection measures, irrigation water and crop harvesting with labour charges. The gross income from the produce was estimated using prevailing market price of crop.The net seasonal income was estimated by subtracting the seasonal cost of production from the gross income of the produce. Benefit: cost ratio was calculated by dividing net return with cost of cultivation of individual treatment.

Statistical analysis

The data obtained for different crop and soil parameters were subjected to statistical investigation following the analysis of variance techniques by using software packages of MS Excel and SPSS 12.0 version. Statistical significance between means of individual treatments was assessed using Fisher's Least Significant Difference (LSD) at 5% level of probability (Gomez and Gomez, 1984).

Results and Discussion

The effects of gravity drip, micro-sprinkler and sub-surface irrigation along with fertigation vis-à-vis conventional surface irrigation with soil fertilization on the crop yield, water utilization pattern, the relationship of water and yield of different crops and the economics of crop cultivation are described in different sub-headwise as below:

Effect of drip nitrogen fertigation and mulch on broccoli

Broccoli is one of the most important winter vegetable crops. It is also called as 'Crown of Jewel Nutrition' because of its richness in dietary fiber, protein, *vitamins*, antioxidants, phytochemicals and minerals like calcium, manganese, iron, magnesium, selenium, zinc and phosphorus (Kumar and Senseba, 2008). Being a shallow rooted crop, broccoli is highly dependent on inputs of irrigation water and nitrogenous fertilizer to achieve optimum production. Short supply of soil moisture and nutrients in any stage of the crop growth cycle reduces productivity (Tiwari *et al.*, 2003). In the recent years commercial cultivation of broccoli is gaining importance in the Indo-Gangetic alluvial plainsdue to its higher nutritive value, palatability, short duration, high productivity and tremendous domestic and foreign market potential. But there is lack of information on optimum water and nutrients requirements of the crop. Drip irrigation and fertigation is the best option to increase the crop and water productivity and fertilizer use efficiency. In this perspective, the present study was initiated to find out the effect of drip nitrogen fertigation vis-à-vis the surface irrigation and soil N fertilization under mulched and unmulched conditions on yield, water productivity and economics of broccoli.

Broccoli (*Brassica oleracea* var. *italica*)cv. green magic seeds were sown on 13 October every year in the nursery bed at a depth of 5 cm with a spacing of 10 cm between the rows. Thirty-day old seedlings were transplanted in the field on 11 November of each experimental year in crop geometry of 45 cm row to row and plant to plant spacing. The area of unit plot size was 12 m^2 (3 m x 4 m). A buffer strip of 1.0 m and 0.5 m was provided between the plots and blocks. Immediately after transplanting a common irrigation equivalent to 20 mm depth was applied uniformly to all treatments. The crop was harvested on 30 January of each experimental year and the curd yield for each treatment was recorded. For mulching, black polythene film of 25μ thickness having small holes of 60 mm diameter at a distance of 45 cm was spread over the prepared field and broccoli seedlings were transplanted in the holes. Paddy straw mulch @ 5 ton/ha on air-dry basis was also spread uniformly at 4-5 cm thickness. The standard agronomic and plant protection measures were adopted uniformly in all plots.The treatments consisted of four levels of irrigation with nitrogen fertigation/fertilization (drip fertigation at 0.6, 0.8 and 1.0 of crop evapotranspiration (ETc) and surface irrigation at 1.0 IW/CPE with soil-fertilization) and three mulches (no-mulch, black polythene mulch and paddy straw mulch). The layout of the experiment was laid out in two factor randomized block design (irrigation with N fertigation/fertilization in main plots x mulch in sub-main plots) with twelve treatments and three replications.

The results showed that the curd yields were significantly influenced by drip N-fertigation over surface irrigation with conventional soil N-fertilization (Table 1). Drip N-fertigation at 0.8 ETc recorded the highest curd yield of 17.80 ton/ha. The effects of drip N-fertigation at 1.0 ETc and 0.6 ETc levels in yield improvement (16.53 and 16.41 ton/ha, respectively) were comparative with each other, and being superior to surface irrigation with N-fertilization (14.69 ton/ha). Drip N-fertigation at 1.0, 0.8 and 0.6 of ETc increased the marketable curd yield by 12.5, 21.2 and 11.7%, respectively over surface irrigation with N-fertilization. The increase in yields under varying levels of drip N-fertigation was ascribed to the improved wetting pattern of soil and effective utilization of water by crop (Raina *et al.*, 1999), higher availability and utilization of fertilizer N in synchrony with crop demand at vital growth period (Rumpel *et al.*, 2003) and better proportion of air-soil-water with plenty oxygen concentration in the root zone which was maintained throughout the life period of crop (Kadam and Karthikeyan, 2006) which might have stimulated better physiological processes, active cell division and photosynthetic activities (Brahma *et al.*, 2010). Surface irrigation with conventional N fertilization, on the other hand, might have resulted in wastage of water in deep percolation, leaching of available plant nutrients and poor aeration resulting in poor yield. There was occurrence of undesirable hazards in relation to soil aeration problem in first few days immediately after irrigation and soil water-nutrient imbalance due to high surface irrigation load (Raina *et al.,* 2011). Besides, the crop also suffered water stress between two irrigations during some parts of physiological growth stages which resulted in reduced crop yield. Mulch treatment had significant effect on curd yield over non-mulch treatment. The increase in curd yield was 19.7% for black polythene mulch and 8.7% for paddy straw mulch. The differences in yield under mulched and unmulched treatments might be due to better availability and utility of fertilizer nutrients, improved soil moisture conditions; favourable hydrothermal regimes within the plant rhizosphere and the minor weed cover (Vankar and Shinde 2007). The interaction effects showed that drip N-fertigation scheduled at 0.8

ETc with black polythene mulch recorded significantly the maximum yield and was superior to other drip N fertigation x mulch treatment combinations. Conversely, surface irrigation with NPK fertilization without mulch significantly produced the lowest yield.

The seasonal consumptive water use by the crop was relatively higher in conventional surface irrigation (209.14 mm) as against drip irrigation showing 126.72, 149.50 and 171.91 mm scheduled at 0.6, 0.8 and 1.0 ETc, respectively (Table 1). Maximum water use efficiency of 129.50 kg/ha-mm was registered in drip irrigation system at 0.6 ETc followed by that of 119.06 kg/ha-mm at 0.8 ETc and 96.15 kg/ha-mm at 1.0 ETc, whereas minimum value of 70.24 kg/ha-mm was obtained in surface irrigation. About 17.80 to 39.41% of water could be saved with overall drip irrigation over the surface irrigation. These results are in accordance with the findings of Rajurkar *et al.* (2012). Mulching had an additional advantage in conserving soil moisture and increasing curd yield over unmulched bare soil. However, black polythene mulch (107.55 kg/ha-mm) was more pronounced in displaying the higher water use efficiency than the paddy straw mulch (99.19 kg/ha-mm).

The additional area brought under cultivation due to saving of water through drip system indicated that drip irrigation scheduled at 0.6 ETc likely to save about 39.41% of water compared to the surface irrigation, which could bring about 65.04% of additional cultivable area under drip irrigated crop. Similarly, the drip irrigation scheduled at 0.8 ETc could save as much as 28.52% of water in comparison with surface irrigation. On utilizing this amount of water saved, it was possible to cultivate an additional area of 39.90% under drip irrigated crop. Likewise, drip irrigation scheduled at 1.0 ETc was found to save about 17.8% of water compared to surface irrigation which could bring an additional area of 21.65% under drip system for crop cultivation.

Fertilizer use efficiency (FUE) was relatively higher in all the drip N-fertigation treatments (82.1-89.0 kg yield/kg NPK) as compared to surface irrigation-N fertilization treatment (73.4 kg yield/kg NPK). This was due to better availability of moisture and nutrients throughout the growth stages in drip N-fertigation system leading to better uptake of nutrients and higher production of curd (Shedeed *et al.*, 2009). FUE was relatively higher in mulch treatments (81.2-89.5 kg yield/kg NPK) than in unmulch treatment (74.8 kg yield/kg NPK), the more so in black polythene mulch than in paddy straw mulch. However, Drip N- fertigation at 0.8 ETC with black polythene mulch exhibited the higher FUE (95.8 kg yield/kg NPK) in comparison to other treatments. These are in accordance with the findings of Vankar and Shinde (2007) who observed that the application of N through the drip irrigation in split doses increased yield and nitrogen economy of tomato and okra crop, respectively.

Table 1: Curd yield, water use and fertilizer use efficiency and economics of broccoli as influenced by drip nitrogen-fertigation and mulches (average of 2010-11 and 2011-12)

Treatment	Curd yield (t/ha)	Water use (mm)	Water use efficiency (kg/ha-mm)	Water saving (%)	Additional cultivated area (%)	Fertilizer use efficiency (Yield/kg NPK)	BCR
Irrigation							
Surface irrigation at IW/CPE 1.0 with N fertilization	14.69	209.14	70.24	-	-	73.4	2.52
Drip N-fertigation at 0.6 ETc	16.41	126.72	129.50	39.41	65.04	82.1	3.09
Drip N-fertigation at 0.8 ETc	17.80	149.50	119.06	28.52	39.90	89.0	3.73
Drip N-fertigation at 1.0 ETc	16.53	171.91	96.15	17.8	21.65	82.7	3.47
CD (0.05)	0.79	-	-	-	-	-	
Mulch							
No-mulch	14.95	162.96	91.74	-	-	74.8	3.20
Black polythene	17.89	166.34	107.55	2.07	-	89.5	3.55
Paddy straw	16.23	163.63	99.19	0.41	-	81.2	3.35
CD (0.05)	0.83	-	-	-	-	-	-

ETc: crop evapotranspiration; IW/CPE: irrigation water to cumulative pan ratio

The economic analysis showed that higher benefit-cost ratio (3.76) was obtained from drip irrigation scheduling at 0.8 ETc with N-fertigation and black polythene mulch followed by drip irrigation scheduling at 0.6 ETc with N-fertigation and black polythene mulch with the corresponding value of 3.54 and drip irrigation at 1.0 ETc with N-fertigation and black polythene mulch with BCR value of 3.36, respectively. Minimum benefit-cost ratio (2.52) was received from the surface irrigation with conventional soil N-fertilization. The higher economic benefit due to drip irrigation with N-fertigation was mainly due to the increase in curd yield on account of higher utilization of water and nitrogen by broccoli plant coupled with the curtailment of labour cost incurred on weeding and fertilizer N application. The economic viability and feasibility of drip fertigation system in different horticultural crops had been communicated earlier (Brahma *et al.*, 2010; Pramanik *et al.*, 2014; Pramanik and Patra, 2014).

It can be concluded from this study that drip N-fertigation with mulch is advantageous in promoting yield and water productivity of broccoli as compared with the conventional surface irrigation and soil N-fertilization. However, drip N-fertigation at 0.8 ETc with black polyethylene mulch proved to be much beneficial with higher yield and water productivity. Under limited water supply condition, drip N-fertigation at 0.6ETc with black polyethylene mulch was also comparative with relatively higher yield and maximum water productivity.

Effect of drip nitrogen fertigation and mulch on okra

Okra (*Abelmoschusesculentum* L Moench) is one of the most important vegetable crops grown throughout the tropical and subtropical regions of the world. It plays a significant role in the human diet by providing carbohydrate, protein, fats, minerals and vitamins that are usually deficient in the staple food. Being a shallow rooted crop, it is highly sensitive to excess or deficit water supply to accomplish the optimum production. Controlled irrigation strategy in any stage of the crop growth is essential to use the limited water judiciously for high yield. The use of plastic mulch in increasing crop yield through their beneficial influence on moisture conservation, soil temperature modulation within the active plant root zone, weed control and reduced disease and pest infestation are well known. The response of okra to drip irrigation was found to be different in different agro-climatic and soil conditions in India.

The seeds were sown on 9 February every year at a depth of 2 cm with crop geometry of 30 cm row to row and plant to plant spacing. Crop after several pod pickings was finally harvested on 6 June of each experimental year and marketable yield for each treatment was recorded. For mulching, black polythene film of 25μ thickness having small holes of 60 mm diameter at a distance of 30 cm was spread over the prepared field and okra seeds were planted in the holes. Paddy straw mulch @ 5 t/ha on air-dry basis was spread uniformly at 4-5 cm thickness. All treatments received a recommended dose of 90, 60 and 40 kg/ha of N, P_2O_5 and K_2O. Full doses of P and K were applied manually as basal during sowing. Nitrogen was applied through drip irrigation in eight equal splits given at weekly intervals; whereas in soil fertilization it was top dressed in two equal splits at basal and 45 days after sowing. The N fertigation was scheduled based on four growth stages of plant *viz*., initial, mid, final and maturity stage.

The results of the study showed that drip irrigation at 0.8 ETc with N fertigation recorded the maximum pod yield of 12.01 t/ha which was at par with drip irrigation at 1.0 ETc with N fertigation displaying pod yield of 11.52 t/ha (Table 2). Significantly minimum average pod yield of 9.16 t/ha was obtained with surface irrigation with soil N fertilization. The overall per cent increase in yield under drip irrigated crop at 1.0, 0.8 and 0.6 ETc with N fertigation was 25.8, 31.1 and 13.9, respectively over surface irrigation with soil N fertilization. These results corroborated with the findings of Tiwari*et al*. (1998) and Abubaker *et al*. (2014) who observed that yield of okra under drip irrigation was significantly higher in comparison with furrow irrigation. The increase in yield under varying drip irrigation levels with N fertigation might be ascribed to improved wetting pattern of soil and better crop water use (Raina *et al.* 1999), higher availability and uptake of fertilizer nutrients in synchrony with crop demand at vital growth stages (Rumpel *et al.,* 2003) and attainment of ambient soil-water-air environment with plenty oxygen supply in the rooting depth, which eventually might have stimulated favourable physiological processes, active cell division and photosynthesis (Brahma *et al*., 2010). The relatively low yield at 1.0 IW/CPE with surface irrigation method could be associated with the tendency to supply water above the moisture content at field capacity. Thus surface irrigation with conventional soil fertilization resulted in excessive wastage of water and available plant nutrients in deep percolation below the crop root zone including higher surface evaporation loss as water spreads over a large area. Addition of mulches as soil cover had significant effect on pod yield over non-mulch treatment. However, black polythene mulch was more pronounced in promoting pod yield. The increase in yield, on an average, was 8.2% for black polythene mulch and 3.7% for paddy straw mulch over unmulched condition. The differences in yield under mulched and unmulched treatments might be due to better availability of plant nutrients, improved moisture conditions and suitable hydrothermal regimes in soil (Vankar and Shinde, 2007).

The interaction effects showed that drip irrigation at 0.8 ETc with N fertigation coupled with black polythene mulch recorded maximum yield of 12.43 ton/ha, but at par with the treatments receiving drip irrigation at 0.8 ETc with N fertigation and paddy straw mulch (11.93 ton/ha), drip irrigation at 1.0 ETc with N fertigation and polythene mulch (11.91ton/ha) and drip irrigation at 1.0 ETc with N fertigation and paddy straw mulch (11.56 ton/ha). On the contrary, surface irrigation with conventional soil fertilization without mulch recorded significantly the lowest pod yield (8.84 ton/ha).

Table 2: Fresh pod yield, water use and fertilizer use efficiency and economics of okra as influenced by drip nitrogen-fertigation and mulches (average of 2010-11 and 2011-12)

Treatment	Pod yield (t/ha)	Total water use(mm)	Water use efficiency (kg/ha-mm)	Water saving (%)	Additional cultivated area (%)	BCR
Irrigation						
Surface irrigation at IW/CPE 1.0 with soil N fertilization	9.16	293.35	32.22	-	-	2.14
Drip N-fertigation at 0.6 ETc	10.43	203.77	51.63	30.54	43.96	2.71
Drip N-fertigation at 0.8 ETc	12.01	222.14	53.48	24.27	32.06	3.28
Drip N-fertigation at 1.0 ETc	11.52	242.52	47.67	17.33	20.96	3.10
CD (0.05)	1.02	-	-	-	-	-
Mulch						
No-mulch	10.37	240.56	43.07	-	-	-
Black polythene	11.22	241.67	49.16	-	-	-
Paddy straw	10.75	239.90	45.02	-	-	-
CD (0.05)	0.34	-	-	-	-	-

ETc: crop evapotranspiration; IW/CPE: irrigation water to cumulative pan ratio

Crop water use was 203.77, 222.14 and 242.52 mm through drip irrigation scheduling at 0.6, 0.8 and 1.0 ETc, respectively and 293.35 mm for surface irrigation. Maximum water use efficiency of 53.48 kg/ha-mm was registered in drip irrigation at 0.8 ETc, followed by that of 51.63 kg/ha-mm in drip irrigation at 0.6 ETc and 47.67 kg/ha-mm in drip irrigation at 1.0 ETc, whereas minimum value of 32.22 kg/ha-mm was obtained in surface irrigation. About 17.33 to 30.54% of water, on an average, could be saved with drip irrigation scheduled at various levels over the surface irrigation. Therefore, by implementing drip system for growing okra, considerable amount of groundwater could be saved due to improved water use efficiency with lower water usage and higher yield. Mulching had an additional advantage in conserving soil moisture over unmulched bare soil. However, black polythene mulch was found to be more effective in increasing the water use efficiency than the paddy straw mulch.

The data on water saving and additional area to be cultivated due to saving of water through drip system is presented in Table 2. The results indicated that the low level of drip irrigation scheduled at 0.6 ETc likely saved about 30.54% of water compared to the surface irrigation, which could bring about 43.96% of additional cultivable area under drip irrigated crop. Similarly, the moderate level of drip irrigation scheduled at 0.8 ETc could save as much as 24.27% of water in comparison with surface irrigation. On utilizing this saved amount of water due to adoption of drip irrigation system, it was possible to cultivate an additional area of 32.06%. Likewise, the higher level of drip irrigation scheduled at 1.0 ETc was found to save about 17.33% of water compared to surface irrigation which could bring an additional area of 20.96% under drip irrigation system for crop cultivation.

The benefit-cost analysis revealed that drip irrigation with N fertigation was more remunerative as compared to surface irrigation and soil N fertilization. Maximum benefit-cost ratio (3.28) was obtained from drip irrigation at 0.8 ETc with N fertigation, closely followed by drip irrigation at 1.0 ETc with N fertigation with BCR of 3.10 and drip irrigation at 0.6 ETc with N fertigation with BCR of 2.71. Minimum benefit-cost ratio (2.14) was obtained from surface irrigation at 1.0 IW/CPE with soil N fertilization. The higher economic benefit due to drip irrigation with N-fertigationwas mainly due to the increase in pod yield on account of higher utilization water and nitrogen by okra plant coupled with the curtailment of labour cost incurred on weeding and fertilizer N application.

Effect of drip fertigation on banana

Banana (*Musa* sp) is one of the most important leading commercial perennial fruit crops in the world. India is the largest producer of banana in the globe. Being a highly exhaustive crop, it requires frequent and liberal supply of water and nutrients at critical physiological stages for proper growth and development and for optimum yield. There are several constraints in increasing the production of banana. Among those constraints, water and nutrients are considered to be the most limiting factors. In the Indo-Gangetic alluvial plains, the banana cultivators usually grow the crop following the traditional flood method of irrigation and conventional soil fertilization. These approaches are quite inefficient due to excessive losses of water and nutrients below root zone and waterlogging and aeration related problems during some parts of crop growth stages leading to decreased yield. Adoption of drip fertigation system may also helpin increasing yields and quality parameters due to improvedirrigation and nutrients use efficiencies.

Banana plant (cv. Martaman, AAB, Silk) consisted of 2.5 month-old healthy sword sucker weighing around 2.0 kg each was planted at a spacing of 2 x 2m (2500 suckers/ha) during 3rd week of January 2008 for plant crop and harvested during 2nd week of March 2009. All the suckers were removed from mother plant and one sucker per mother clump was allowed as follower for ratoon crop.These suckers were cut back at the height of 20 cm from the ground level to get the uniform cropping and were maintained from 2nd week of March 2009 and harvested during 3rd week of April 2010. A buffer strip of 2 m was maintained between the two plots. The treatments consisted of three drip irrigation levels at 50 (I_1), 60 (I_2) and 70 (I_3) per cent of cumulative pan evaporation (CPE) and three fertigation levels at 50 (F_1), 60 (F_2) and 80 (F_3) per cent of recommended dose of fertilizers (RDF) including a control treatment which includes surface irrigation at 100 per cent CPE with conventional soil fertilization at 100 per cent RDF. The recommended dose of fertilizers for banana is 250:50:300 g NPK/ plant. The fertilizers were scheduled in 18 equal splits for N, 2 splits for P and 9 splits for K apportioned in different growth stages of crop as tabulated below.

Week after planting	N (%)	P (%)	K (%)
9 to 16	20 (4 splits)	50 (1 split)	20 (2 split)
17 to 28	40 (6 splits)	50 (1 split)	40 (3 split)
29 to 40	30 (6 splits)	-	30 (3 split)
41 to 42	10 (2 splits)	-	10 (1 split)

The water requirement of banana was computed on the basis of CPE, pan factor, crop coefficient, canopy area factor and wetted area factor. The drip was scheduled once in three days in summer and five days in winter based on evaporation replenishment splitting 80 irrigations in each plant and ratoon. In conventional soil fertilization, the RDF was applied in 4 splits at 2, 5, 7 and 9 months after planting of plant and ratoon crop. Banana cv. Martaman, AAB, Silk was used as the test crop. All the suckers were removed near mother plant till flowering and one sucker was retained per plant for the ratoon crop.

A perusal of the study revealed that different levels of drip irrigation and drip-fertigation had significant effects on the fruit yield of main and ratoon crop (Table 3). Among the irrigation levels, drip irrigation at 60% of CPE registered the highest yield of 42.86 t/hafor main crop and 39.48 t/hafor ratoon crop. Further increase in evaporation replenishment resulted in a significant decrease in yield of both crops. Similarly, among the levels of drip-fertigation, fruit yield consistently and significantly increased up to 80% of recommended NPK fertilizers. This increase in yield was largely due to higher nutrients uptake of crop due to timely application of judicious amounts of nutrients directly to the root zone with simultaneous decrease in N and K losses through leaching and deep percolation (Pawar and Dingre, 2013), which in turn influenced the yields. The interaction between drip irrigation and NPK fertigation on fruit yields were significant. Maximum yield of 49.14 t/ha for main crop and 44.04 t/ha for ratoon crop was recorded with drip irrigation at 60% of CPE with 80% of recommended NPK fertilizers and were found significantly superior to the remaining levels of irrigation-fertilizers combinations. This higher yield might be due to optimum supply of irrigation and nutrients through drip system matching with crop water-nutrient requirements maintained throughout the crop growing period. The traditional surface irrigation, on the other hand, resulted in considerable wastage of water in runoff and deep percolation below root zone and might invite a chain of undesirable hazards such as leaching loss of available plant nutrients, water congestion with poor soil aeration and weed infestation leading to the declined fruit yield (Patel and Rajput, 2004). Another possible reason of low yield in surface irrigation might be that crop had to undergo water stress between two irrigations (Pawar and Dingre, 2013). It is interesting to note that the fruit yield of ratoon crop was relatively lower than that of main crop regardless of different irrigation and fertigation levels.

During the cropping seasons, total water use by the main and ratoon crop was 1135.1, 1190.4, 1245.9 mm and 859.3, 892.1, 924.8 mm through drip irrigation at 50, 60 and 70% of CPE, respectively and the corresponding figures for surface irrigation was 1928.1 and 1525.5 mm. It is evident from Table 3 that drip irrigation at 60% of CPE recorded the higher WUE as compared to that of drip irrigation at 50 and 70% of CPE. Similarly, WUE increased progressively with the concomitant increase in drip-fertigation levels and accordingly, the higher value of WUE was registered with 80% of RDF administered through drip irrigation. The interactions between drip irrigation and fertigation revealed that maximum water use efficiency was registered in drip irrigation at 60% of CPE with 80% of RDF (0.41 and 0.49 t/ha/mm for main and ratoon crop, respectively) followed by drip irrigation at 60% of CPE with 60% of RDF (0.38 and 0.47 t/ha/mm for main and ratoon crop, respectively). The savings of water combined with higher yield under drip irrigation are the reasons for increased WUE. On the contrary, the plants under surface irrigation showed low WUE, because the increased amount of water applied did not result in corresponding increase in yield.

Table 3: Different levels of irrigation and drip fertigation on fruit yields and water use efficiency (WUE) of banana (pooled data of 2008-09 and 2009-10)

Treatment	Fruit yield (t/ha)		Water use (cm)		WUE (t/ha-cm)		BCR
	Main	Ratoon	Main	Ratoon	Main	Ratoon	
Irrigation							
I_1 (CPE 50)	35.33	33.00	113.51	85.93	0.31	0.38	1.65
I_2 (CPE 60)	42.86	39.48	119.04	89.21	0.36	0.44	1.98
I_3 (CPE 70)	41.40	38.07	124.59	92.48	0.33	0.41	1.91
CD (0.05)	0.85	1.22	-	-	-	-	-
Fertilizer							
F_1 (50% RDF)	33.75	31.21	119.04	89.21	0.28	0.35	1.68
F_2 (60% RDF)	41.71	38.91	119.04	89.21	0.35	0.43	1.98
F_3 (80% RDF)	44.13	40.43	119.04	89.21	0.37	0.45	1.88
CD (0.05)	0.85	1.22	-	-	-	-	-

Recommended dose of fertilizers (RDF): N:P:K :: 250:50:300 g/plant/year

The fruit quality parameters were influenced by the drip irrigation and fertigation levels (Table 4). Relatively higher TSS, non-reducing sugar, total sugar, and higher acidity was observed in drip irrigation at 70% of CPE, whereas higher reducing sugar and ascorbic acid with lower acidity was recorded in drip irrigation at 60% of CPE. Similarly, application of 80% of RDF promoted all these quality parameters as compared with other fertigation levels. The combination of drip irrigation at 60% of CPE with 80% of RDF showed maximum peel content (30.63 g), TSS (23.75), reducing sugar (6.59) and ascorbic acid, whereas drip irrigation at 70% CPE with 80% RDF recorded maximum non-reducing sugar (9.72), total sugar (16.10) and higher acidity (0.43).

The benefit-cost analysis showed that drip irrigation with NPKfertigation was more remunerative as compared to surface irrigation and soil N fertilization. Maximum benefit-cost ratio (5.55) was obtained from drip irrigation at 60% CPE, closely followed by drip irrigation at 70% CPE with BCR of 5.40. Similarly, maximum BCR of 5.61 was obtained from drip fertigation at 80% of RDF followed by 5.45 from drip fertigation at 60% of RDF. The interaction study recorded maximum BCR of 2014 obtained from drip irrigation at 60% CPE in conjunction with drip fertigation at 60% RDF. Minimum benefit-cost ratio (1.49) was obtained from surface irrigation with conventional soil NPK fertilization.

Table 4: Fruit quality of banana as influenced by irrigation methods and drip fertigation levels (pooled data of 2008-09 and 2009-10)

Treatment	TSS (°brix)	Reducing sugar (%)	Non-reducing sugar (%)	Total sugar (%)	Acidity (%)	Ascorbic acid (mg/100 g pulp)
Irrigation						
I_1 (CPE 50)	21.11	5.73	9.10	14.83	0.36	5.04
I_2 (CPE 60)	22.54	6.22	9.07	15.29	0.39	5.55
I_3 (CPE 70)	22.88	6.15	9.35	15.50	0.41	5.40
CD (0.05)	0.20	0.06	0.12	0.12	0.01	0.13
Fertilizer						
F_1 (50% RDF)	20.71	5.60	9.09	14.70	0.35	4.93
F_2 (60% RDF)	22.56	6.16	9.09	15.25	0.39	5.45
F_3 (80% RDF)	23.27	6.35	9.33	15.68	0.42	5.61
CD (0.05)	0.20	0.06	0.12	0.12	0.01	0.13

The study suggests the advantage of drip fertigation with applying precise amounts of water and fertilizer nutrients over surface irrigation and conventional soil fertilization for banana production. The drip irrigation at 60% CPE with fertigation at 60% RDF through drip may be recommended for obtaining yield, fruit quality, water use efficiency and economics from banana.

Effect of drip irrigation and nitrogen fertigation on guava

Guava (*Psidium guajava* L.) is one of the leading fruit crops in India due to wide adaptability to varying soil and climatic conditions. In the Indo-Gangetic alluvial soils of West Bengal, the crop has immense potential in increasing productivity and yield sustainability. However, limited availability of irrigation water during the dry season is a major constraint in increasing area under guava cultivation. Even the unscientific water management practices coupled with lack of proper water saving technologies can lead to the reduction in crop yield. Judicious application of water and plant nutrients in guava is prerequisite to achieve the targeted yield and quality of fruits. Drip fertigation (application of fertilizer with drip irrigation) offers its superiority over surface irrigation and conventional soil fertilization to ensuring the right amounts of water and nutrients available at the root zone and nourishes the crop requirements for stabilizing yield and quality of produce. It also increases the nutrient use efficiency of crop by permitting timely application of fertilizers in small quantities in the vicinity of root zone matching with the plants' nutrient need, besides substantial saving in fertilizer usage and reducing nutrient losses.

The treatments consisted of four irrigation (surface irrigation at IW/CPE 1.0, drip irrigation at 60, 80 and 100% ETc) and three N-fertigation (80, 100 and 120% of recommended dose of nitrogen) schedules. Healthy, vigorous and disease free seedlings of guava (cv. Khaja) were planted on 23rd June 2007 with a spacing of 5 m x 5 m. Every plant received about 2 kg wood ash, 500g bone meal and 15 kg farm yard manure during planting. The recommended dose of fertilizers perplant was 200 g N, 160 g P_2O_5 and 260 g K_2O. Phosphorus and potassium were applied in two equal splits broadcasted in mid-January and mid-August everyyear. Nitrogen as per treatment was applied through drip fertigation in 10 equal splits, whereas in surface irrigation, nitrogen was top-dressed in three equal splits. Crop was harvested in several pickings between July to November each year and data were added to calculate the total fruit yield.

The results showed that different levels of drip irrigation and nitrogen-fertigation had pronounced effects on the fruit yield of guava (Table 5). Drip irrigation at 100% of crop evapotranspiration (ETc), recorded significantly the highest fruit yield (19.7 t/ha) and was found superior to the conventional surface irrigation at IW/CPE 1.0 and remaining drip irrigation schedules. The lowest fruit yield was observed at drip irrigation at 60% ETc. Drip irrigation at 100% ETc increased the fruit yield by about 62.3, 16.5 and 11.2 % over drip irrigation at 60% ETc, 80% ETc and surface irrigation, respectively. These variations in yields might be ascribed to the differences in wetting pattern and relative water uses by the crop as influenced by different irrigation schedules and amounts of water application. The results are in agreement with findings with Ramniwas *et al.* (2012) who observed maximum fruit yield of guava at 100% irrigation water through drip system. Maximum yield under drip irrigation at 100% ETc was might be due to the attainment of favourable soil environment which likely to have facilitated the better water utilization (Raina *et al.*, 1999) and higher uptake of nutrients by crop (Rumpel*et al.*, 2003) and excellent balance of soil-water-air with plenty oxygen concentration in the root zone (Bangar and Chaudhary, 2004). The lowest fruit yield in surface irrigation, on the contrary, is likely to be water-stress situation during critical period coupled with choked aeration in first few days immediately after irrigation and losses of water and fertilizer nutrients in deep percolation below the crop root zone on heavy load of irrigation water application (Raina *et al.*, 2011). Similarly, fruit yield was increased significantly with increase in levels of nitrogen fertigation/fertilization. Nitrogen fertigation at 120% RDN showed maximum yield (17.6 t/ha) which was superior to 80% of RDN, but at par with 100% of RDN. This reveals to the fact that optimal dose of fertilizer nitrogen applied through drip system is necessary for achieving higher fruit yield. The interaction effects between irrigation method and nitrogen level recorded that drip irrigation at 100% ETc with 120% of RDN documented maximum fruit yield (21.6 t/ha) and was at par with drip irrigation at 100% ETc with 100% of RDN (21.2 t/ha). This might be ascribed to the timely supply of judicious amounts of water and nitrogen in the root zone matching with the crop demands, which thereby helps in promoting fruit yield of crop.

The amounts of water applied during the cropping season by drip irrigation at 0.6, 0.8, 1.0 ETc and surface irrigation at IW/CPE 1.0 were 432, 576, 720 and 850 mm, respectively. The corresponding figures for crop water use were 933, 1074, 1215 and 1344 mm, respectively. The results showed that maximum water use efficiency of 16.2 kg/ha-mm was obtained from drip irrigation system at 1.0 ETc, followed by that of 15.1 kg/ha-mm at 0.8 ETc and 12.0 kg/ha-mm at 0.6 ETc. Water use efficiency of 13.5 kg/ha-mm was observed in surface irrigation. About 9.6 to 30.6 % of water could be saved with various drip irrigation schedules over surface irrigation. The higher water use efficiency and water saving in drip irrigation as compared to surface irrigation was the results of precise amount of water delivery directly in the crop root zone at right time and better water utilization coupled with minimization of water losses in deep percolation and surface runoff (Dubey *et al.*, 2002).

Maximum BCR of 3.79 was observed in drip irrigation at 1.0 ETc and minimum BCR of 1.82 in drip irrigation at 0.6 ETc (Table 6). Surface irrigation at IW/CPE 1.0 recorded BCR of 3.36. In terms of economic benefit, drip irrigation at 0.8 ETc exhibited BCR of 3.01. The higher economic benefit in drip irrigation at 1.0 ETc was found to be more profitable compared to surface irrigation. This was mainly due to increase in fruit yield coupled with reduction on labour cost incurred on weeding and fertilizer application.

It may be advocated that drip irrigation at 1.0 ETc with full recommended dose of N-fertigation could be recommended for higher fruit production and economics of guava. Alternatively, surface irrigation scheduling at IW/CPE 1.0 could be advocated if the initial investment for laying the drip irrigation system is an impediment for the guava growers.

Table 5: Interaction effects of different irrigation schedules and nitrogen fertilization/fertigation on fruit yield of guava (pooled data of 2009-10, 2010-11 and 2011-12)

Irrigation	Fruit yield (t/ha) Recommended dose of nitrogen			
	80%	100%	120%	Mean
Surface at IW/CPE 1.0	15.9	19.1	19.4	18.1
Drip at 0.6 ETc	9.6	11.8	12.2	11.2
Drip at 0.8 ETc	14.9	16.7	17.1	16.2
Drip at 1.0 ETc	16.3	21.2	21.6	19.7
Mean	14.2	17.2	17.6	-
	I	N	I x N	-
CD (0.05)	1.7	1.1	1.4	

ETc: Crop evapotranspiration; RDN: N:P:K:: 200:70:215 g/plant/year; N applied in soil in surface irrigation and fertigated in drip system

Table 6: Fruit yield, water use efficiency and economics of guava as influenced by drip irrigation schedules and nitrogen fertigation(average data of 2009-10, 2010-11 and 2011-12)

Treatment	Fruit yield (t/ha)	Irrigation water (mm)	Total water use(mm)	Water use efficiency (kg/ha-mm)	Water saving (%)	Benefit-cost ratio (BCR)
Surface irrigation @ IW/CPE 1.0	18.1	850	1344	13.5	-	3.36
Drip at 0.6 ETc	11.2	432	933	12.0	30.6	1.82
Drip at 0.8 ETc	16.2	576	1074	15.1	20.1	3.01
Drip at 1.0 ETc	19.7	720	1215	16.2	9.6	3.79

ETc: crop evapotranspiration; RDN: N:P:K::200:70:215 g/plant/year; N applied in soil in surface irrigation and fertigated in drip system

Effect of micro- and sub-surface irrigation on pineapple

Pineapple (*Ananascomosus* Merr.) is an important and popular fruit crop in India. It is extensively grown in different parts of the country is due to its wide adaptability to varying soil and climatic conditions. The area and production of this crop in the state of West Bengal is around 13.4 thousand ha and 379.2 thousand ton, respectively. It has immense potential in increasing the productivity and yield sustainability with assured supply of irrigation water especially during the lean dry season. The curtailment in the liberal supply of water in the critical physiological stages of crop limits its growth and productivity. The pineapple growers in this region generally follow the traditional surface irrigation which is quite inefficient leading to excessive losses of water and nutrients through runoff and deep percolation beyond the root rhizosphere leading to low yield. Different types of microirrigation systems such as drip, microsprinkler and subsurface have the potential in enhancing crop productivity and quality of produce. It also provides opportunity of judicious use of water, fertilizer and other inputs, and requires less energy for irrigation.The information on the efficacy of micro irrigation system in pineapple is limited, thus the present study was initiated.

The experiment consisted of eight treatments viz., drip irrigation at 100% Eo,drip irrigation at 80% Eo,drip irrigation at 60% Eo, microsprinkler irrigation at 100% Eo, microsprinkler irrigation at 80% Eo, microsprinkler irrigation at 60% Eo, subsurface irrigation at 100% Eo and conventional surface irrigation. Thirty day-old healthy suckers of pineapple cv. Kew were planted at a spacing of 30 cm x 50 cm x 70 cm in paired row system for a density of 44,440 plants/ha. Crop was planted on 19th November of 2005, 2007 and 2009 and harvested in several pickings in June-July of 2007, 2009 and 2011, respectively. The recommended fertilization rate for pineapple was 600-400-600 kg/haofN, P_2O_5 and K_2O. Farm yard manure @ 10 t/ha along with full P and K and one-fourth of N were given at planting and the remaining N was top-dressed in three equal splits at 15, 50 and 65 weeks after planting.

The analysis of data showed that different levels of drip, microsprinkler and subsurface irrigation registered significantly the higher fruit yield over surface irrigation (Table 7). Being a shallow rooted crop, pineapple appeared to be sensitive to the curtailment of water application. Significantly the highest fruit yield (56.86 t/ha) was obtained with drip irrigation at 100% of evaporation replenishment (E_0). The increase in yield over surface irrigation was 8.6, 10.9 and 15.4% in drip irrigation at 60, 80 and 100% E_0, respectively. The improvement in yield under drip irrigation might be ascribed to better water utilization, higher uptake of plant nutrients and excellent maintenance of soil-water-air relationship with higher oxygen concentration in the root zone by way of timely and precise application of water directly to the crop root zone with concomitant reduction of nutrients losses through runoff, leaching and deep percolation (Tiwari *et al.*, 2005). In addition, drip irrigation might facilitate to maintain the soil moisture at field capacity in the crop root zone which might have influenced the root CEC and thereby increased the water and nutrient uptake for increasing fruit production (Bangar *et al.*, 2004). The Improvement of soil water distribution around the root rhizosphere under drip irrigation as compared to conventional surface irrigation system might also result in larger and heavier fruits of pineapple

(Paoli *et al.*, 2006). In the present study, the individual weight of the fruit without crown under drip system regardless of irrigation levels varied from 1.4 kg to 1.6 kg.

Similarly, among the three levels of microsprinkler irrigation system, application of irrigation at 100% of E_0 recorded maximum fruit yield over 80 and 60 % of E_0. The increase in yield at highest irrigation level of microsprinkler system was only 2.7 % in comparison to surface irrigation method. These indicate that microsprinkler irrigation in promotion of fruit yield was not as effective as in drip irrigation. The effect was almost comparable with surface irrigation, mainly due to unbalanced soil moisture distribution pattern around the root zone caused by the extended wetted front and considerable evapotranspiration loss of applied water. The subsurface irrigation also gave significantly the higher fruit yield as compared to surface irrigation and the increase in yield was only 5.5 %. The performances of subsurface irrigation in enhancing the fruit yield were relatively higher than the microsprinkler irrigation. The results are in agreement with the findings of Matouk *et al.* (2000) who observed the highest fruit yield of grape under subsurface irrigation as compared to surface irrigation system. The conventional surface irrigation, on the other hand, resulted in water stress during critical period coupled with aeration problem in first few days immediately after irrigation. There was considerable wastage of water and available plant nutrients in runoff and deep percolation below root zone and appearance of a chain of undesirable hazards such as water logging, weed infestation and soil-water-nutrient imbalance due to excess load of water application which was reflected in declined fruit yields (Hegde and Srinivas, 1990).

During the cropping season, the water use by the crop was 954, 859 and 762 mm for drip irrigation at 100, 80 and 60 % of E_0, respectively and the corresponding data for microsprinkler irrigation was 987, 883 and 777 mm. This figure was 997 mm for subsurface irrigation and 1164 mm for surface irrigation (Table 7). The results showed that water use efficiency was different in different micro-irrigation systems. However, the higher crop water use efficiency was recorded in drip irrigation immediately followed by microsprinkler and subsurface irrigation and the least was observed in surface method of irrigation. Relatively higher water use efficiency in drip irrigation in comparison to microsprinkler, subsurface and surface irrigation was the result of better utilization of irrigation water as precise amounts of water was delivered directly into crop root zone. Water savings under micro-irrigation irrespective of irrigation regimes was found to vary from 14.3 to 34.5% as compared to surface irrigation. Maximum water saving of 34.5% followed by to 33.2% was obtained by drip and microsprinkler irrigation at 60% of E_0, respectively. The water savings for subsurface irrigation system was only 14.3%.

Table 7: Effects of drip, microsprinkler, subsurface drip and surface irrigation on fruit yield, water use efficiency and economics of pineapple (pooled data of 2005-2007 to 2009-2011)

Irrigation regime	Fruit yield (t/ha)	Water use (mm)	WUE (kg/ha-mm)	Water saving (%)	BCR
Drip at 100% E_0	56.86	954	59.63	18.0	3.15
Drip at 80% E_0	54.64	859	63.63	26.2	3.20
Drip at 60% E_0	53.51	762	70.22	34.5	3.32
Microsprinkler at 100% E_0	50.58	987	51.25	15.1	2.86
Microsprinkler at 80% E_0	47.90	883	54.23	24.1	2.85
Microsprinkler at 60% E_0	46.23	777	59.47	33.2	2.91
Subsurface at 100% E_0	51.97	997	52.13	14.3	2.84
Surface at 50 mm depth	49.26	1164	42.34	-	2.55
CD (0.05)	1.98	-	-	-	-

Eo: evaporation replenishment

The economic feasibility of different micro-irrigation systems on pineapple production showed that higher benefit-cost ratio was obtained from drip irrigation, followed by microsprinkler and subsurface irrigation, respectively. Higher levels of irrigation through drip or microsprinkler resulted in higher monetary return, but lesser BCR and *vice-versa* for the crop. BCR of subsurface irrigation was 2.84. Lowest BCR was obtained from crop adopting conventional surface irrigation.

It is concluded that drip irrigation technology at 60% of evaporation replenishment for pineapple was much remunerative in terms of higher fruit yield, water use efficiency, water savings and benefit-cost ratio and thus recommended to the pineapple growers in the region.

Effect of drip irrigation and integrated nitrogen management on gladiolus

Gladiolus (*Gladiolus grandiflorus* L.) is one of the most popular ornamental bulbous plants grown in many parts of the world for its dazzling florets colour, sturdy spike, size, attractive appearance and keeping quality. It has high economic value because it is one of the most important cut flowers grown in India. The plant is very responsive to the optimum use of irrigation water with adequate and balanced use of inorganic fertilizers and organic manures which can sustain the growth and yield of flower. In the Gangetic alluvial plain the flower yield is adversely affected due to inappropriate irrigation and nutrient management. Thus the field trial was conducted.

There were four irrigation treatments viz., conventional surface irrigation and drip irrigation at 100, 80 and 60% of evaporation replenishment and four nutrition treatments *viz.*, control (without fertilizers and manure), 100:60:60 kg NPK/ha,

50:30:30 kg NPK/ha + vermicompost @ 2.5 t/ha and vermicompost @ 5 t/ha. Vermicompostwas incorporated in the soil during the final land preparation. Full P and K and 1/3 N was applied as basal. The remaining 2/3 N was applied in two equal splits at 30 and 60 days interval after basal application. Medium sized corms were planted during second week of November each year in lines maintaining row and plant spacing of 30 and 25 cm, respectively at a depth of 5 cm. Flower was harvested 5-times between mid January and first week of February.

The results of the study demonstrated that the flower yield of gladiolus was significantly influenced by the levels of gravity drip irrigation, integrated nitrogen application and their interactions (Table 8). Maximum spike yield of 9.30 t/ha was obtained with drip irrigation at 1.0 Eo (evaporation replenishment) which was competitive with drip irrigation at 0.8 Eo. The traditional surface irrigation method registered significantly the lowest yield. The marked reduction in yield improvement parameters and spike yield in drip irrigation at 60% of evaporation replenishment (0.6 ETo) was particularly due to water stress which might have failed to fulfill the water requirement of the plant and consequently resulted in lower yield characters and thereby spike yield (Begum *et al.*, 2007). The water stress decreased the mobilizing ability of inflorescence and increased that of corms, reduced $^{14}CO_2$ fixation and slightly delayed assimilate translocation from the source leaves. The conjunctive use of 50% inorganic fertilizer N plus 50% organic vermicompost N in the recommended fertilizer nitrogen recorded maximum spike yield (9.25 kg/ha). The interaction between irrigation and nitrogen source revealed that the increased flower yield (10.13 kg/ha) was obtained with gravity drip irrigation at 1.0 Eo with 50% inorganic N plus 50% vermicompost N, which was competitive with 0.8 Eo coupled with 50% inorganic N plus 50% vermicompost N (9.86 kg/ha).

The water use efficiency by plant varied with the amount of water application and addition of inorganic and organic nitrogen sources either in single or combination. The interaction effect between irrigations and nitrogen sources revealed that drip irrigation schedule of 0.6 Eo with 50% inorganic N plus 50% organic N recorded the maximum water use efficiency of 65.63 kg/ha-mm.

The relationship for spike yield with amount of irrigation water applied is illustrated in Fig 1. A second degree polynomial equation is fitted to the data of spike yield and amount of irrigation water regardless of methods of water application. Coefficient of determination (R^2) value was found to be 0.99. Spike yield reached a maximum at 118 .8 mm water application with a yield of 9.4 t/ ha. This is interesting from irrigation management point of view. This focuses that higher amount of water application by conventional surface irrigation is not conducive for higher marketable flower production. On the contrary, precise amount of water application in several fractions based on evaporation replenishment through low-cost gravity drip irrigation is much advantageous in promotion of higher flower yield of gladiolus.

Table 8: Interaction effects of irrigation schedules and nitrogen sources on spike yield, water use, water use efficiency and economics of gladiolus (pooled over 2011-12, 2012-13 and 2013-14)

Irrigation x nitrogen	Spike yield (t/ha)	Irrigation (mm)	Total water use (mm)	WUE (kg/ha-mm)	Benefit: cost ratio
$I_1 N_1$	7.35	200	220.78	33.24	2.11
$I_1 N_2$	8.42	200	220.62	37.46	2.46
$I_1 N_3$	8.00	200	220.91	36.06	2.35
$I_2 N_1$	9.31	143.5	164.86	56.25	2.66
$I_2 N_2$	10.13	143.5	164.62	60.05	2.83
$I_2 N_3$	9.75	143.5	164.93	58.31	2.74
$I_3 N_1$	8.74	114.8	137.44	62.94	2.55
$I_3 N_2$	9.86	114.8	137.25	71.23	2.78
$I_3 N_3$	9.46	114.8	137.42	68.2	2.70
$I_4 N_1$	7.55	86.1	109.83	68.68	2.15
$I_4 N_2$	8.60	86.1	109.97	79.04	2.49
$I_4 N_3$	8.55	86.1	110.21	77.23	2.37
CD (0.05)	0.65	-	-	-	-

I_1: surface irrigation, I_2: drip irrigation at 1.0 Eo, I_3: drip irrigation at 0.8 Eo, I_4: drip irrigation at 0.6 Eo, N_1: 100% N as vermicompost, N_2: 50% N as vermicompost + 50% N as inorganic, N_3: 100% N as inorganic

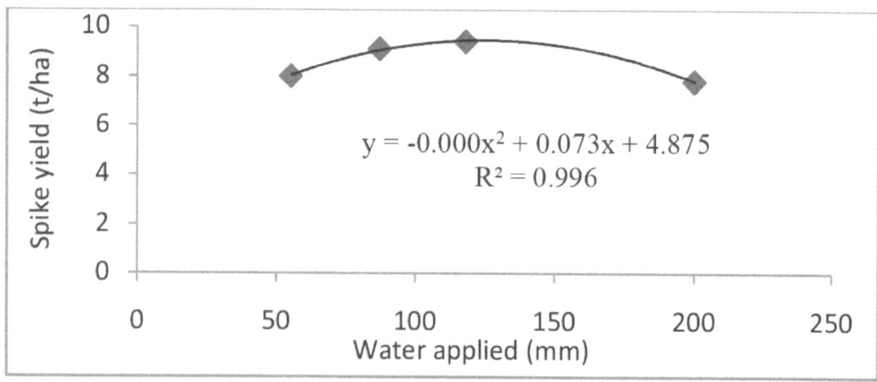

Fig 1: Relations of spike yield vs. amount of water

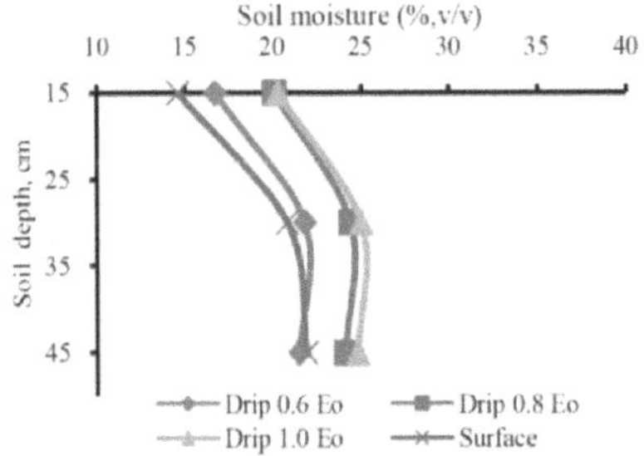

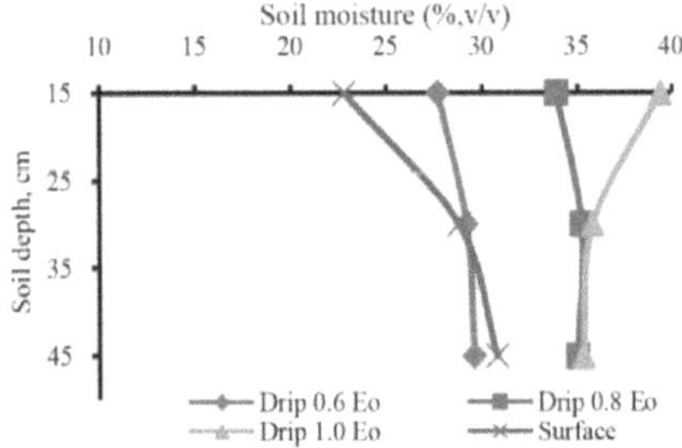

Fig. 2: Distribution of soil moisture before and after drip and surface irrigation

Table 9: Average soil moisture storage (cm) prior to the next irrigation under drip and surface irrigation methods at two depths of active crop root zone

Irrigation level	Soil moisture storage (cm)				
	Soil depth (cm)			Rooting depth (cm)	
	0-15	15-30	30-45	0-30	0-45
Drip at 0.6 E_0	2.5	1.9	1.9	4.4	6.3
Drip at 0.8 E_0n	3.1	2.7	2.6	5.8	8.4
Drip at 1.0 E_0an	4.3	2.7	2.5	7.0	9.5
Surface at 40 mm	1.8	2.0	2.1	3.8	5.9
				(50.9)	**(36.7)**

Figure in the parenthesis indicate the excess amount of moisture (%) stored within crop root zone (up to 30 cm and 45 cm soil depth) for overall drip irrigation as compared with conventional surface irrigation.

Soil moisture distribution under three gravity drip irrigation levels as compared with conventional surface irrigation was computed from the moisture data collected along the vertical distance of 0-15, 15-30 and 30-45 cm (Fig 2). The results indicate that the soil moisture contents decreased progressively with the increasing soil depth and the percentage variation was relatively more when higher amount of water was applied through drip irrigation. In case of surface irrigation, the variation in soil moisture was comparatively less than drip irrigation and consistently increasing with increasing depth of soil profile. It is conspicuous that maximum accumulation of soil moisture content was observed in 0-30 cm soil depth.

The results also show that the soil moisture storage gradually decreasing with increasing soil depth in drip system and the quantum of decrease was more in higher drip irrigation level than in lower level (Table 9). The trend was reverse in surface irrigation where soil moisture storage consistently increased with increasing depth of soil profile, but of lower magnitude. There were 50.9 and

36.7% more moisture storage at 0-30 cm and 0-45 cm depth considering active rooting zones under the overall drip system as compared to the surface irrigation. This reveals to the fact that under drip irrigation system there is maintenance of sufficient soil moisture within the root zone checking the deep percolation loss of soil moisture to deeper layers. The results also revealed to the fact that drip irrigation at 0.6 Eo was not sufficient to meet the crop evapotranspiration demand and the plants might have experienced soil moisture stress temporarily during some periods of their life cycles. Surface irrigation is indicative of soil moisture losses through deep percolation, seepage and evaporation.

It is concluded that drip irrigation at 80% of evaporation replenishment along with the conjunctive use of 50-30-30 kg NPK/ha and vermicompost @ 2.5 t/ha may be recommended for improving the quality and productivity of gladiolus.

Effect of drip irrigation and integrated nitrogen management on sweet corn

Sweet corn (*Zea mays* L var. *Saccarata*) is a popular multi-purpose cereal crop. In the lower Indo-Gangetic plain of West Bengal, the sweet corn has a high demand during the winter season, but its supply chain is inadequate to satisfy the consumers' requirements and retail market need. Some progressive farmers of the state has grown the crop commercially in patches during the *rabi* season with the conventional surface irrigation and imbalanced mineral nutrition especially the nitrogenous fertilization resulting in low yield and quality of crop. It is a deep rooted vegetable crop requiring high demand water and supply of plant nutrients especially the nitrogenous fertilizer throughout the growth stages. The cultivation of the crop under proper irrigation and right amount of nitrogen application may be encouraging for achieving higher yield and quality of produce. The lack of judicious irrigation and nitrogen management may cause low crop yield and poor return from investment. In this context, the present study was under taken.

The field experiment consisted of four irrigation viz., drip irrigation at 1.0, 0.8, 0.6 ETc and surface irrigation at 50 mm depth and four N nutrition viz., 100% inorganic N, 75% inorganic N + 25% inorganic N as vermicompost, 75% inorganic N + 25% inorganic N as FYM and 75% inorganic N + 25% inorganic N as mustard oil cake. The recommended dose of N was 120 kg/ha.

The marketable cob yield and quality parameters such as reducing sugar, total sugar and total soluble solids of corn were significantly influenced by the schedules of drip and conventional surface irrigation method and the integrated nitrogen management (Table 10). The data showed that main effect of both gravity drip irrigation at 1.0 ETc and surface irrigation resulted the highest cob yield of 6.91 t/ha which were superior to drip irrigation scheduling at 0.8 ETc (6.22 t/ha) and

0.6 ETc (5.41 t/ha). The higher amount of water application at desired crop physiological stages thus helps for keeping the plant photosynthetically active which might have resulted into proper growth and development and ultimately maximum possible yield (Rathore and Singh, 2009; Acharya *et al.*, 2013). Water stress, on the other hand, restricts transpiration, stomatal opening and reduced $^{14}CO2$ fixation resulting into low photosynthetic activity and retarded growth and expansion of leaf area and eventually the depressed yield (Karam *et al.*, 2002). The drip irrigation at 1.0 ETc recorded the higher values of quality parameters like reducing sugar (6.42%), total sugar (20.7%) and total soluble solids (7.4%). Surface irrigation demonstrated the lowest quality attributes in terms of total sugar and total soluble solids. Similarly, the main effect of integrated nitrogen management revealed that maximum cob yield (6.96 t/ha) was obtained with the conjunctive use of 75% inorganic N and 25% organic N as vermicompost and was significantly superior to that with the combined application of 75% inorganic N plus 25% organic N as FYM (6.38 t/ha) or mustard cake (5.88 t/ ha) and 100% inorganic N as fertilizer (6.23 t/ha). However, in respect with the improvement of quality parameters, the inclusion of 25% N through different organic sources in the fixed inorganic N fertilizer resulted in significant variations. The highest values of reducing sugar (7.14%) and total sugar (21.1%) were obtained with 75% inorganic N plus 25% organic N as mustard cake and that of total soluble solids (7.31%) with 75% inorganic N plus 25% organic N as vermicompost. Conversely, 100% inorganic N as fertilizer showed the least values of all quality parameters. The interactions recorded that significantly the mean maximum cob yield of 8.43 t/ha was obtained with surface irrigation provided with 75% inorganic N as fertilizer and 25% organic N as vermicompost. Significantly the minimum cob yield of 4.91 t/ha was recorded with drip irrigation schedule at 0.6 ETc coupled with 75% inorganic N as fertilizer and 25% organic N as mustard cake.

The highest water use (342.18 mm) was recorded in the farmers' practice of conventional surface irrigation and the lowest (168.25 mm) in drip irrigation schedule at 0.6 ETc. Accordingly, the average highest water use efficiency (32.15 kg/ha-mm) was obtained with drip irrigation scheduling at 0.6 ETc, whereas the average lowest WUE (20.19 kg/ha-mm) was recorded with conventional surface irrigation.As regards to the main effect of nitrogen management on water use efficiency, the conjunctive use of 75% inorganic N plus 25% organic N through vermicompost recorded maximum water use efficiency (29.90 kg/ha-mm) and the minimum WUE was registered in 75% inorganic N plus 25% organic N as mustard cake with a value of 26.01 kg/ha-mm. The interaction between irrigation and N management revealed that drip irrigation at 0.6 ETc with 100% N as inorganic fertilizer recorded maximum WUE of 36.01 kg/ha-mm.

Table 10: Effect of different irrigation schedules and nitrogen management on cob yield and quality, water use efficiency and economics of sweet corn (pooled over 2013-14 to 2015-16)

Treatments	Cob yield (t/ha)	Reducing sugar (%)	Total sugar (%)	TSS (%)	Water use (mm)	WUE(kg/ha-mm)
Irrigation schedule						
Drip 1.0 ETc	6.91	6.42	20.7	7.4	245.56	28.15
Drip 0.8 ET)	6.22	6.25	20.1	6.7	203.16	30.63
Drip 0.6 ETc	5.41	5.71	19.6	7.1	168.25	32.15
Surface irrigation	6.91	5.92	18.9	6.5	342.18	20.19
CD (0.05)	0.27	0.73	0.82	0.92	—	—
Nitrogen management						
100% inorganic N	6.23	4.77	17.7	5.96	240.39	27.75
75% inorganic N + 25% vermicompost N	6.96	6.29	20.3	7.31	239.58	29.90
75% inorganic N + 25% FYM N	6.38	5.96	20.8	6.79	239.62	27.45
75% N inorganic N + 25% mustard cake N	5.88	7.14	21.1	7.17	239.56	26.01
CD (0.05)	0.32	1.15	1.41	1.36	—	—

**including 20 mm pre-sowing irrigation for better seed germination and crop establishment*

Conclusion

It may be concluded that gravity drip-fertigation can be used as a potential tool to increase the marketable yield and quality of crops with augmenting water and nutrient use efficiency. The studies clearly suggest the multiple advantages of drip fertigation with applying precise amounts of water and fertilizer nutrients over surface irrigation and conventional soil fertilization. With the increase of coverage of area under drip irrigation, more area can be brought under drip fertigation programme. However, the farmers need to be educated and motivated to use fertilizer and water efficiently through fertigation which can considerably enhance the economy of farmers through more yields with less water and fertilizer consumption.

References

Abubaker, B.M.A., Alhadi, Mohammad, Shuang-En, Yu and Guang-Cheng, Shao. 2014. Different irrigation methods for okra crop production under semi-arid conditions. *International Journal of Engineering Research and Technology*, 3(4): 787-794.

Acharya, S.K., Shukla, Y.R. and Khatik, P.C. 2013.Effect of water regime on growth and yield of lettuce (*lactuca sativa* L.).*The Bioscan*, 8(1): 201-206.

Allen, R. G., Pereira, L. S., Raes, D. and Smith, M. 1998. Crop evapotranspiration. Guidelines for computing crop water requirements. FAO irrigation and Drainage 56, FAO, Rome, Italy, pp 300.

Ayas, S., Orta, H. and Yazgan, S. 2011. Deficit irrigation effects on broccoli (*Brassica oleracea*L. var. *Monet*) yield in unheated greenhouse condition. *Bulgarian Journal of Agricultural Science,* 17(4): 551-559.

Bangar, A.R. and Chaudhary, B.C. 2004.Nutrient mobility in soil, uptake, quality and yield of Suru sugarcane as influenced by drip fertigation in medium vertisols. *Journal of the Indian Society of Soil Science,* 52(2): 164-171.

Begum, R.A.; Rahman, M.N., Mondol, A.T. M.A.I., Rahman, M.J. and Khan, F.N. 2007. Effect of different moisture regimes on the growth and quality of gladiolus.*International Journal of Sustainable Crop Production*, 2(5): 43-45.

Brahma, S., Phookan, D.B., Kachari, M., Hazarika, T.K. and Das, K. 2010.Growth, yield and economics of broccoli under different levels of nitrogen fertigation. *Indian Journal of Horticulture,* 67: 279-282.

Deshmukh, M.M. and Sen, N.L. 2000. Studies on irrigation scheduling and point of application through drip irrigation to guava (*Psidiumguajava* L.) variety–Allahbadsafeda.*Haryana Journal of Horticultural Sciences,*29(3/4): 156-158.

Dubey, A.K., Pathak, R.A. and Pathak, R.K. 2002.Effect of drip irrigation on guava (*Psidiumguajava* L.) on plant growth and nutrient status of leaves.*Progressive Horticulture*, 34(1): 56-59.

Fanish, S. Anitta and Muthukrishnan, P. 2013. Nutrient distribution Uuder drip fertigation systems. *World Journal of Agricultural Sciences,* 9(3): 277-283.

Gomez, K. A. and Gomez, A. A. 1984. Statistical procedure for agricultural research, 2nd Edition, International Rice Research Institute, John Wiley and Sons, New York, pp 1-340.

Hagin, J., Sneh, M. and Lowengart, A. 2002.Fertigation: Fertilization through irrigation research Topic No. 23. International Potash Institute, Basel, Switzerland, pp 81.

Hedge, D.M. and Srinivas, K. 1990. Growth, productivity and water use of banana under drip and basin irrigation in relation to evaporation replenishment.*Indian Journal of Agron*omy, 35: 106-112.

Kadam, J.R. and Karthikeyan, S. 2006. Effect of soluble NPK fertilizers on the nutrient balance, water use efficiency, fertilizer use efficiency of drip system in a tomato. *International Journal of Plant. Science,* 1: 92-94.

Karam, F., Mounzer, O., Sarkis, F. and Lahoud, R. 2002.Yield and nitrogen recovery of lettuce under different irrigation regimes.*Journal of Applied Horticulture,* 4(2):70-76.

Kumar, J. L. G. and Senseba, T. 2008. Yield irrigation production efficiency and economic return of broccoli (*Brassica oleracea* var. *italica*) under different irrigation method and schedules. *Journal of Environmental Research and Development* 2(4): 513-522.

Matouk, A.M., El-Gindy, A.M., El-Adl, M.A. and Bondok, M. Y. 2000.Grape yield response to subsurface drip irrigation in old valley. *Egyptian journal of Agricultural Res*earch, 18: 1411-1433.

Mohammad, M.J. and Said, Z. 2003. Enhancement of yield and nitrogen and water use efficiencies by nitrogen drip-fertigation of garlic. *Journal of Plant Nutrition,* 26(9): 1749-1766.

Panigrahi, P., Sahu, N.N. and Pradhan, S. 2011. Evaluating partial root-zone irrigation and mulching in okra (*AbelmoschusesculentusL.*) under a sub-humid tropical climate. *Journal Agriculture and Rural Development in Tropics and Subtropics,*112(2): 169-175.

Paoli, E.R., Perez, F.R. and Echevarria, J.Z. 2006.Increasing pineapple productivity with improved management practices. In: Proceedings of the 18th World Congress of Soil Science, Philadelphia, Pennsylvania, USA, pp159.

Patel, N. and Rajput, T.B.S.2004. Fertigation - A technique for efficient use of granular fertilizer through drip irrigation. *IEI Journal,* 85: 50-54.

Patel, N. and Rajput, T.B.S. 2010.Use of simulation modeling for enhancing potato production using sub-surface drip. *Potato Journal,* 37: 21-27.

Pattanaik, S.K., Sahu, N.N., Pradhan, P. C. and Mohanty, M. K. 2003. Response of banana to drip irrigation under different irrigation designs. *Journal of Agricultural Engineering* ISAE, 40(3): 29-34.

Pawar, D.D. and Dingre, S.K. 2013. Influence of fertigation scheduling through drip on growth and yield of banana in western Maharashtra. *Indian Journal of Horticulture,* 70: 200-205.

Pramanik, S. and Patra, S.K. 2014. Influence of fertigation on fruit yield, water use and distribution efficiency and economics of guava (*Psidiumguajava* L.). *Journal Soil and Water Conservation,* 13(2): 164-169.

Pramanik, S., Tripathi, S.K., Ray, R. and Banerjee, H. 2014.Economic Evaluation of Drip-fertigation System in Banana cv. Martaman (AAB, Silk) Cultivation in New Alluvium Zone of West Bengal. *Agricultural Economics Research Review,*27(1): 103-109.

Raina, J.N., Sharma, T. and Suman, S.2011. Effect of drip fertigation with different fertilizers on nutrient distribution in soil, leaf nutrient content and yield of apricot (*Prunusaremeniaca* L.). *Journal of the Indian Society of Soil Science,* 59: 268-277.

Raina, J.N., Thakur, B.C. and Verma, M.L. 1999. Effect of drip irrigation and polythene mulch on yield, quality and wateruse efficiency of tomato (*Lycopersiconesculentum*). *Indian Journal Agricultural Science,* 69 (6): 430-433.

Rajurkar, G., Patel, N. and Rajput, T.B.S. and Varghese, C. 2012. Soil water and nitrate dynamics under drip irrigated cabbage. *Journal of Soil and Water Conservation,*11(3): 196-204.

Ramniwas, R.A., Kaushik, D.K. Sarolia, S.P. and Singh, V. 2012.Effect of irrigation and fertigation scheduling on growth and yield of guava (*PsidiumguajavaL.*) under meadow orcharding.*African Journal of Agricultural Research,*7(47): 6350-6356.

Rathore, A.C. and Singh, J.N. 2009.Optimization of nitrogen application and irrigation for improved growth and spike production of tuberose (*Polianthus tuberose* L.).*Indian Journal of Soil Conservation,* 37(1): 45-49.

Rumpel, J., Kaniszewski, S. and Dysko, J. 2003. Effect of drip irrigation and fertilization timing on yield of onion.*Journal of Vegetable Crop Production,* 9(2): 65-73.

Sandal, S. K. and Kapoor, R. 2015.Fertigation technology for enhancing nutrient use and crop productivity: An overview. *Himachal Journal of Agricultural Research*, 41(2): 114-121.

Sharma, I.P. and Kumar, S. 2007. Effect of drip irrigation and N and P application on soil moisture distribution and water uses efficiency of chrysanthemum (*Dendranthemagrandiflorum*) in Inceptisols. *Indian Journal of Soil Conservation,* 35(1) 50-53.

Shedeed, Shaymaa I., Zaghloul, Sahar M. and Yassen, A. A. 2009.Effect of method and rate of fertilizer application under drip irrigation on yield and nutrient uptake by tomato.*Ozean Journal of Applied Sciences,* 2(2): 139-147.

Singandhupe, R. B., Rao, G. G. S. N., Patil, N. G. and Brahmanand, P. S. 2003. Fertigation studies and irrigation scheduling in drip irrigation system in tomato crop (*Lycopersiconesculentum* L.) *European Journal of Agronomy,* 19: 327-340.

Tarawalie, I. F., Wengang, X., Guangcheng, S. and Chunli, H. 2012. Effect of water use efficiency on growth and yield of hot pepper under partial root-zone drip irrigation condition. *International Journal of Scientific and Engineering Research,*3(1): 1-13.

Tiwari, K. N, Singh, A. and Mal, P. K. 2003.Effect of drip irrigation on yield of cabbage (*Brassica oleracea* L. *var. capitata*) under mulch and non-mulch conditions. *Agricultural Water Management* 58:19–28.

Tiwari, K.N., Mal, P.K., Singh, R.M. and Chattopadhya, A. 1998. Response of okra to drip irrigation under mulch and no-mulch conditions.*Agricultural Water Management,* 38: 91-102.

Tiwari, K.N., Singh, B.K., Chourasia, S.K., Mandal, S. 2005. Yield response of pineapple (*Ananascomosus*Merr.) cv. Kew under drip irrigation and plastic mulch. International conference on plasticulture and precision farming, Ashoka, New Delhi, 17-21 November.

Vankar, R.R and Shinde, P.P. 2007. Effect of micro irrigation systems and polythene mulches on yield and yield attributes of okra. *The Orissa Journal of Horticulture,* 35(1): 22-26.

9

Bio-drainage for Improving Agricultural Productivity in Low Lying Areas

Sagar Maitra[1] *and A. Zaman*[2]

[1&2]*MSSSoA, Centurion University of Technology and Management*
Paralakhemundi-761 211, Odisha, India
Email: sagar.maitra@cutm.ac.in

Abstract

The bio-drainage system used to control the water logging hazards and restore the ground water table depth at a level favourable for other crops to grow. The water utilization efficiency per unit area of land improved through bio-drainage and associated cropping activities. The huge quantity of shallow and good quality groundwater that remains perpetually unutilized by the prevalent lowland cultivation system was effectively used by the biological drainage system. Thus the bio-drainage is expected to alleviate the land degradation problem due to high depth of water stagnant, to restore the crop cultivation programme on a sustainable way, to improve the water management practices, to improve the economic conditions of the lowland farmers and finally, to conserve the ecological values of this fragile agro-ecosystem.The objectives of efficient bio-drainage system are to evaluate ET and depended on growth characteristics of different trees and vegetation species having bio-drainage potential in the waterlogged land thereby to study the water table behaviour in the land under plantation and in the surrounding region and use the result in developing suitable plantation geometry to ensure desired water table draw-down, to assess interaction between bio-drainage plantation and co-existing crops and/or companion crops, to assess the value of the tree species through fodder, fuel wood and timber use, and to development of a new technology for land reclamation and evaluating its impact on soil, water and plant bodies in the basin.

Keywords: Bio-drainage;waterlogged areas; crop productivity; sustainable agriculture

Introduction

Bio-drainage technology helps in reclamation of salt affected waterlogged areas. Benefits include increased cropping intensity, increased nutrient use efficiency, growing arable crops including pulses and oilseed, which otherwise is not possible on waterlogged soils, and increased employment generation.The driving force behind the bio drainage concept is the consumptive water use of plants. The deep-rooting characteristics of trees make them extremely efficient users of water whereas shallow-rooted grasses and crops have limited access to underlying water tables, deep-rooted trees can access water tables down to several metre.

One of the major factors determining the sustainability of plant water use issalt balance. If the salts moving into the rootzone are not either (i) taken up by the vegetation and harvested or (ii) removed from the rootzone by leaching, the vegetation is doomed to succumb to salinity. Where bio drainage results in salt accumulation, engineering assistance is needed to make the system sustainable. Where the groundwater table is sufficiently declined due to high groundwater withdrawal by the trees, the land between the rows may even be put under field crops to give additional return to the land holders. The root system of herbaceous annuals, including majority of crop plants, has little contact and impact on water table levels. Farmers of disadvantageous areas lose precious cultivable land due to sustained waterlogging and salinity. Attempts to reclaim this land by providing conventional subsurface drainage through perforated pipes have not always proved effective. Bio-drainage presents itself as a cost-effective feasible and environment-friendly option that farmers could adopt to reclaim their land. It is based on the ability of plants and trees to transpire water luxuriantly and thus remove excess water and salinity from the soil system for restoring crop cultivation.

Scientific basis of bio-drainage

In natural environments the components of the hydrological system, i.e. rainfall, evapotranspiration, change in soil-water storage and drainage is in balance. Periods of high rainfall might temporarily result in increased drainage flows, a rise in the groundwater table and/or soil moisture storage, then over a period of about 5-10 years equilibrium is established. Vegetation plays a vital role in the evapotranspiration and soil-water storage components of water balance. The natural vegetation is cleared and replaced by crops or tree plantations, the seepage losses to the groundwater table under the new land use system are either higher or lower than under the pre-clearing situation. The increased seepage scenario prevails and the development of land for either rainfed or irrigated agriculture generally increases groundwater recharge rates. Where this increased recharge results in shallow water tables, evaporation at the soil surface causes groundwater to move upwards through the soil, and consequently salts accumulate in the rootzone, limiting plant growth. Plantations of fast growing tree species such as eucalyptus, when grown on land, could also result in strongly reduced accessions to the groundwater and in the drying up of wells and springs.

The driving force behind the bio-drainage concept is the consumptive water use of plants. The exemplary observations over a period of time shown the consumptive water eater use of the different plantation trees were given below in Table 1 and Table 2. The consequent decline in the depth of groundwater table was given in Table 3. One of the major factors determining the sustainability of plant productivity (and thus plant water use) processes is salt balance. If the salts moving into the root zone are not either (1) taken up by the vegetation and harvested or (2) removed from the root zone by leaching, the vegetation is doomed to succumb to salinity. Therefore, salt balance considerations are considered to be most crucial to the viability of bio-drainage. The deep-rooting characteristics of trees make them extremely efficient users of water. The shallow-rooted grasses and crops have limited access to underlying water tables wherein deep-rooted trees can access water tables down to several meters.

In view of considering their long-term viability, bio-drainage systems have been to be subjected to the some scrutiny as other plant-based biological systems with regards to nutrient and salt-balance, soil texture and other important parameters. Moreover, as large-scale adoption of bio-drainage crops could include considerable tracts of land, socio-economic considerations have to be taken into account when considering the adoption of this technique.

Table 1: Consumptive use of water for plant species: Luxury water consumption

Plant species	Consumptive use of water (m^3/ha/annum)
Willow	7700 – 13500
Poplar	6500 – 23500*
Alder	16800
Tamarix	22000 – 28000**

* 3-year old plant consumes 45 litres /day; ** 3-year old plant consumes 45 litres /day

Table 2: Daily water consumption by plant species (two and half month old)

Plant species	Water consumption (liters)/day
Acacia arabica	129
Eucalyptus tereticornis	137
Eucalyptus camaldulensis	143
Moris alba	153
Dalbergiasissoo	156
Tamarixarticulata	174

Table 3: Efficacy of Eucalyptus and Bamboo in controlling watertable (initial depth: 1.5 m)

Year after plantation	Watertable decline (m)	
	Eucalyptus	Bamboo
1.	1.95	1.09
2.	3.48	1.86
3.	3.76	2.46
4.	3.64	2.96

Bio-drainage controls water table

Shallow water table levels pose a threat to agricultural crops as they often result in salinization of the plant root zone. The management of irrigation areas often aims to keep water tables below the critical depth, which is defined as the depth at which capillary salinization is negligible. Sustainability of irrigation is determined by the leaching capability of soils. To avoid salinity problems, the salts present in the irrigation water have to be removed from the rootzone by leaching them either laterally to adjoining non-irrigated areas or streams or vertically down to levels below the vegetation root zone.

Plants can remove water from the soil either directly from the saturated zone below the water table, from the unsaturated capillary fringe above the water table, both resulting in water table control and leaching becomes restricted and salt accumulation processes begin to occur, especially where water tables are shallow, as is often the case in irrigation areas or from unsaturated topsoil layers after rainfall or irrigation that control groundwater recharge, leaching unimpeded when water application exceeds plant water demand, leaching takes place.

Seepage interception

Introduction of canal irrigation in arid zones underlain with brackish groundwater has led to rising water table and consequent waterlogging and salinity problems in the adjoining land. Under such a situation bio-drainage is envisaged as a benign and cost effective technology to lower the rising water table so as to take it well below 1.5 meter of the crop root zone of crop plants. Conventional surface and subsurface drainage techniques that are not cost-effective require periodic maintenance and have the problem of effluent management. On the other hand, it has been demonstrated that under ideal conditions, a tree canopy may lower water tables by 1-2 meter over a time period of 3-5 years.

Principles of bio-drainage planning

The aim of bio-drainage is to remove excess groundwater through the process of transpiration by vegetation. This is achieved by enhancing the transpiration capacity of the landscape by introducing high-water use vegetation types in large enough areas to balance recharge/discharge processes to maintain groundwater balances below the root zone of the agriculture crops. The following issues are considered in the development of bio-drainage systems:

1. **Water balance**: Bio-drainage plantations should be able to extract groundwater volumes equal to the net recharge. The water balance is to be maintained such that the water table is kept below the root zone.
2. **Plantation area**: The bio-drainage plantation area should be kept as small as possible. Irrigated agriculture is practiced primarily to produce high-value crops. Conversion of high-value cropping land to relatively low-return forestry may be difficult.
3. **Salt tolerance:**Bio-drainage crops need to be salt tolerant. The water use capacity of trees and other crops decreases with increase in water salinity. In the case of Eucalypt species, it reduces to about one-half of potential when the water salinity increases to about 8 dS/m.
4. **Drawdown of water table**: Crops, including trees, act as bio-pumps; they depress the water table directly underneath plantation areas and consequently lower the water table in the surrounding area. The drawdown effect under trees/crops depends on the tree/crop's water use, the rate of recharge in the surrounding area, the hydraulic conductivity of substrata and the depth to deeper barrier layers. Bio-drainage plantings should be established in blocks or strips and spaced to keep water table levels in the irrigated farmland in between the plantings below the rootzone. The harvesting of the bio-drainage plantations would need to be planned in such a manner that the "drainage" function is not lost.
5. **Salt balance:**The introduction of irrigation always upsets the salt balance. Although irrigation supplies often have relatively low salinities, the large volumes of water that are introduced in the landscape increase salt imports significantly. To achieve salt balance without conventional drainage, the irrigated crops, along with interspersed bio-drainage plantings, would have to accumulate the salts introduced by irrigation, and would subsequently have to be harvested and removed from the region. This is only achievable in situations where very low-salinity water is available to the plants.
6. **Economic aspects:**The growing of bio-drainage trees and crops requires a different operational management approach than the growing of agricultural crops. Costs associated with planting and maintenance and the income from harvesting may be considered to make the system acceptable to landholders.

7. **Social acceptance:**The introduction of new crops such as tree plantations affects rural societies. New markets might have to be developed. Active participation of local communities in the development of tree plantation-based bio-drainage systems is extremely important to overcome problems and to reap the benefits of the bio-drainage systems to the maximum extent.

Eco-physiological traits

The clonal efficiency to fit for bio-drainage suitability was evaluated by measuring the bio-drainage related eco-physiological traits *viz.*, transpiration rate, stomatal conductance, CINT and photosynthesis rate. The unique process of transpiration which helped in the loss of water to the extent of 360 lit/daycould be achieved only by tree crop viz., *Eucalyptus* spp. (Singh *et al.*, 1998). Bhutta and Chaudhry (2000) had studied the average daily water consumption by individual trees of *E. tereticornis* and *E. camaldulensis* were 137 lit/day and 143 lit/day respectively. Kumar *et al.*, (2006) also conducted an experiment and identified that *Eucalyptus* clones had high transpiration rate. The studies carried out by Pereira and Kozlowski (1976) on the stomatal conductance of *Eucalyptus camaldulensis* reported that high level of stomatal conductance of 0.67cm/sec could be effectively used as a bio-drainage agent for reclaiming the waterlogged area. Similarly, studies of Clemens *et al.* (1978) reported that the eucalyptus species had no stomatal control and used large quantities of water for transpiration, when abundant water was available in the root zone. The stomatal conductance of the *Eucalyptus* was higher and ultimately helped in draining of water by way of increased transpiration rate. Stomatal conductance is also an important trait for deciding the bio-drainage efficiency of clones. The modulation of photosynthetic efficiency of the clones helped to identify the efficient clones for bio-drainage. The genetic makeup of *Eucalyptus* and other physiological and biochemical characters might have played a vital role in making the clones fit for high bio-drainage efficiency. It was also important to note that the rooting behaviour, the root spread in the subsurface layer of the soil also helped *Eucalyptus* variety to absorb excess water.

Rainfed systems

A major problem with bio-drainage in rainfed conditions is that plant water requirement is generally low during cooler winter periods with high rainfall. So there is soil reservoir filling with winter rainfall and being depleted by vegetation water use over summer, thus creating a storage buffer to accommodate the next rainfall season.

The use of bio-drainage in waterlogged discharge areas is based on the concept of enhanced evapotranspiration. One special application of the bio-drainage concept is the amelioration of waterlogged soils with vegetation species having vigorous, deep and extensive root system. It is suggested that bio-drainage could be considered for waterlogged landscape depressions and canal seepage interception, and could be applied in 'parallel field drainage' arrangements as an alternative to conventional field drainage systems. The scope of bio-drainage seems to be more favourable in arid zone where drainage surpluses are small in relation to evapotranspiration rates. In Australia it is experienced that in discharge situations, enhanced evapotranspiration bio-drainage sites will lead to salinity, unless some form of conventional drainage is installed to control salt balance to the vegetation's root zone by removal of saline drainage effluent

Plants can use water both from the unsaturated part of the soil profile above the water table and from the saturated part below the water table. Plants in the latter category are called phreatophytes. They often grow in semi-arid climates where they tap deep water tables. Prosopis seedlings growing in desert in the southeastern United States where the groundwater is sufficiently shallow can occasionally produce deep enough roots to reach the water table in wet years. In the same area the introduced *Phreatophyte* and *Tamarix* has lowered the water table to such a low level that other species with shallower rooting depths are being eliminated.

Irrigated systems

In landscapes with undulating topography, recharge and discharge areas are often relatively easy to delineate. Recharge occurs at the higher parts of the landscape and discharge lower down the slope. In irrigation areas, with their flat topography and shallow water tables, the distinction between recharge and discharge is less clearly delineated and frequently areas that are discharging groundwater by evapotranspiration between irrigation events temporarily turn into recharge areas during and immediately after irrigation.

'Walking' plantations

Where trees, after harvesting, are replaced by shallow-rooted irrigated crops, water table control could be provided by a new plantation established on an adjacent site. This second tree crop should be planted about five years before the planned harvesting date of the original plantation so that the new trees would be old enough to have an impact on the water table and protect the adjacent paddocks from shallow water table induced salinity problems. The tree plantation width would depend on its drawdown impact. Under this system tree plantations 'walk' through the landscape, each new planting providing water table and salinity protection to the preceding site.

Benefits of bio-drainage system

- Involving certain categories of plants that habitually draw water from ground water
- Control waterlogging and Salinization of irrigated lands
- Evacuates undesirable and undrainable water with the help of transpiration from plantation trees
- Controlling of depth of groundwater table
- Help establish water and salt balance in the groundwater
- Acts as sub-surface drainage/ shallow tube-well/bio-pump
- Cost-effective and socially acceptable
- Environment friendly

Conclusion

In the arid and semi-arid regions, in addition to waterlogging the major problem associated with irrigation was salinization, which caused land degradation and consequently yields reduction. The solution to this problem was the installation of subsurface drainage measures that keeps the water table down and allows for leaching of salts from the crop root zone. Bio-drainage system summarily delay the salinization process and enhance agricultural sustainability in water-logged areas of the country.

References

Bhutta, M.N., Chaudhry, M.R. 2000. Biological control of waterlogging. In: Proceedings Eighth ICID international drainage workshop, New Delhi, India, vol IV, Group V: 33–45.

Clemens, J., Kirk, A.M., Mills, P.D. 1978. The resistant to waterlogging of three *Eucalyptus* species. effect of waterlogging and an ethylene-releasing growth substances on *E.grandis, E. Robusta* and *E. saligna*. *Oecologia* 34(2):125–131.

Dubey Kumud. 2012. SWOT analysis for the application of biodrainage technology to phytoremidiate water logged sites. International Journal of Social Forestry. 5(2): 47-59

Kumar, R., Pandey, S. and Pandey, A. 2006. Plant roots and carsequestration. *Current Science*. 91: 885–890.

Pereira, J.S. and Kozlowski, T.T. 1976. Leaf anatomy and water relations of *Eucalyptus camaldulensis* and *E. globulus* seedlings. *Can J Bot* 54: 2868–2880

Singh, B., Singh, V., Singh, R.P. and Srivastava, B.K. 1998.Economic prospects of vegetable intercropping in young Eucalypts plantation. *Annals of Agricultural Research.* 19(4): 470-474.

Toky, O.P., Angrish, R., Datta, K.S., Arora, V., Rani, C., Vasudevan, P. and Harris, P.J.C. 2011. Bio-drainage for preventing waterlogging and concomitant wood yields in arid agro-ecosystem in North Western India.*Journal of Scientific and Industrial Research.* 70: 639-644.

Zaman, A. 2012.Water management technologies to increase crop and income per drop with reference to effective utilization of low land eco-system in eastern region. *J. Water, Energy and Food Security*, 1(1):1-9.

10

Bio-remediation of Arsenic Problem for Sustainable Agriculture

A. Zaman[1] *and Md. Hedayetullah*[2]

[1]*Emeritus Professor, Centurion University of Technology and Management*
MS Swaminathan School of Agriculture Paralakhemundi-761211, Odisha, India
E-mail: profazaman@gmail.com
[2]*Assistant Professor, Department of Agronomy, Directorate of Research*
Bidhan Chandra KrishiViswavidyalaya, Kalyani-741235
West Bengal, India
Email: hedaye.bckv@gmail.com

Abstract

Adoption of agricultural practices is cost effective bioremediation techniques to mitigate arsenic problems and can contribute substantially to the cause of removal of arsenic and other heavy metal contamination of soils and water thereby quality enrichment in marketable agricultural produce leading to increased crop productivity in the area of intensive cropping under irrigated eco system. Though the farmers are already practicing some traditional techniques to mitigate the arsenic problems in soils and water, like cultivation of crops which are less affected to arsenic and heavy metals pollution, growing hyper accumulating crops and so many practices wherein the modern methods are not prevalently used by the farmers due to the lack of scientific validity of these techniques, particularly in the context of prevailing socio-economic conditions. Moreover, acceptance of these improved practices by the small farmers of this region is an evolutionary process and should be intensively proved. Hence, extensive studies need for bio or phyto-remediation of heavy metals including arsenic contamination for quality marketable agricultural produce, which ultimately can realize better crop productivity.

The general objective and goal of such study aims to evaluate the effectiveness of advanced bioremediation techniques in comparison with traditional system towards improving the quality of water for irrigation vis-à-vis soil and crop quality improvement in diverse soil types through on-farm field experiments.

Keywords: Arsenic; mitigation; bio-remediation; agriculture; sustainability

Introduction

Arsenic in elemental form is insoluble in water and it is soluble in oxidized.WHO's maximum permissible limit is 0.05 mg/L in India and 0.01 mg/L in USA. All the tube wells in 17 blocks of Nadia in West Bengal had been analyzed and 89% tube wells contained arsenic above 0.05 mg/L as reported. Average arsenic concentration in different water bodies other than groundwater is shown in Table 1 and spread of arsenic contamination in water/soil in Indian subcontinent is shown in Table 2.

Table 1: Arsenic concentration in water other than groundwater

Source	Arsenic (μg As/L)
Rainwater and snow	<0.002-0.59
Rivers	0.20-264
Lakes	0.38-1.00
Sea water	0.15-6.00
Ponds (West Bengal, India)	4-70
Canals (West Bengal, India)	40-150

Source: Welch *et al.* (1988) and ICAR (2003).

The level of arsenic in edible plants is generally low, often being close to the limit of detection, even when the crops are grown on contaminated land. The data indicated that when each soil type contains similar As concentrations, lower levels are found in plants grown on clays and silts, with their higher clay mineral and Fe/Al oxide content, than in plants grown on lighter soils, e.g. sands or sandy loams. This reflects the sorption characteristics of soils. In this context, the proposed study carried out (i) to develop more efficient hyper accumulator from the selected species using biotechnological tools and (ii) to select the cultivated crops and cultivars of local origin having resistant to heavy metal accumulation without impairing crop yield and economy of the farmers;

Table 2: Groundwater arsenic contamination in the Indian Subcontinent

State	Coverage	Level of contamination	Citation
West Bengal	**12 districts** (Mushidabad, Maldah, Nadia, North 24 Parganas, South 24 Parganas, Bardhaman, Howrah, Hooghly, Kolkata, Coochbehar, North Dinajpur and South Dinajpur), **111 blocks**	50-3700 µg/l	http://www.soesju.org/arsenic wb.htm
Assam	18 districts, 72 blocks **5 districts** (Barpeta, Dhemaji, Dhubari, Darrang and Golaghat) **4 districts** (Jorhat, Lakhimpur, Nalbari and Nagoan)	>50 µg/l100 - 200 µg / l228–657 µg /l	Singh, A. K. (2007) Current SCIENCE 92 (11); 1506-1515
Bihar	**12 districts** (Bhagalpur, Khagaria, Munger, Begusarai, Lakhisarai, amastipur, Patna, Baishali, Saran, Bhojpur, Buxar, and Katihar), **32 blocks and 201 villages.**	> 50 µg/l	Acharya, S.K. and Shah, B.A. (2004) *Environmental Health Perspective* 112 (1); 19- 20
Jharkhand	1 district (Sahibgunj)	> 50 µg/l	http://www.soesju.org/arsenic/ jharkhand.htm
Uttar Pradesh	**21 districts** (Ballia, Lakhimpur, Kheri, Bahraich, Chandauli, Ghazipur, Gorakhpur, Basti, Siddarthnagar, Balrampur, Sant Kabir Nagar, Unnao, Bareilly, Moradabad, Rae Bareli, Mirzapur, Bijnore, Meerut, SantRavidas Nagar, Shahjahanpur and Gonda)	> 50 µg/l	http://www.nerve.in/ news:253500133730
Madhya Pradesh	**1 district** (Rajnandgaon)	52-88 µg/l	*Press Trust of India, September 4, 1999*
Manipur	**1 district** (Thoubal)	798-986 µg /l	Singh, A. K. (2007) Current Science 92 (11); 1506-1515

Source: Adriano, 1986.

Soil plant relationships on arsenic

In common with most trace elements, the degree of uptake varies widely from species to species. Unlike some marine and fresh water organisms where very high concentrations (over 1000 mg As/kg fresh weight, similar to those in the sediments, have been found in some macrophytes), the levels in terrestrial plants remain well below the level in the soil. In general roots contain higher levels than stems, leaves or fruit. Whilst it is rare to obtain levels of As approaching 1 mg/kg fresh weight in food crops, grasses growing on spoil tips from old As mines in SW England contained upto 3460 mg/kg dry weight with spoil material levels up to 26530 mg/kg. Similar grasses growing in urban areas on soil containing 20 mg As/kg were found to have a maximum of 3 mg As/kg dry weight.

When a number of crops were grown on dredged spoil (35-108 mg As/kg) from the rivers As levels were shown by radishes with 0.8-2.1 mg As /kg dry weight. The order of decreasing As accumulation is: radish (1)> grass (0.33) > lettuce (0.26) > carrot (0.17) > potato tuber (0.07) > spring wheat grain (0.04).

Pot and field trials on the effect of applying sewage sludge, containing arsenate, to a sandy loam (pH 6) and to a calcareous loam (pH 8) indicated enriched factors that were 3-4 times greater in the lettuces and rye grass grown on the calcareous loam. The highest level of As reached in the crops was 2 mg/kg dry weight in lettuce. The increases in plant tissue As concentration were significantly correlated ($p < 0.01$) with soil As for both types of soil.

There are some reporting from the research findings that application of FYM and P_2O_5 could resist the release of native arsenic and could be lowering down to moderate toxic effect in soil-plant system. Application of Zn has reportedly been reduced the concentration of arsenic in summer rice. A large number of other significant findings have ensued from the extensive research at Bidhan Chandra KrishiViswavidyalaya. The application of FYM and phosphate was found to have opposing effect on release of native and applied As in the contaminated soils, with FYM reducing such release, thereby tending to moderate the toxic effect of As in the soil–plant system. Organic manures, namely compost, poultry manure, neem cake, etc., have also been examined as ameliorants, not only for rice, but for other crops as well (e.g., potato, elephant foot yam), with encouraging results. The crop with less water requirement also might be helpful to lowering down the effect of arsenic and its severity. In these backdrops, water productivity of summer rice with differential sowing and transplanting dates as its relation with exploitation of groundwater with arsenic contamination are to be quantified and to be established. The crop those require less water are considered as substitute for the dominant rice-

rice-fallow sequence in the affected areas to minimize the arsenic uptake by food crops where signified the importance of crop diversification during summer season. This apart, inclusion of pulses /other legumes /green manure crops in the cropping sequence, coupled with organic manure addition, was found helpful in moderating arsenic build-up in soil and plant parts. Pond-stored water contaminated groundwater was also found to get partially decontaminated over time by sedimentation-cum-dilution due to rainfall, and conjunctive irrigation using such water bodies reduced the dependence on groundwater irrigation. The research results also showed that the incorporation of pulse crop(s)/ legume(s)/green manure coupled with organic/compost in crop sequence (s) would also be helpful in moderating the effect of arsenic. BGA (*Anabaena* sp. and *Nostoc* sp.) and some bacteria were shown the effect having decontamination ability. The tube wells with activated alumina are being used to get arsenic free drinking water to mitigate the severity of the problem in these areas. Surface water is free from arsenic contamination and safe for use in crop cultivation. The efficient of judicious application of intermittent irrigation water in summer rice was found effective as integral part of water saving technologies.

Phyto-bio-remediation is defined as the use of vegetation or other macro- and microscopic biota for *insitu* or *exsitu* treatment of contaminated soils, sediments and water where green plants degrade, assimilate, metabolise or detoxify inorganic and organic chemicals in soils to environmentally acceptable levels (Thuraisingham *et al.*, 2004). In this context, phytoremediation may tend to offer a potentially useful avenue to address the problem of arsenic contamination of soil-crop system. Phytoremediation may also prove attractive due to its relative cost-effectiveness, coupled with its asthetic nature, namely the use of plants for 'clean-up' activities. Indeed, a number of plant species, including the dominant wild species, growing in the arsenic-affected areas of West Bengal, have the capability to accumulate arsenic.

The reported hyperaccumulation of arsenic (As) from the contaminated soils by the brake-fern, *Pterisvittata,* and its subsequent translocation into the above-ground biomass suggests that the plant-accumulated arsenic was present almost entirely in the toxic inorganic forms with the proportion of highly toxic As (III) being in fact much greater in the plant body than that of the less toxic As (V) form, as compared to the distribution of these two forms in the contaminated soil in which the fern grows (Ma *et al.*, 2001). Conversion of the plant-accumulated inorganic forms of the toxin to non (or less) – toxic organometallic forms by plant metabolism would certainly aid the detoxification process (Sanyal and Nasar, 2002). Such detoxification within the plant body assumes importance, particularly in view of the report that arsenic in plant residues may be mobilized by unicellular algae, namely *Polyphysapeniculas* resident in normal (moist) agriculturalsoils. The hyper accumulating crop species and its comparative sensitivityis shown in Table 3 and arsenic cotent in different economic crop produce is given in Table 4.

Table 3: Comparative sensitivity of various crops to arsenic

Crops	Tolerant	Moderately Tolerant	Low Tolerant
Fruit Crops	Apples, Grapes, Rasberries, Dewberris	Cherries, Strawberries	Peaches, Appricot
Vegetable Crops	Cabbage, Carrots, Tomato, Potato,	Beets, Squash, Turnips, Radish	Peas, Onion, Cucumber, Snap beans, Lima beans, Spinach
Field Crops	Rye, Mint, Asparagus, Wheat, Oats, Cotton, Peanut, Tobacco	Corn	Soybean, Rice
Forage Crops	Sudan, Blue, Italian Ryegrasses,	Crested wheat grasses, Timothy	Alfaalfa, Bromegrass, Clover, Vetch

Source: Adriano, 1986.

Table 4: Arsenic content in economic produce of crop

Crops	Arsenic content (mg/kg)
Rice	0.1-0.89
Potato	0.03-1.02
Wheat	0.45-1.08
Mustard	1.37-1.59
Sesame	0.31-1.02
Okra	0.91-4.2
Aurum	11.6-33.2
Banana	1.2-3.7
Papaya	1.8-3.6

Source: Sanyal, 1999; Sanyal and Dhillon, 2005

A scan of literature reveals a number of plant/microbial species, known for arsenic accumulation/or as bio-indicator, which can effectively remove arsenic (and other heavy metals) from the aquatic system, for instance, to the tune of 170 and 340 mg As/g dry weight of water hyacinth in its stem and leaves, respectively (Chigbo *et al.*, 1982), when grown in a pond containing 10 mg As/dm^3. However, such accumulated arsenic in water hyacinth (*Eichornia-crassipes*) is also liable to leaching out in the water body, particularly so on decomposition of such aquatic weed. Consequently, appropriate precaution has to be exercised while interpreting the arsenic status of aquatic environment by water hyacinth accumulation.

A number of microbial species (e.g. the bacterial species, namely *Proteus* sp., *Escherichia coli*, *Flavobacterium* sp.; *Corynebacterium* sp. and *Pseudomonas* sp., the fungus, namely *Candida humicola;* the freshwater algae, namely *Chlorella ovalis. Phaepdactuylum tricornutum. Oscillatoria rubescens*) has been reported to possess varying degrees of arsenic accumulating abilities (Sanyal and Nasar, 2002).

Several weed species, identified by the present authors, normally found along with crops like rice, potato, jute, mustard, etc., growing on arsenic contaminated soils (2-14 mg As/kg soil) and subjected to irrigation (given to the desired crops) with arsenic contaminated groundwater, were noted to accumulate considerable amounts of arsenic in their biomass.

However, it is worthy to note in this context that such accumulation of arsenic does *not* necessarily lead to its detoxification *per* se (Sanyal, 1999; Sanyal and Dhillon, 2005) unless the plant- accumulated toxin is effectively detoxified within the plant body by its metabolic processes. Hence, the success of such phytoremediation technique would depend on establishing a desirable plant community. A systematic search for phytoaccumulating or phytoexcluding plant species is necessary for this. Some of the techniques in this field, that may prove useful, include phytodegradation (the process in which the given plant species are capable of metabolizing and destroying the contaminant within the plant tissues), phytovolatilization or biotransformation (the process employing microorganisms and plants to transform the toxicants into the volatile forms to escape to the atmosphere), rhizofiltration (the process in which plants, used for clean-up, accumulate the toxicants), etc. (Das *et al*., 2005). This would certainly be a useful tool for remediating arsenic contamination in an effective manner.

Adoption of agricultural practices to mitigate arsenic problems

- Vermicompost application reduced the arsenic accumulation in root, shoot and grain of rice by 30%, 30% and 75%, respectively, whereas the same application reduced the arsenic load in root, shoot and seed of mustard at the rate of 20-25%. $FeSO_4$ (30 kg/ha) or $ZnSO_4$ (20 kg/ha) had significant effect on reducing arsenic accumulation in rice or mustard.
- Arsenic uptake *vis-à-vis* translocation factor were substantially reduced with higher level of phosphorus in the growing media.
- Low arsenic accumulating-crops like potato, pumpkin, sesame can be included in rice- based cropping system in arsenic exposed areas to bring down the arsenic load in food-chain. Replacement of 25% N-fertilizer by FYM along with increased P-application reduced arsenic accumulation in the economic produce of wheat, pea, potato and French bean.

- Arsenic content in rice significantly varies among the genotypes. Most of the high yielding rice cultivars accumulate greater than 500μg/kg of rice. Arsenic accumulation ability in rice grain is a polygenic trait.
- Possibilities of using pure cultures of bacteria and cyanobacteria capable of sustaining high arsenic concentration have been explored as soil ameliorating microorganism.
- Several weed species, identified by earlier research project, normally found along with crops like rice, potato, jute, mustard, etc., growing on arsenic contaminated soils (2-14 mg As/kg soil) and subjected to irrigation (given to the desired crops) with arsenic contaminated groundwater, were noted to accumulate considerable amounts of arsenic in their biomass. Table 4 demonstrates the extent of such arsenic accumulation by these weed species.
- Irrigation from adjacent pond or permanent water body, recharged by arsenic contaminated shallow tube well, significantly reduce the arsenic load of the irrigation water.

Remedial measures suggested

i) Identification/ development of genotypes/crops:

    a) which uptake less arsenic from the soil.

    b) where ratio of organic and inorganic form of arsenic is high.

ii) Efficient agro techniques:

    a. Increased use of FYM and other manures and green manure crops, inclusion of pulses/other legumes as well as application of appropriate amendments (Zn/Fe salts as and where applicable)

    b. Irrigation with pond-stored groundwater–decontamination facilitated by rainfall and sedimentation

    c. Recharge groundwater resource

    d. Prefer low-water requiring farmer-attractive cropping sequences (especially for the lean period).

iii) Identification of suitable soil ameliorating substances including organic matter and microbes which will reduce entry of arsenic into the plant.

iv) Suitable cost-effective phytoremediation options.

v) Creation of general awareness: Mass campaigning, holding of farmers' day, field demonstrations.

Conclusion

Arsenic toxicity of contaminated drinking water in humans in many parts of the world, including West Bengal, India, has deservedly received increasing concern. Arsenic toxicity in livestock and other animal species like poultry, fish and sea food has been given comparatively much less attention. On the other hand, arsenic contamination in agro-ecosystem, which acts as a conduit for the passage of the toxicant to human population via food web, came under serious consideration only recently.

The complex aspects of accumulation of arsenic in the food web *vis-à-vis* that in drinking water aloneis necessary for understanding, and its ultimate passage to the human populations. As vast arsenic affected area of Nadia and North 24 Paraganas is considered as vegetables export zone so immediate attention is also required towards meaningful mitigation options. A comprehensive approach is necessary to combat the present arsenic crisis. Some of the potential remedial options aimed at reducing the toxic effect of arsenic in agricultural systems include developing low arsenic accumulating genotypes, optimum conjunctive use of ground and surface water (e.g., use of harvested rainwater) irrigation with pond-stored excess rainwater and groundwater, recharge of the groundwater resource with rainwater, and enhancing water use efficiency for the agricultural crops by way of adopting appropriate soil and irrigation water management strategies.

Acknowledgement

The author carried out research on water management in the state of West Bengal and abroad having in depth collaboration with Central Ground Water Board and Bidhan Chandra KrishiViswavidyalaya and expressed in debt and acknowledges the resources collected from several sources as cited.

References

Chigbo, F.E., Smith, R.W. and Shore, F.I. 1982.*Environ. Pollut. Ser. A.,* 27, 31 – 36.

Das, D. Chatterjee, A., Mandal, B.K., Samanta, G., Chakraborty, D. and Chanda, B. 1995.*Analyst,* 120, 917-924.

Das, I. Ghosh, K. and Sanyal, S.K. 2005.*Everyman's Science,*40 (No. 2), 115 – 123.

Ghosh, K., Das, I., Das, D.K. and Sanyal, S.K. 2003.*Proc. National Seminar on Development in Soil Science:* 2003, 68th Annual Convention of Indian Soc. Soil Sci. held at C.S. Azad Univ. Agril. Tech., Kanpur, November 4-8, Abstr., 34.

Ghosh, K., Das, I., Saha, S., Banik, G.C., Ghosh, S., Maji, N.C. and Sanyal, S.K. 2004. *J. Indian Chem. Soc.*, 81, 1063 – 1072.

Ma, L.Q., Komar, K.M., Tu, C., Zhang, W., Cai, Y. and Kennelley, E.D. 2001. *Nature,* 409, 579.

Sanyal, S.K. and Dhillon, K.S. 2005.In: *Proceedings of the International Conference on Soil, Water and Environmental Quality: Issues and Strategies,* New Delhi, India, January 28 to February 1, 2005, p. 239 – 263.

Sanyal, S.K. and Nasar, S.K.T. 2002.In :*Analysis and Practice in Water Resources Engineering for Disaster Mitigation,* New Age International (P) Limited, New Delhi, 216 – 222.
Thuraisingham, R., Muller, R. and Arguello, R. 2004.*The Association for Environmental Health and Sciences*.

Section 3

Soil Resource Management

11

Soil Health: An Imperative for Efficient Nutrient Management for Sustainable Crop Production

Dilip Kumar Das

Former Professor & Head, Department of Agricultural Chemistry and Soil Science, BCKV & Emeritus Fellow (UGC), Institute of Agricultural Science University of Calcutta, 51/2, Hazra Road, Kolkata-700019, India
Email: dkdas1231@gmail.com

Abstract

In a developing country like India with exploding population pressure and shrinking availability of resources per capita, agricultural intensification maintain soil health is the only way to solve hunger and poverty. Soil is a fundamental resource base for agricultural production systems. Besides being primary medium for crop growth, soil functions to sustain crop productivity, maintain environmental quality and provide for plant, animal and human health. Soil organic matter (SOM) content is frequently identified as a primary attributes of soil quality assessment. It is evident that SOM influences various soil physical,chemical and biological properties. Besides, SOM serves as a slow release reserviour for macro-, secondary and micronutrients to plants. The impaired soil health due to imbalanced fertilizer use coupled with low use of organic manures is much blamed for declining fertilizer responses and crop productivity. Site specific integrated nutrient practices and site specific soil quality indices including conservation agriculture (CA) improve soil health and so under CA system, the establishment and maintenance of healthy soil condition is extricably linked to the achievement of effective nutrient management. Maintenance of soil health with respect to physical, chemical and biological indices through different management practices is of utmost importance to ensure food and nutritional security of the country. Therefore, it is felt necessary for having a relook at the state and appraisal of soil quality, impact on agricultural intensification, industrialization /urbanization and other anthropogenic activities on soil resource quality, issues involved and strategies required to fulfill the commitments of sustaining and aggrading soil health optimizing the effective utilization of agricultural inputs.

Keywords: Nutrient management; organic matter; soil health; soil quality indices

Introduction

Healthy soils are vital to a sustainable environment. They store carbon, produce food and timber, filter water and support wildlife and landscapes. However, there are increasing signs that the condition of soils has been neglected and that soils may not recover following loss and damage. Soil quality indices (SQIs) should be based on soil function, with the most important functions being those of environmental interaction, food and fibre production, providing a building platform, support for ecological habitat and biodiversity, the provision of raw materials and protection of cultural heritage.

Out of 67 potential indicators for all the functions were identified and subsequently classified on the basis of these soil functions.

Classification fell into three broad groupings:

- Those for which there was significant data and experience in interpretation;
- Those for which there was less data, but which could nevertheless be interpretedwith a good measure of confidence;

Those for which there was little data or which represented emerging science, butwhich might make useful indicators if more research were undertaken.

SQIs are not just useful for assessing the state or condition of soil; they canalso help shape soil and land use policies.Therefore, in order to identify an effect of land management practices or environmental stressor through the use of a soil quality index (SQI) an interpretative framework needs to be established. Interpretation of change in an indicator and a consistent understanding of what change means, especially in relation to direction and magnitude within this framework is most often effected through the use of 'trigger values', indices, thresholds and target ranges. There is a lack of scientific information on likely trigger values for indicators for the soil function of environmental interaction. Indicators of agricultural status of soils are almost exclusively chemical and often have well-established trigger values that reflect a soils state in regard to production. The application of such triggers to soils not under intensive agriculture, such as those under semi-natural habitats, which have often developed in conditions of low nutrient availability, is clearly inappropriate.

Soil Health

Soil health is defined as the continued capacity of soil to function as a vital living system, by recognizing that it contains biological elements that are key to ecosystem function within land-use boundaries (Doran and Zeiss, 2000; Karlen *et al.*, 2001). These functions are able to sustain biological productivity of soil, maintain the quality of surrounding air and water environments, as well as promote plant, animal, and human health (Doran *et al.,* 1996).

Soil quality was been defined as "the capacity of a reference soil to function, within natural or managed ecosystem boundaries, to sustain plant and animal productivity, maintain or enhance water and air quality, and support human health and habitation." Subsequently the two terms are used interchangeably (Karlen*et al.*, 2001) although it is important to distinguish that, soil quality is related to soil function (Karlen *et al.*, 2003; Letey *et al,* 2003), whereas soil health presents the soil as a finite non-renewable and dynamic living resource. For the purpose of this review, we will preferably rely on the soil health concept, since it directly includes interactions between plant inputs and soil in creating a healthy environment. Because of the numerous alternative uses of soil as a living resource, the meaning of the terms soil health and soil quality depend on the defined purpose such as for agricultural use (Andrews and Carroll, 2001; Doran and Parkin, 1994). In agriculture, we mainly pay attention to plant and animal productivity as these would be of greatest importance in cultivated soils (Doran *et al.*, 1996a) as opposed to urban soils (Idowu *et al.,* 2007). More comments on the meaning of these terminologies and the context in which they are applied can be found by examining the contributions of Doran and Zeiss, (2000). Protection of soil quality under intensive land use and fast economic development is a major challenge for sustainable resource use in the developing world (Doran *et al.*, 1996b). The basic assessment of soil health and soil quality is necessary to evaluate the degradation status and changing trends following different land use and smallholder management interventions (Lal and Stewart, 1995).

In Asia, adverse effects on soil health and soil quality arise from nutrient imbalance in soil, excessive fertilization, soil pollution and soil loss processes (Zhang *et al.,* 1996; Hedlund *et al.*, 2003).

Soil quality index is a function of the following parameters:

Soil Quality Index (SQI) = f (SP, P, E, H, ER, BD, FQ, MI)
Where, SP= Soil properties, P=Potential productivity, E=Environmental factors, H=Health of animal & human, ER=Erodibility, BD=Biological diversity, FQ=Food quality safety, MI=Management inputs.

Soil quality is a measure of the condition of soil relative to the requirements of one or more biotic species and or to any human need or purpose. According to the United States Department of Agriculture Natural Resources Conservation Service, "Soil quality is the capacity of a specific kind of soil to function, within natural or managed ecosystem boundaries, to sustain plant and animal productivity, maintain or enhance water and air quality, and support human health and habitation." The European Commission's Joint Research Centre proposed a definition, stating that "Soil quality is an account of the soil's ability to provide ecosystem and social services through its capacities to perform its functions under changing conditions."

The soil health refers to self-regulation, stability, resilience, and lack of stress symptoms in a soil as an ecosystem. Soil health describes the biological integrity of the soil community-the balance among organisms within a soil and between soil organisms and their environment."

Physical attributes can be as basic as color. It can also be used to describe more complex soil characteristics such as soil organic matter, nutrient amounts, soil structure, etc. These attributes can all be influenced by management practices and have the capability to enhance or diminish soil health.

The concept of soil health and soil quality has consistently evolved with an increase in the understanding of soils and soil quality attributes. Soil quality cannot be measured directly, but soil properties that are sensitive to changes in management can be used as indicators. Soil health indicators are needed that help smallholder farmers understand the chain of cause and effect that links farm decisions to ultimate productivity and health of plants and animals. The soil health approach is better applied when specific goals are defined for a desired outcome from a set of decisions. Therefore, the soil health and soil quality may be considered as an evaluation process which consists of a series of actions:

i) Selection of soil health indicators, preferably site specific based on local soil conditions

ii) Determination of a minimum data set (MDS)

iii) Development of an interpretation scheme of indices

iv) On-farm assessment and validation

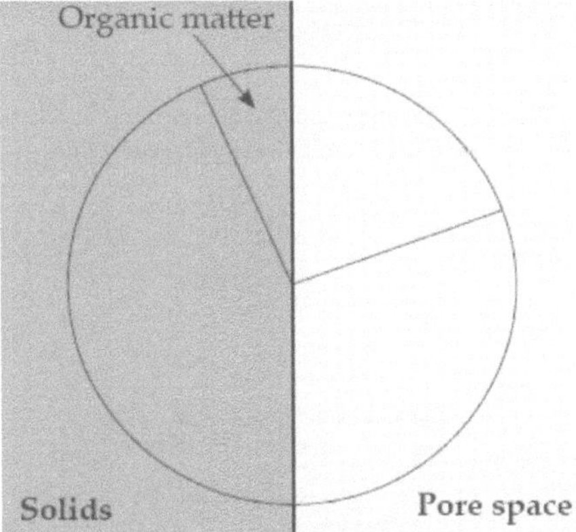

Fig 1. Soil is composed of four basic components: mineral solids, water, air and organic matter (including living biota) *Source*: Das, 2015

Concepts of Soil health

- v Soil health is the capacity of the soil to function as a living system in which soil biological processes or the endogenous inputs are utilized alongside any exogenous inputs required to achieve the desired level of agricultural production that is economically and environmentally sustainable
- v Thus with CA systems, the establishment and maintenance of healthy soil condition is inextricably linked to the achievement of effective nutrient management

Soil health indices

Soil quality is defined as " the capacity of a soil to function within ecosystem and land use boundaries to sustain productivity , maintain environmental quality and promote plant and animal health". Soil health is a concept that deals with the integration and optimization of the physical, chemical and biological properties of soil for improved productivity and environmental quality.

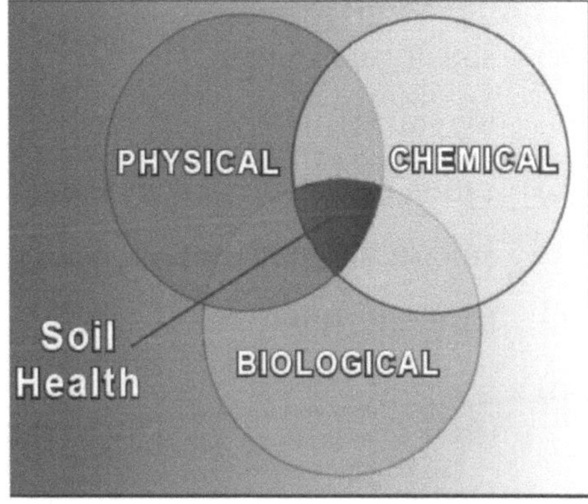

Fig. 2. Soil health attributes. (Adapted from Rodale Institute & cited in Das,2015)

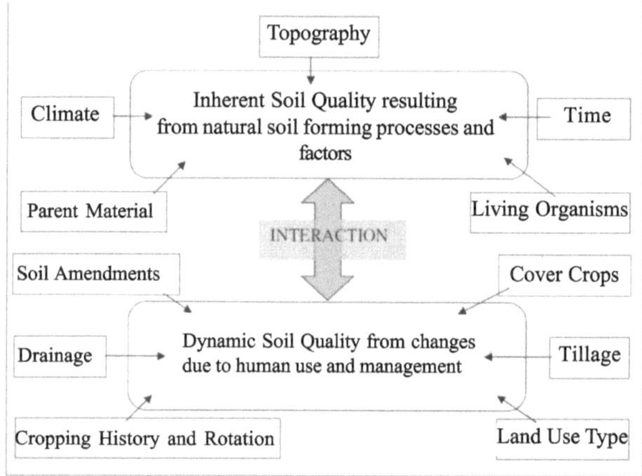

Fig 3: Inherent and dynamic soil quality and various factors affecting them (*Source*: Karlen *et al.*, 2003)

Soil health deals with both inherent and dynamic soil quality. The inherent quality relates to the natural or genetic characteristics of the soil (e.g texture).They are usually represented in soil surveys and hence can not be amended while that of the dynamic quality relates to the levels of compaction, biological functioning, root proliferation etc. and is affected by management practices. At the heart of the soil health is the integration of the soil physical, chemical and biological processes and function. A healthy soil will be balanced for all the three components.

Soil Health Indicators

A. Chemical

i) soil organic carbon

ii) top soil pH

iii) cation exchange capacity (CEC) to 1 m depth

iv) anion adsorption capacity in topsoil

v) base saturation

vi) concentration of potential pollutant elements/organic micropollutants (POPs).

B. Biological

i) SOC(soil organic carbon)

ii) microbial biomass carbon/SOC

iii) soil biomass

iv) Biological score

v) DNA-based microbial diversity index and enzyme assays.

C. Physical

i) integrated air capacity to 1 m depth

ii) number of locations with erosion features

iii) soil organic carbon

iv) topsoil surface condition

v) aggregate stability

vi) soil bulk density

vii) topsoil plastic limit to a depth of 1 m

viii) time to ponding

ix) water dispersible clay.

Others

x) catchment hydrograph

xi) surface water turbidity

xii) biological status of rivers with and without sewage treatment works

xiii) number of eutrophication incidents per year.

The quality of soil is rather dynamic and can affect the sustainability and productivity of land use. It is the end product of soil degradative or conserving processes and is controlled by chemical, physical and biological components of a soil and their interactions. Indicators, however, will vary according to the location, and the level of sophistication at which measurements are likely to be made. Therefore, it is not possible to develop a single short list which is suitable for all purposes (Table 1).

Table 1: Summary of soil health indicators used to asses soil function

Indicator	Soil function
Soil organic matter (SOM) soil erosion (Carter, 2002)	Soil structure, stability, nutrient retention;
Physical: soil aggregate stability, infiltration and bulk density	Retention and mobility of water and nutrients; habitat for macro and micro fauna Swift *et al.*, 2004)

Chemical: pH, extractable soil nutrients, N-P-K and base cationsCa Mg & K loss of Ca, Mg &K (Drinkwater *et al.*,1996).	Soil biological and chemical activity thresholds; plant available nutrients and potential for N and P
Biological: microbial biomass C and N; Potentially mineralizable N	Microbial catalytic potential and repository for C and N productivity and N supplying potential

1. Biological

Identification of biological indicators of soil quality is reported as critically important by several authors because soil quality is strongly influenced by microbiological mediated processes (nutrient cycling, nutrient capacity, aggregate stability). Of particular importance is to identify those components that rapidly respond to changes in soil quality. Biological indicators of soil quality that are commonly measured include soil organic matter, respiration, microbial biomass (total bacteria and fungi,) and mineralizable nitrogen. Soil organic matter plays a key role in soil function, determining soil quality, water holding capacity and susceptibility of soil to degradation . In addition, soil organic matter may serve as a source or sink to atmospheric CO2 and an increase in the soil C content is indicated by a higher microbial biomass and elevated respiration . It is also the principal reserve of nutrients such as N in the soil and some tropical soils may contain large quantities of mineral N in the top 2 m depth (Das, 2015; Havlin *et al.*, 2005).

2. Chemical

In order to achieve high crop yields smallholder farmers have to provide soil nutrients in large quantities (Sanchez and Swaminathan, 2005). Therefore it is possible to alter the pool of available nutrients by adding inorganic fertilizers, incorporating cover crops, and using other organic materials in form of manures and composts (Stocking, 2003). Results of chemical tests are soil quality indicators which provide information on the capacity of soil to supply mineral nutrients, which is dependent on the soil pH. Soil pH is an estimate of the activity of hydrogen ions in the soil solution. It is also an indicator of plant available nutrients. High activity is not desirable and the soil may require liming with base cations Ca or Mg in order to bring the solution back to neutral.

3. Physical

Soil physical properties are estimated from the soil's texture, bulk density (a measure of compaction), porosity, water-holding capacity (Hillel, 1982). The presence or absence of hard pans usually presents barriers to rooting depth. These properties are all improved through additions of organic matter to soils. Therefore, the suitability of soil for sustaining plant growth and biological activity is a function of its physical properties (porosity, water holding capacity, structure, and tilth).

Aggregates are soil particles that are composed of smaller soil particles, which may themselves be aggregates. Aggregates range in size from microns to millimetres . A measure of the mean aggregate size that is present in a soil is termed Mean Weight Diameter (MWD) and a MWD greater than 2 mm is optimal for soil sustainability, both in terms of production and environmental well being (Sparling *et al.*, 2003). The stability of aggregates and the pore spaces between them affects the movement and storage of water, aeration, erosion susceptibility, microbial activity and the growth of crops. Thus, good aggregate stability is important to a wide range of physical and bio-chemical processes in natural and agricultural environments.

Soil bulk density (BD) is the single most useful parameter of soil physical structure. It is a direct measure of soil compaction (or loosening) and is essential to assess total available pore space within a soil (that is, total porosity) . Bulk density is an excellent measure of a most important contemporary form of soil degradation: that which occurs due to ill-timed cultivation, trafficking and stocking.

Packing density (PD) - a derivative of BD calculated as PD = BD*0.009(% clay) - is a better parameter than BD for comparison of physical structure between different soils and, based on derived statistics, affords an excellent indirect estimation of soil porosity.

Soil quality indicators: metals and organic micropollutants

Soil needs to be protected as a limited resource for production (food and agricultural products), as a habitat for plants and animals, and for the numerous other functions it performs, such as neutralising contaminants and acting as a base for parks and buildings, sinks for carbon dioxide and reservoirs for water. The aim of this report is to present changes in concentrations of metals and organic contaminants as potential indicators of soil quality. The presence and fluctuation of metals and contaminants in soil may indicate stress, particularly for soil microbial communities; however, this may be difficult to relate to the soil functions of environmental interaction. Metals and organic micro-pollutants (OMPs) may enter the soil through natural processes, such as weathering of parent materials for metals or burning of vegetation for polycyclic aromatic hydrocarbons (PAHs), and through human activities, such as atmospheric deposition from industrial and vehicle emission or direct application of waste (sludge/slurry, landfills), pesticide and fertiliser. Levels of metals and organic contaminants must therefore be viewed in relation to soil types, land uses and other external influences.

Criteria of Selecting Soil Health Indicators

There are several criteria to consider when selecting soil health and soil quality indicators. In general, appropriate indicators should be:

- easy to asses.
- able to measure changes in soil function both at plot and landscape scales.
- assessed in time to make management decisions.
- accessible to many farmers.
- sensitive to variations in agro-ecological zone.
- representative of physical, biological or chemical properties of soil.
- assessed by both qualitative and/or quantitative approaches.

Minimum Data Set: Concepts and Applications

Tired approach

The use of a MDS is often criticised because of its use of a limited number of indicators and its justification in terms of the wider soil data needs of a broad range of interest groups. However, the use of a MDS can be supported through the use of an interpretative framework in which other SQIs are not excluded, but can be used for specific purposes at different, lower tiers than the MDS. Such an approach is advocated for this project and has a number of advantages over single tier schemes, the primary one being that levels of risk can be balanced against the effort involved. This approach is not new. Indeed, recently European working groups on soils have suggested three levels for a soil monitoring strategy:

- Level 1: basic monitoring
- Level 2: reference sites
- Level 3; special sites (for example, sites with unusual climatic conditions or localised issues).

A modification of this approach is suggested here, whereby Tier 1 represents broadscale monitoring with data collected using the MDS. Tier 2 is more specific monitoring, using a more targeted set of indicators for sites identified through the use of the triggers or through the risk-based approach. This type of sampling could, for example, be spatially defined by a 'gross' soil type and land use combination resulting in specific threat of capping and monitored through the use of 'time to ponding' indicator. Tier 3 monitoring refines the information collected to answer localised issues. The type and magnitude of the issue would dictate the type of monitoring carried out. At Tier 1, generic information is collected about soils of the area to identify headline trends and areas at risk. At this level, existing data on soil properties can be used to broaden the dataset and enable early trend analysis to be undertaken. Transition from Tier 1 to Tier 2 and then Tier 3 occurs through the identification of increasing levels of risk, such as the exceedance of a trigger or the identification of a scenario that presents an obvious risk. Changing management practices or land use may modify the level of risk, where sampling at the site(s) could revert to Tier 1 or Tier 3. The MDS is also collected at Tier 2 and 3 (in addition to any extra/alternative sampling) and includes site characteristics, land use history, soil classification and a profile description. Tier 3 sites represent highly localised issues or threats requiring localised management responses. These are likely to be significantly fewer in number and scale compared with Tier 2 sites.

A minimum data set (MDS) was proposed to measure soil quality and its changes due to management practices through selection of key indicators such as soil texture, organic matter, pH, nutrient status, bulk density, electrical conductivity and rooting depth (Larson and Pierce, 1994). Collecting a minimum data set helps to identify locally relevant soil indicators and to evaluate the link between selected indicators and significant soil and plant properties (Arshad and Martin, 2002). It is a minimum set of indicators required to obtain a comprehensive understanding of the soil attributes evaluated (Figure 1). More importantly they serve as a useful tool for screening the condition, quality, and health of soil (Doran *et al.*, 1996; Larson and Pierce, 1994; and Doran and Parkin, 1994). For researchers, there is need to conduct sufficiently detailed tests while controlling for variation in order to develop meaningful assessments of soil status often expressed as an index of soil quality (Kang *et al.*, 2005).

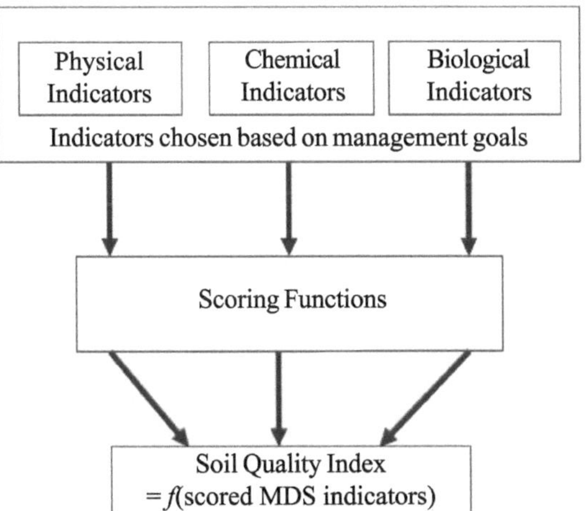

Fig. 4: Adapted from Karlen *et al.*, 2003

Characteristics of a healthy soil

- Good soil tilth
- Depth of the soil
- Sufficient but not excess supply of nutrients
- Small population of plant pathogens and insect pests
- Good soil drainage
- Large population of beneficial organisms
- Low weed pressure
- Free of chemicals and toxins that may harm the crop
- Resistant to degradation
- Resilience when unfavourable conditions occur

Soil constraints

i) Soil compaction: it can happen from soil wetness/excessive moisture/ heavy traffic, tillage under wet soil conditions. Due to such compaction, it can cause reduced root growth, poor infiltration, poor aeration, runoff and erosion, stagnation of water,drought sensitivity, inceased cost of tillage, low crop yields.

ii) Poor aggregation and soil crusting: arsing from poor aggregate stability, very low / no application of organic matter, intensive tillage, limited use of soil building crops. All these can result poor seedling emergence as well as poor establishment of crops, poor infiltration causing soil erosion, poor aeration, reduced root growth, low microbial activity.

iii) Weed pressure: it happens from from poor crop rotations, resistance to herbicides, poor weed management practices etc. which can cause poor crop growth, reduced yield with poor quality,increased pests and disease infestations, increased cost of weed control, affect cultural practices and harvest.

iv) Soil borne pathogens and root disesases: poor crop rotations, poor sanitary practices, ineffective/poor timing of management, low organic matter, poor soil physical quality, low microbial diversity etc. can cause all these. However, all these can result damaged and diseased roots,uneven and poor crop growth and hence reduced yield of crops.

v) Low water and nutrient retention: low organic matter, poor retention/ recycling of nutrients, poor structure, excess tillage, low water holding capacity are the contributing factors. All these can result ground water pollution, reduced microbial activity,nutrient deficiencies and poor plant growth, drought stress etc.

There is need to develop diagnostic measures and indicators of soil health and soil quality changes in order to derive classifications from the minimum data set due to these complexities, (Table 2) that can better assist farmers and inform research and extension to target solutions at farm level (Murage *et al.*, 2000).

Table 2: Proposed minimum data set (MDS) of physical, chemical, and biologicalindicators for assessing soil quality

Indicator	Rationale for assessment
Biological	
Microbial biomass C and N	Describes microbial catalytic potential and repository for carbon and nitrogen.
Potentially mineralizable N	Provides an early warning of management
effects on organic matter.	Describes soil productivity and nitrogen supplying potential.
Provides an estimate of biomass.	
Soil respiration	Defines a level of microbial activity.
Provides an estimate of biomass activity.	
Chemical	
Soil organic matter (SOM)	As a proxy for soil fertility and nutrient availability
pH	Biological and chemical activity thresholds.
Electrical conductivity (EC)	Plant and microbial activity thresholds.
Extractable N, P and K	Describes plant-available nutrients and potential for N loss. Indicates productivity and environmental quality.
Physical	
Soil texture	Indicates how well water and chemicals are retained and Transported. Provides an estimate of soil erosion and variability.
Soil depth and rooting	Indicates productivity potential.
Infiltration and bulk density	Describes potential for leaching, productivity and erosion.
Water holding capacity	Describes water retention, transport and erosion. Available water is used for the calculation of bulk density and organic matter.

Source: Das, 2015

Approaches for the improvement of soil health

- identify the vital processes and functions of the soil required for soil health assessment in relation to agronomic land use.
- test different soil properties that can serve as potential soil quality indicators

- develop appropriate sampling and measurement protocol for soil health which can compliment existing chemical laboratory and can be offered on free for service basis
- develop criteria for interpreting soil health indicator in an agronomically meaningful way
- develop and evaluate accessible data bases as repositories for high quality, reliable soil information
- recommend improved soil management practices based on soil health assessment that will ensure economic viability , environmental safety and social acceptability

On-farm soil health assessment

On-farm assessment of soil quality and health is recommended to assist farmers evaluate the effects of their management decisions on soil productivity (Andrews and Carroll, 2001). Also, this approach permits interaction between researchers, extension and policy personnel when providing interpretation to link on-farm knowledge to soil health information. The main challenge is to develop soil quality and soil health standards to assess changes which are practical and useful to farmers. For instance, linking soil health measurements with farmer perceptions of soil quality can bridge the gap in interpretation of complex data sets.

1. **Scorecards:** The use of scorecards for on-farm soil quality assessment is emphasized where qualitative observations of soil health are scored to obtain an overall measure of soil quality soil health (Romig et al., 1995). These cards may be developed to evaluate soil health through farmer observations of soil physical, chemical and biological properties (Romig et al., 1996). These soil characteristics are classified in terms of descriptive indicators which are interpreted on a graded scale. Examples such as observations on earthworm numbers can yield a general index of biological activity in the soil.

Scoring function

An overall soil quality score is computed from the sum of all the individual indicator scores and is expressed on a percentage scale. The overall classifiction of the soil based on the percentage score is given as;

i. 55-70% Medium
ii. > 85% Very high
iii. 70-85% High
iv. 40-55% Low
vi) <40% Very low

2. **Soil quality test kits:** Assessment tools such as soil quality test kits (Liebig et al., 1996) focus on farmer-based evaluations and education regarding various soil and smallholder management practices. Further, they aim to produce an educational tool to increase public awareness of the importance of soil quality. Such kits, which are commercially available, provide for measurements of soil respiration, infiltration, bulk density, water content, electric conductivity (EC), pH, soil nitrates, aggregate stability, slake test and earthworm numbers.

To enable a more direct interpretation of the soil health test report, a four step procedure are necessary;

1) Identify and prioritize soil health constraints
2) List soil health management options that can address the constraints
3) Compile information on site history and farm background to help decide a management strategies
4) Select appropriate management strategies

3. **Soil quality indices:** Finally, various soil quality indexing approaches are available and can be applied to derive a range of critical test values within which soil quality and soil health assessments can be defined. Soil quality indices should be assessed based on site specific soil constraints which will be not only more appropriate but also cost effective for an efficient nutrient management.

Management of soil health and balancing nutrient contents in soil

i) Cover crops are to be included during crop cultivation with a view to enhance soil organic matter, reduce soil compaction, improvement of soil nutrient regimes, suppression of weeds, reduce soil erosion etc. However, type of cover crops should be selected based on the climate, seasons, current production systems etc,. Winter cover crops, summer fallow cover crops and season long cover crops should be grown keeping in mind that there will be no competition with the main crop for water and nutrients as well as useful for the soil having low fertility and very limited scope for the application of organic manures.

ii) Application of organic manures like, animal manures, composts, green manure crops, crop residues, vermicomposts and others locally available low cost organic resource materials at an regular interval are to be incorporated to the soil primarily for the augmentation of soil organic carbon contents and microbial activity (most important biological indicators) and soil physical properties namely, structure, porosity, aeration, bulk density, water holding capacity etc, (physical indicators) and improves soil nutrient status (chemical indicators). All these practices are sustainable for long term crop cultivation through an overall improvement of soil health. Crop residues left on the surface soil will protect against erosion and improve surface aggregation and hence reducing crusting and compaction. Different kinds of green manure crops are grown for the purpose of improving soil fertility and soil health by increasing microbial diversity and organic matter content in general.

iii) Minimum tillage operation can also improves soil health by maintaining organic carbon content in soil. Extensive tillage reduces soil aggregation resulting in crusting and soil compaction and at the same time reducing organic matter content of the soil through stimulation of microbial activity. Therefore, type and timing of tillage are often site specific and dependent on the cropping systems and availability of equipments. Reducing both tillage frequency and intensity will reduce the burning of organic matter and lead to improved soil tilth and microbial proliferation resulting in soils that are less susceptible to compaction and more resilient and healthy.

iv) Appropriate crop rotation such as inclusion of leguminous crops in a crop sequence also improves soil health by enhancing nutrient contents especially nitrogen content in soil. Of late, crop rotation is also an important component of soil and pest management in many agricultural crop production systems. Crop rotation can also help in nutrient management through incorporation of crop residues and improve soil resiliency after a root crop. Therefore, rotating with a diversity of root structures from tap root to fibrous rooted crops will also improve physical, chemical and biological soil qualities.

Some important case studies are appended below which have improved different soil health indices like physical, chemical and biological indices through different management practices.

Table 3: Effect of different rice residue management practices on soil physical indicators in 0-15 cm soil layer in a sandy loam soil under rice wheat system after 5 cropping cycles

Treatment	Organic C(%)	Mean Weight Diameter (mm)	Bulk Density (g cm)
Residue removed	0.44	0.35	1.68
Residue burnt	0.51	0.53	1.65
Residue incorporated*	0.63	0.60	1.51
LSD(p= 0.05)	0.02	0.07	0.03

Kang *et.al.* (2005)

Table 4: Effect of Recycling of cowpea or pearl millet crop residues (Equivalent to 20kg N ha^{-1}) on chemical properties of two Alfisols in Hyderabad , India

Soil property	Loamy sand Alfisol		Sandy loam Alfisol	
	Initial	After 5 Years	Initial	After 5 Years
Organic C (%)	0.34	0.39	0.65	0.62
Available N(kg ha^{-1})	235	286	284	316
Available P(kg ha^{-1})	2.9	5.6	2.4	6.4
Available K(kg ha^{-1})	224	214	213	215

Kang *et.al* (2005)

Table 5: Effect of continuous application of chemical fertilizers and organic manures on soil quality indicators of a sandy loam soil under rice- wheat system for 17 years in a semi – arid subtropical environment

Chemical Indicators

Treatment	OC(%)	Nutrient (kg ha^{-1})			
		Total N	NO_3^- - N	Av.P	Av.K
Control	0.47	1292	26.7	33.3	83.6
100% NPK	0.50	1299	32.0	34.4	93.4
50% NPK 50% N (FYM)	0.63	1508	30.1	150.7	115.0
50% NPK 50% N (WCS)	0.52	1236	12.2	45.3	79.2
50% NPK 50% N (GM)	0.68	1442	46.0	62.3	90.4
CD(P=0.05)	0.05	NS	10.4	39.8	17.8

Kang *et al.* (2005)

100% NPK – 120 kg N, 60 kg P_2O_5& 30 kg K_2O ha^{-1}

WCS - Wheat cut straw @ 6 Mg ha^{-1}

GM -*Sesbaniacannabina* green manure

FYM - Farm yard manure

OC -Organic carbon

Table 6: Changes in some biological indicators affected by management practices

Treatment	Biomass C (mg kg^{-1})	Biomass N (mg kg^{-1})	Phosphatise (mg PNP g^{-1} h^{-1})	Dehydrogenase (mg TPF g^{-1} 24 h^{-1})
Control	141.1	48.3	15.1	105.6
100% NPK	157.0	50.7	18.3	82.2
50% NPK 50% N (FYM)	163,0	55.6	23.2	123.9
50% NPK 50% N (WCS)	153.3	52.1	22.9	105.6
50% NPK 50% N (GM)	172.8	63.1	27.6	119.0
CD(P=0.05)	31.8	7.1	6.6	34.4

Kang *et.al* (2005)

100% NPK – 120 kg N, 60 kg P_2O_5& 30 kg K_2O ha^{-1}

WCS - Wheat cut straw @ 6 Mg ha^{-1}

GM -*Sesbaniacannabina* green manure; FYM - Farm yard manure; OC -Organic carbon

Improvement of soil quality

- Improved soil and water management practices along with integrated nutrient managements
- Use of covercrops , increasing the N use efficiency and site specific management of plant nutrients
- There is a need for future research to generate information on the long term effects of farming practices on soil and water quality .

Emerging issues

Sensing soil health

Assessments of soil chemical properties normally rely on laboratory data with large numbers of samples required to adequately characterize spatial variability at farm scales have developed a promising approach that directly estimates several soil properties simultaneously from diffuse reflectance spectra in rapid non destructive ways. The soil reflectance spectral data successfully predicted crop yields of an 18-year field experiment testing different levels of fertilizer, manure and crop residue management. Since spectral techniques allow large numbers of samples to be rapidly analyzed, resources can be directed towards thorough characterization of the soil and its spatial variability within a target region.

Soil and human health

Hunger in Africa has been linked to unhealthy people and unhealthy soils (Sanchez and Swaminathan, 2005). Currently, the concept of soil quality is in the process of evolution and progressively moving from a concept focused on yield potential and nutrient levels to one of environmental quality, food safety and human health (Karlen *et al.*, 1997).

Soil and climate change

Global climate change is dramatically increasing the variability of weather conditions worldwide and soil is a critical buffer medium for hydrologic and biogeochemical processes and therefore can mitigate the effects of extreme weather conditions and uncertainty in the availability of water resources (Larson and Pierce 1994). The present threats of global climate change and ozone depletion, through elevated levels of certain atmospheric gases and altered hydrological cycles, necessitate a better understanding of the influence of land management on soil quality (Doran and Safley 1997). Management systems need to be further fine tuned in order to balance the need and priorities for food production with those for a safe and clean environment.

Conclusions

Maintenance or enhancement of soil quality is a common criterion when assessing long-term sustainability of agricultural ecosystems. However, the task of establishing a specific criterion for soil quality is challenging because functions and subsequent values provided by agricultural production ecosystems are variable and rely on the interplay of soil physical, chemical, and biological properties and processes which often differ significantly across spatial and temporal scales. Choice of a standard set of specific soil properties as indicators of soil quality can be complex and may vary among agricultural production systems. Despite these challenges, development of soil quality indices is progressing and minimum data sets (MDS) have been proposed which recognize soil properties or processes that are likely to be sensitive to management perturbations and are related to agricultural productivity and soil health. These lists commonly include properties such as organic matter content, nutrient supplying capacity, acidity, bulk density, porosity, and available water holding capacity. Any of these soil properties may be relevant to several soil functions simultaneously and will have varying levels of influence which can be weighted accordingly in soil quality index models. Indices of soil quality which incorporate chemical and physical soil properties will be most readily adopted if they are: (i) sensitive to management induced changes; (ii) easily measured; (iii) relevant across sites or over time; (iv) inexpensive; (v) closely linked to measurement of desired values such as productivity or biodiversity; and (6) adaptable for specific ecosystems. Indices of soil quality which meet these criteria must be developed based on our current knowledge and must be adaptable for sustainable agricultural crop production.

Soil resource assessment and monitoring is a new era in terms of quality information produced by new information technologies through innovative use of GIS and remote sensing significantly improving the acceptance and use of soil health information. Application of organic manures and conservation agriculture ensures not only improvements in soil fertility but also better biological and physical quality of soils.Tillage also produces a counterproductive effect on the accumulation of soil organic matter. Green manure with very low C/N ratio behaving similarly to mineral fertilizer was unable to immobilize N. Development of cost effective protocol for assessing the health status of soils are necessary. Maintenance of soil health is of vital importance to ensure food and nutritional security of the country. The integrated nutrient management envisaging conjunctive use of chemical fertilizers including secondary and micronutrients, organic manures, biofertilizers and soil amendments for ameliorating acid / saline /sodic soils is the most novel way of nutrient management. The 4 R concepts of nutrient management should be followed for an efficient and cost effective nutrient management by maintain soil health.

Soil testing services vis-à-vis providing soil health card to the farmers should be ensured for judicious fertilizer application. Judicious crop rotation, use of cover crops etc. may be useful strategy for increasing short term SOM and for establishing healthy, fertile and productive soils. In years ahead, major emphasis has to be given on research–extension linkages, skill upgradation, educational programmes involving self-help groups, NGOs, Use of mass media in transfer of location specific low cost technologies ensuring farmers participation.

Research Needed

Rapidly increasing population in countries that can least afford it have made them food-insecure. With inadequate inputs in agriculture, developing countries are degrading their lands rapidly and destroying ecosystems. Currently, soil resource assessment and monitoring is entering a new era, in terms of quality of information produced by new information technologies through the innovative use of Geographic Information Systems and remote sensing and will significantly improving the acceptance and use of soil health information. The use of electronic technology will significantly increase the demand for and ability to process more data. Further innovations will result in model approaches in soil genetic studies that will demonstrate the integral role of soils in ecosystems. Particularly the use of information and communications technology (ICT) including on-line delivery content through wikis, blogs and social book marking is set to push sharing of soil health information to a new level. At the more fundamental level, basic research will be urgently required in order to select and develop proper indicators based on a site specific soil quality indices and also applicable at different farmer scales. Soil health cards (SHCs) are to be prepared based on soil test values and the same are required to be delivered to the farmers with appropriate recommendations. Innovations will also be needed in setting up effective study programs, which would guarantee the accumulation of the necessary baseline soil data in order to develop appropriate minimum datasets (MDS).

References

Andrews, S.S., Carroll, C.R., 2001. Designing a soil quality assessment tool for sustainableagroecosystem management. *Ecological Applications* 11, 1573–1585.

Arshad, M.A. and Martin, S. 2002. Identifying critical limits for soil quality indicators in agro-ecosystems. *Agriculture, Ecosystems and Environment* 88: 153–160.

Carter, Martin, R. 2002. Soil quality for sustainable land management: Organic matter and aggregation interactions that maintain soil functions. *Agron J.* 94: 38-47.

Das, D.K. 2015. In: *Introductory Soil Science*, 4[th]edn. Published by Kalyani Publishers, New Delhi, India.

Doran, J. W. and Parkin, T. B. 1994. Defining and assessing soil quality. Pages 3-21 in J. W. Doran et al., (eds.) Defining soil quality for a sustainable environment. Soil Science Society of America Special Publication no. 35, Madison, WI.

Doran, J.W., Sarrantonio, M., Liebig, M., 1996b. Soil health and sustainability. In: Sparks, D.L. (Ed.), Advances in Agronomy, Vol. 56. Academic Press, San Diego, pp. 1–54.

Doran, J.W., Jones, A.J. (Eds.), 1996a. Methods for Assessing Soil Quality. Soil Science Society of America Special Publication, vol. 49. Soil Science Society of America, Madison, Wisconsin.

Doran, J.W., Parkin, T.B., 1994. Defining and assessing soil quality. In: Doran, J.W., Coleman, D.C., Bezdicek, D.F., Stewart, B.A. (Eds.), Defining Soil Quality for a Sustainable Environment. Soil Science Society of America Special Publication, vol. 35. Soil Science Society of America, Madison, Wisconsin, pp. 3–21.

Doran, J.W., Zeiss, M.R., 2000. Soil health and sustainability: managing the biotic component of soil quality. *Applied Soil Ecology* 15: 3–11.

Doran, J.W. and Safley, M. 1997. Defining and assessing soil health and sustainableproductivity. In: Pankhurst, C. et al. (eds.). Biological indicators of soil health. Wallingford, UK: CAB International. p. 1–28.

Drinkwater, L.E., Cambardella, C.A., Reeder, J.D. and Rice, C.W. 1996. Potentially mineralizable nitrogen as an indicator of biologically active soil nitrogen. In: Doran, J.W., Jones, A.J. (Eds.), Methods for Assessing Soil Quality. Soil Science Society of America, Special Publication 49, Madison, WI, pp. 217–229.

Havlin, J.L., J.D. Beaton, S.L. Tisdale, and W.L. Nelson. 2005. Soil fertility and nutrient management: An introduction to nutrient management. 7th Edition.515 p.

Pearson/Prentice Hall. Upper Saddle River, N.J. Hedlund, A., Witter, E., An, B.X., 2003. Assessment of N.P. and K. management by nutrient balances and flows on peri-urban smallholder farms in southern Vietnam. *Eur. J. Agron*. 20, 71–87.

Hillel, D. 1982. Introduction to soil physics.Academic Press, San Diego, CA. http://soilhealth.cals.cornell.edu/ http://soils.usda.gov/. USDA, Natural Resources Conservation Service.

Idowu, O.J. van Es, H. Schindelbeck, R., Abawi, G. Wolfe, D. Thies, J. Gugino, B. Moebius, B. and Clune, D. 2007. The new Cornell Soil Health Test: Protocols and interpretation. What's Cropping Up? 17, (1) 6-7 (http://www.css.cornell.edu/extension/)

Kang, G.S., Beri, V., Sidhu, B.S., Rupela, O.P., 2005. A new index to assess soil quality and sustainability of wheat-based cropping systems. *Biology and Fertility of Soils* 41: 389–398.

Karlen, D. L., Mausbach, M. J., Doran, J. W., Cline, R.G., Harris, R. F. and Schuman, G. E. 1997. Soil quality: a concept, definition, and framework for evaluation. *Soil Science Society of America Journal*. 61: 4-10.

Karlen, D. L., Andrews, S. S. and Doran, J. W. 2001. Soil quality: Current concepts and applications. *Advances in Agronomy,* 74: 1-40.

Karlen, D. L., Doran, J. W., Weinhold, B.J. and Andrews, S. S. 2003. Soil quality: Humankind's foundation for survival. *Journal of Soil and Water Conservation,* 58.

Larson, W.E. and Pierce, F.J. 1994. The dynamics of soil quality as a measure of sustainablemangement. Pages 37 - 51 in J. W. Doran, D. C. Coleman, D. F. Bezdicek and B.A. Stewart, eds. Defining soil quality for a sustainableenvironment. SSSA- Special Publication 35. Soil Science Society of America,Madison , WI, USA.

Lal, R. and Stewart, B.A. 1995. Soil Management: Experimental Basis for Sustainability and Environmental Quality. Advances in Soil Science, CRC Press, Boca Raton, Florida.

Letey, J., Sojka, R. E., Upchurch, D.R., Cassel, D.K., Olson, K. R., Payne, W. A., Petrie, S. E., Price, G.H., Reginato, R.J., Scott, H.D., Smethurst, P.J. and Triplett, G.B. 2003. Deficiencies in the soil quality concept and its application. *Journal of Soil and Water Conservation*. 58:180-187.

Murage E.W., Karanja N.K., Smithson P.C. and Woomer P.L. 2000. Diagnosticindicators of soil quality in productive and non-productive smallholders fields of Kenya's Central highlands. *Agriculture, Ecosystems and Environment*, 79: 1–8.

Romig, D.E., Garlynd, M.J., Harris, R.F. and McSweeney, K. 1995. How farmers assess soil health and quality. *Journal of Soil and Water Conservation,* 50: 229–236.

Romig, D.E., Garlynd, M.J., Harris, R.F., 1996. Farmer-based assessment of soil quality: A soil health scorecard. In: Doran, J.W., Jones, A.J. (Eds.), Methods for Assessing Soil Quality. Soil Science Society of America, Special Publication 49,Madison, WI, pp. 39–60.

Sanchez, P. A., Swaminathan M. S., 2005. Hunger in Africa: the link between unhealthypeople and unhealthy soils. *The Lancet,* 365: 442-444.

Sparling, G., Parfitt, R. L., Hewitt, A. E., Schipper, L. A. 2003. Three approaches todefine desired soil organic matter contents. *J. Environ Qual,* 32: 760-766.

Stocking, M.A., 2003. Tropical soils and food security: The next 50 years. *Science,* 302: 356-1359.

Swift, M.J., A. Izac, and M. Van Noordwijk. 2004. Biodiversity and ecosystem servicesin agricultural landscapes—are we asking the right questions? *Agriculture, Ecosystems and Environment,* 104(1): 113-134.

Zhang, W.L.,Tian, Z.X., Zhang, N., Li, X.Q. 1996. Nitrate pollution of groundwater in northern. China. *Agricult, Ecosyst Environ,* 59:223–231.

12

Processes and Mineralogy of Arsenic in Soils: An Overview

Siddhartha S. Mukhopadhyay

M S Swaminathan School of Agriculture, CUTM, Paralakhemundi-761211 Odisha, India
E-mail siddhartha_soil@yahoo.co.in

Abstract

Arsenic, although a non-metal, is very often referred to as a heavy metal. Its provenance and contents in recently formed or least weathered soils depend on the type of rocks and intensity of geochemical processes. The variation is exemplified in shale, where As content varies from 1 to 900 mg kg^{-1}. In highly weathered soils, where pedochemical processes exceed geochemical processes, the content is generally below the levels that are toxic to life forms. In the Bengal basin, As content is low in soils, but high in the ground water (beneath the vadose zone). This suggests that As has been sufficiently leached from soils. The solubility of most of the As-bearing soil minerals is high, and it occurs as anions (AsO_3^{3-}, or AsO_4^{3-}) in soil solution. The soils of the Bengal basin are alkaline in reaction, mostly fine textured with almost even distribution of particle size fractions, and moderately drained. Minerals like micas and feldspars dominate their coarse fractions, whereas illite (40-47%) and smectite (24-30%) are dominant in clay fractions. Both inorganic and organic colloids are negatively charged that cannot retain anions. In many soils in the world, leaching of As is well documented. Therefore, it appears that the problem of As in human food and drink chains in the Bengal basin arises, because leached As is recycled back through lifting of ground water for drinking and irrigation. There could be other sources like deposition from atmospheric emission, pesticide, and mine wastes, but the accretions might be very small. One root of the problem of As toxicity in the Bengal basin is that population and anthropogenic biotic pressure exceed the supporting capacity of natural resources. This may be tackled by surface storing of rainwater, which is abundant in the region during monsoon, for off-season use. Arsenic is known to replace Si^{4+} in Si-crystal but there is little information available to suggest that this type of substitution is occurring in the soil minerals that are common in the soils of the Bengal basin.

Keywords: Arsenic; soil contents; soil mineralogy arsenic relationship; soil processes; arsenic distribution; soil properties-arsenic

Introduction

Arsenic in rocks: Arsenic, although a non-metal, is very often referred to as a heavy metal. Its provenance and contents in recently formed or least weathered soils depend on the type of rocks and intensity of geochemical processes. It belongs to lithophile element (ionize readily or form stable oxyanions, but bonding changes with temperature, pressure, and oxidation or reduction state), chalcophile element (form covalent bond with sulfide), and biophile element (associated with organisms) groups. The most of the earth crust is made-up of lithophile elements (Jackson, 1964). The content of As varies widely in different rocks; as well as within the same rock type. For example, its content in shale varies from 1 to 900 mg kg^{-1} (Table1, Alloway 1990).

Table 1: Mean content of As in major rock types (mg kg^{-1})

Igneous rocks			Sedimentary rocks		
Ultra-mafic	Mafic	Felsic	Limestone	Sandstone	Shales
1.0	1.5	1.5	0.12	0.25	13.0 (Range 1-900)

Source: Alloway (1990, p.31).

Arsenic in soils

The content of As in soils varies widely. There are several soils in the world, which are free from As. Many soils, on the other hand were believed to be free from As from natural sources, but now got contaminated due to anthropogenic reasons. These soils perhaps accrued As through pesticides. Walsh and Keeney (1975) reported 13 to 2553 mg kg^{-1} of As in some contaminated soils in North America. Incidentally, land use of these soils is orchard.

Amongst the naturally-occurring (non-anthropogenic) soils (apart from the soils of Bengal basin), acid sulfate soils developed on pyretic parent materials are known to contain high amounts of As. Dudas (1987) attributed weathering of pyrites as a major contributor of As to some acid sulfate soils of Canada (8 to 40 mg kg^{-1}). It appears that As concentration in the pedosphere is governed by natural processes of soil formation, intensity of weathering, and as also anthropogenic factors. Readers may see the review of Smith *et al.* (1998) for details.

Geo-chemical and pedochemical processes of Arsenic

The distribution of elements in the pedosphere has long been perceived as a function of rocks from which the soil parent material was derived, and on the geochemical and pedochemical processes of weathering. Geochemical processes primarily determine elemental composition in recently formed or least weathered soils, and thereby elements show strong provenance. In highly weathered soils, where pedochemical processes exceed geochemical processes, elemental composition is an exclusive domain of pedosphere characteristics (Mitchell, 1964). Although doubts were expressed about the soundness of the hypothesis on the ground of lack of sufficient data (Chesworth, 1991), some of the recent works on the distribution of a few micronutrients implicitly showed that pedogenic processes control the distribution of total Zn, Cu, Mn and Fe in the soils of the Indo-Gangetic region (Sharma *et al.*, 2000), in the Entisols of Punjab (Sharma *et al.* 2003), and in the Alfisols of Punjab (Sharma *et al.*, 2005a, Table 2). The pedogenic processes vis-à-vis hydrologic cycles governed the distribution of some elements in the benchmark profiles of the Vertisols of India (Sharma *et al.*, 2005b), in spite of the fact that it is a soil group, which is remarkably homogeneous due to continuous churning and mixing (Blokhuis, 1982; Soil Survey Staff, 1999). These publications established that the role played by soil process often gets masked due to drawing of samples from arbitrarily decided depths rather than from genetically meaningful horizons.

Peryea and Creger (1994) observed considerable movement of As to depths greater than 120 cm in soils with lead arsenate pesticide residues. Similarly, McLaren *et al.* (1998) reported substantial downward movement of as in soils at cattle dip sites in Australia. In studied with undisturbed soil columns contaminated with timber preservative containing As, Carey *et al.* (1996) leaching of substantial amount of arsenate. These studies confirm that Ascan not be present in soils for long irrespective of its provenance or anthropogenic accumulation.

This appears to be a fact, when the data on the distribution of As in the soils of the Bengal basin are examined. These soils are low in As content, and is generally below the levels that are toxic to life forms. But ground water (beneath the vadose zone) contains high amount of As, which implies that As has been sufficiently leached from soils, and concentrated in the ground water. Moreover, the solubility of most of the As-bearing soil minerals is high, and it occurs as anions (AsO_3^{3-}, or AsO_4^{3-}) in soil solution.These facts, perhaps emphasize that the problem of As toxicity in human food and drink chains in the Bengal basin must be looked from a holistic angle, taking into account the entire eco-system that includes human factors.

Table 2: Total contents of Zn, Cu, Mn, and Fe in the diagnostic horizons in the Alfisols of Punjab

Land form unit	Diagnostic Horizon	Total-Zn	Total-Cumg kg^{-1}	Total-Mn	Total-Fe%
Hill	Eluvial	61	26	470	3.11
	Illuvial	75	34	543	3.33
Piedmont	Eluvial	51	23	363	2.43
	Illuvial	80	29	508	3.95
Plain	Eluvial	48	16	267	2.33
	Illuvial	58	25	347	3.29
All together	Eluvial	52	21	353	2.57
	Illuvial	70	28	453	3.57

Source: Sharma *et al.*, 2005a

Mechanism of as retention in soils of Bengal Basin

Arsenic is present in soil solution and on exchange sites as anions. Both inorganic and organic colloids that are present in the soils of the Bengal Basin are negatively charged, and thereby they cannot retain As-anions. In many soils in the world, leaching of as is well documented. As source of arsenic in the Bengal Basin is primarily ground water, it could be discerned that as has been leached from the surface soils, and is left below the vadose zone.

The soils of the Bengal basin

The soils of the Bengal basin are alkaline in reaction, mostly fine textured with almost even distribution of particle size fractions, and moderately drained. Minerals like micas and feldspars dominate their coarse fractions. Illite (40-47 %) and smectite (24-30 %) together constitute 70% of minerals in the clay fraction. Other minerals in clay fraction are vermiculite, chlorite, interstratified minerals, and kaolinite (Ghosh and Dutta, 1974, Mukhopadhyay and Datta, 2001).

The problem of As toxicity in the Bengal basin

It appears that the problem of As in human food and drink chains in the Bengal basin arises, because leached As is recycled back through lifting of ground water for drinking and irrigation. There could be other sources like deposition from atmospheric emission, pesticide, and mine wastes, but the accretions might be very small. One root of the problem of as toxicity in the Bengal basin is that population and anthropogenic biotic pressure exceed the supporting capacity of natural resources.

Tackling the problem of as toxicity in the Bengal basin

The severe problems of as toxicity shows that anthropogenic activities have exceeded the carrying capacity of the eco-system. For the eco-system, to come back to thc healthy and sustainable to the human life, it must be allowed to be under forest land-use and human population must be rehabilitated elsewhere. For short-term measure, rainwater can be stored during monsoon, and then used for drinking and irrigation during water-deficit periods. Incidentally, the region falls in high rainfall isohyets.

Arsenic is known to replace Si^{4+} in Si-crystal (Evans 1964), but there is little information available to suggest that this type of substitution is occurring in the soil minerals that are common in the soils of the Bengal basin.

References

Alloway, BJ. 1990. Heavy metals in Soils. Blackie Academic and Professional, New York.

Blokhuis, W.A. 1982. Morphology and genesis of Vertisols.In Vertisols and rice soils of the tropics, 23-47. New Delhi, India: 12th International Congress of Soil Science.

Carey, P.L., McLaren, R.G. and Adams, J.A. 1996. Sorption of cupric dichromate and arsenate ions in some New Zealand soils.*Water, Air, and Soil Pollution* 87:189-203.

Chesworth, W.1991. Geochemistry of micronutrients. InMicronutrients in agriculture, 2nd edition, eds. J.J. Mortvedt, F.R.Fox, L.M.Shuman, and R.M.Welch. Madison, Wisconsin: Soil Science Society of America.

Dudas, M.J. and S.Pawluk. 1980. Natural abundances and mineralogical partitioning of trace elements in selected Alberta soils. *Can. J. Soil Sci.* 60: 763-771.

Evans, R.C. 1964. An Introduction to Crystal Chemistry. Cambridge Univ. Press, Cambridge.

Ghosh,S.K., and N.P. Dutta 1974. X-ray investigations of the soils of West Bengal. *Proc. Indian Nat. Sci. Acad.* 40B:138-143.

Jackson, M.L. (1964). Chemical composition of soils. p. 71-141. *In* F.E.Bear (ed.) Chemistry of Soils.Oxford & IBH, Calcutta.

McLaren, RG. 2003. Micronutrients and toxic elements. Chapter 18. p.589- 626. *In* Benbi, D.K. and Nieder, R. (ed.) Handbook of Processes and Modeling in Soil-Plant System. Haworth Press, New York.

Mitchell, R.L. 1964. Trace elements in soils, p.320-368.In Chemistry of the Soil. Bear, F.E. (ed.) Oxford & IBH, Calcutta.

Mukhopadhyay, S.S., andS.C.Dutta. 2001. Potassium availability in relation to mineralogy in the Indo-Gangetic Plains. p.33-44. Importance of Potassium in Nutrient Management for Sustainable Crop Production in India. International Potash Institute, Berne/Potash Research Institute of India, Gurgaon, India.

Peryea,F.J., and T.L.Creger. 1994. Vertical distribution of lead and arsenic from soils contaminated with lead arsenate pesticide residues. *Wat.Air Soil Poll.* 78:297-306.

Sharma, B. D., S.S.Mukhopadhyay, and P.S.Sidhu.1998.Micro topographic controls on soil formation in the Punjab region, India. *Geoderma*, 81:357-368.

Sharma, B. D., S.S.Mukhopadhyay, P. S. Sidhu, and J. C. Katyal. 2000. Pedospheric attributes in distribution of total and DTPA- extractable Zn, Cu, Mn and Fe in Indo-Gangetic plains. *Geoderma*, 56: 31-35.

Sharma, B.D., S.S. Mukhopadhyay, and Harsh-Arora.2005a. Total and DTPA-extractable micronutrients in relation to pedogenesis in some Alfisols of Punjab, India. *Soil Science* 170: 559-572.

Sharma, B. D., S. S. Mukhopadhyay, and J. C. Katyal. 2005b. Distribution of Total and DTPA-extractable Zinc, Copper, Manganese, and Iron in Vertisols of India. Communications in Soil Science and Plant Analysis.

Sharma, B. D., V. K. Aggarwal, S. S. Mukhopadhyay, and Harsh-Arora. 2002. Micronutrient distribution and their association with soil properties in Entisols of Punjab. *Ind. J. Agric. Sci.* 72:334-340.

Smith,E., R.Naidu, and A.M.Alston. 1998. Arsenic in the soil environment: A review. *Adv. Agron.* 64:149-195.

Soil Survey Staff, 1999. Soil Taxonomy, 2nd Edition. Soil Conservation Service, U.S.D.A., Washington, D.C.

Walsh,L.M., and D.R. Keeney. 1975. Behaviour and phytotoxicity of inorganic arsenicals in soils. p. 35-53. *In* E.A.Woolson (ed.) Arsenical Pesticides. *Am. Chem. Soc.*, Washington DC.

13

Management of Acid Soils for Agricultural Sustainability in Eastern India

A. Zaman[1] and Parveen Zaman[2]

[1]Emeritus Professor, Centurion University of Technology and Management
M S Swaminathan School of Agriculture
Paralakhemundi-761 211, Odisha, India
Email: profazaman@gmail.com
[2]Assistant Director of Agriculture, Pulse and Oilseed Research, Sub-Station
Beldanga-742133, Murshidabad, West Bengal, India
Email: parveenzaman1989@gmail.com

Abstract

Soil acidity poses a great problem for crop production due to very poor fertility status and nutrient imbalances in eastern and north eastern region and hence, it needs a special attention to address such acidity problems in relation to crop production. The integrated nutrient management system for the maintenance of soil fertility, sustainable agricultural productivity and improving profitability through the judicious and efficient use of mineral fertilizers especially secondary and micronutrients, organic matters, green manures, bio-fertilizers in acid soils. Rainwater harvesting, efficient storage of excess rainwater and subsequent uses as supplemental irrigation especially in the areas of rainfed situations as well as selection of crops and cropping systems in different acid soils region need to be developed. The management of secondary and micronutrients in acid soils can be made either by the addition of cost-effective lime or liming materials to counteract the abnormalities of soils by raising pH as the first step and subsequently by modifying agricultural practices and applying balanced fertilizers including limiting nutrients *viz.* S, B, Mo and Zn at optimal levels based on soil test values for maximum crop production. Poor growth of crops on acid soils is attributed to the presence of toxic amounts of Fe, Mn and Al and deficiency of B, Mo and Zn and less activity of soil microorganisms. Acid soils have been formed due to drastic weathering associated with hot humid and high rainfall situations. Hence, the ways and means of amelioration and management of acid soils are paramount importance for sustainable crop prop production.

***Keywords*:** Acid soils; its spread, crop productivity; amelioration; constraints

Introduction

Eastern India comprising North Eastern Hill states, West Bengal, Bihar, Jharkhand, Orissa, eastern Uttar Pradesh and eastern Madhya Pradesh (Chatisgarh) has the geographical area of 91.04 m ha (28% of India) with cultivated area of 34.47 m ha (24.3% of India), gross cropped area of 48.48 m ha and cropping intensity of 141% (Table 1). About 40% of the total cultivated area is irrigated. In India, acid soils are about 100 million hectares out of the total geographical area of 328 million hectares, of which, 51 m ha comes under forest and 49 m ha under cultivation. About 25.9 m ha of land has a pH below 5.6 termed as strongly acid soils and 23.1 m ha has pH value ranging between 5.6 and 6.5 called as slightly acid soils. Out of total acid soils in the country around 45 m ha is located in the eastern states constituting 49% of the geographical area. About 80% soils (12.5 m ha) of Orissa, 76% (20.0 m ha) of NEH states, 39% (3.5 m ha) of West Bengal, 30% (5.2 m ha) of Bihar, 20% (2.9 m ha) of eastern MP and 10% (0.86 m ha) of eastern UP are acidic (Table 2). About 24.2 m ha constituting 49.4 per cent of 49 m hacultivated acid soils in India are found in Eastern India. Out of this roughly 15.2 m ha have pH less than 5.5 and 9.0 m ha between 5.5 and 6.5 (Table 3). Most of the soils in acid soil regions are sedentary and mostly found in hilly terrains. A major portion of the acid soils are grouped into red, laterite as well as mixed red and black or yellow soils which come under the modern soil classification in the order of *Alfisols, Oxisols, Spodosols, Ultisols, Mollisols, Entisols and Inceptisols.* Some classified acid soils into several groups amongst which (1) Laterites, (2) Laterite and lateritic red soils; (3) Brown forest soils, (4) Acid sulphate soils, (5) Foot hill soils, (6) Coastal alluvial Soils and (7) Marshy soils are predominant in the eastern regions. Increased solubility and toxicity of Al, Mn and Fe; deficiency of Ca and Mg; reduced availability of P, Mo, B and Si; reduced microbial activity with decreasing organic matter and pH are characteristic features of acid soils. The fertility status of acid soils is very poor and under strongly to moderately acidic soils the plant growth and development are affected. Crops grown on these problematic soils do not give remunerative return rather it lowers down the yield to a great extent. In the north-eastern states of India, more than 90 % of the cultivated area is affected by varying degrees of acidity, which has restricted the crop choice in more than 70% of the gross cropped area of this region to cereals, particularly to rice with meager area under pulses and oilseeds with poor productivity level (Table 4). There is a positive correlation between acid soils regions and poverty where the resource poor farmers use to grow crop like wheat, potato, pulses and oilseeds as winter crops during dry season. The productivity of such crops are below average, even growing rice in *kharif* or *boro* season is far from satisfactory. These areas, mostly inhabited by tribal accustomed with shifting cultivation and denudation of forest trees creating tremendous soil productivity losses. The limited land resource it needs judicious management practices of soil-water-nutrient-plant continuum so that the yield of different crops can be increased substantially.

Table 1: Land use pattern in Eastern India

State	Total geographical area (m ha)	Net cropped area (m ha)	Gross cropped area (m ha)	Cropping intensity (%)	NCA under % of irrigation
NEH States	26.22	3.82	5.17	135	21.7
Bihar and Jharkhand	17.39	7.60	10.52	138	50.3
West Bengal	8.88	5.33	8.30	156	35.0
Orissa	15.57	6.28	9.17	146	33.7
Eastern UP	8.58	5.05	7.51	149	67.0
Eastern MP	14.40	6.39	8.20	126	30.0
Total	91.04	34.47	48.48	141	39.6
India	328.0	143.0	186.56	131	37.6

Table 2: Extent of occurrence of acid soil regions (ASR) in eastern India

S.No.	States	Geographical area (m ha)	Acid soil (m ha)	ASR (%)
1.	Assam and hilly tract of North East states	26.2	20.0	76.0
2.	Orissa (eastern ghats region)	15.6	12.5	80.0
3.	West Bengal (hilly terrains, red and lateritic, coastal alluvial regions)	8.9	3.5	39.0
4.	Bihar (Chottonagpur plateau) and Jharkhand	17.4	5.2	30.0
5.	Eastern MP	14.4	2.9	20.0
6.	Eastern UP	8.6	0.86	10.0

Table 3: pH ranges of acid soils in Eastern India

Soil group	pH ranges	States of eastern India
Laterite	4.8 - 7.0	Orissa, West Bengal, Assam, Jharkhand
Laterite and red lateritic	5.0 - 7.0	Orissa, West Bengal, Assam & NEH Jharkhand
Mixed red and black yellow	5.5 - 6.5	Bihar
Ferruginous red	5.0 - 6.8	Orissa, Bihar, West Bengal, Assam
Podzolic and brown forest	4.5 - 6.5	Assam, West Bengal, Orissa
Foot hill soils	4.5 - 6.5	West Bengal, Bihar
Acid sulphate soils	3.3 - 5.9	West Bengal

Table 4: Expected loss of productivity due to acidification

Class	pH	Degree	Loss in productivity (%)
0	>6.5	Nil	Nil
1	5.5-6.5	Slight	Up to 10
2	4.5-5.5	Moderate	10-25
3	3.5-4.5	Strong	25-50
4	<3.5	Extreme	>50

Source: Velayutham *et.al.*, 1998

Causal factors for formation of acid soils

Climate plays a major role in the formation of acid soils. Rainfall, temperature, vegetation and above all, parent material are the dominant factors influencing development of acid soils. In regions of high rainfall and high temperature, acid soils are exclusively found. The following factors are responsible for the formation of acid soils:

a) Hot and humid climate with high temperature and rainfall,

b) Leaching due to high rainfall,

c) Origin of soils from acid parent materials,

d) Origin of soils from non-acid parent materials,

e) Vegetation species

f) Use of acid formation fertilizers

g) Microbial action, and

h) Cultivation practices

Formation and occurrence of acid soils

Acid soils occur in the tropics and subtropics as well as in moderate climate areas while acid sulphate soils occur in the low inundated lands and in the mangrove swamps of coastal region. The formation and distribution of acid soils depend upon several factors like temperature, vegetation, parent material, hydrologic conditions, etc. Its formation is generally associated with drastic weather condition under hot humid climate and heavy rainfall. The processes involved in the formation of acid soils in the tropical zones are *laterization* of varying degrees, *podzolization*in temperate and sub-temperate zones, or intense base leaching in light alluvial soils. In acid soil region, soil leaching is predominant because precipitation exceeds the evaporation, thereby causing removal of soluble salts, readily soluble soil minerals and bases resulting in the acidity of surface soils. When the process of weathering is drastic, the sub-soil and in many situations the whole soil profile becomes acidic. Acid sulfate soils are also formed under marshy conditions with presence of considerable amount of partly decomposed organic matter under anaerobic condition with subsequent formation of sulphuric acid due to oxidation of accumulated sulphides under aerobic condition. Application of selective fertilizer like ammonium sulphate and water soluble phosphatic fertilizer superphosphate causes acidity. In addition, decomposing plant residues and organic wastes particularly in the forest soils contribute to acidity of soil. Crops suffer usually from physiological disorders under water logged condition which induced low nutrients transformations due to poor drainage particularly during rainy seasons. Heavy downpour left very little scope of percolation conducive to sheet erosion in undulating and rolling topography in hilly and sloping terrains. Hill soils of NE regions used to vary in texture sandy loam prevalence of shifting cultivation.

Spread of acid soils regions in eastern India

The eastern states of Assam, Bihar, West Bengal and Jharkhand and north-eastern states of Arunachal Pradesh, Nagaland, Manipur, Meghalaya, Sikkim and Tripura situated between latitude of 17° and 29° N and longitude of 80° and 97° E constitute the eastern region of India having a tropical monsoon climate with an average annual rainfall of 1544 mm and mean monthly temperature of 9.3 to 39.1°C. India is divided into 21 agro-ecological regions and the eastern states comprise 4 regions, *viz*:

i) Eastern plateau (Chhatisgarh, Bihar, Jharkhand): Hot sub-humid eco-region with red and yellow soils having neutral to slightly acidic;

ii) Eastern plateau and Eastern Ghats (Chhottanagpur plateau of Jharkhand and Bihar and western parts of West Bengal): Hot sub-humid eco-region with red loamy soils having slightly to moderately acidic;

iii) Assam and Bengal plains: Hot humid eco-region with alluvium derived soils moderately to strongly acidic; and

iv) Eastern coastal plains: Hot sub humid eco-region with alluvium derived acid sulphate soils.

Distribution of acid soils in West Bengal

In West Bengal, different types of acid soils are found due to peculiar topography and land situation accompanied by high precipitation. Laterite, red gravelly soils, alluvial terai, coastal and foothill soils are the most dominant acid soils. Laterite and red gravelly type of acid soils are normally found in the districts of Midnapur (W), Bankura, Birbhum, north-western part of Burdwan, Puruliya, whereas alluvial and terai soils are found in Jalpaiguri and Coochbehar. In addition, some acreage of acid soils is found in foot hill soils of Darjeeling district. Red and lateritic soils of West Bengal mainly having coarse and medium textured with low water retentive capacity mainly concentrated in the regions of high rainfall and temperature. In the deltaic area of Sundarbans in coastal plains, some soils are acid sulphate in nature. The acid soil coverage in different districts of West Bengal is depicted in Table 5 as below.

Table 5: District wise distribution of acid Soils in West Bengal

District	Area (m ha)	District	Area (m ha)
Bankura	0.104	Jalpaiguri	0.369
Birbhum	0.086	Midnapur (E and W)	0.306
Burdwan	0.620	Puruliya	0.235
Coochbehar	0.115	Parts of Murshidabad, Malda, Dinajpur (N) and S) and 24 Parganas (S)	0.200
Darjeeling	0.235		

Source: Das (2004)

Crop production constraints in acid soils

1. Physical constraints

The upland and lowland acid soils have distinct physical properties which are given below -

a. Upland acid soils

i) Single grained to granular or crumb structure

ii) Coarse textured, highly porous, usually compact at soil surface

iii) Low organic matter content

iv) High infiltration rate and hydraulic conductivity

v) Low water holding capacity with high permeability

vi) High bulk density and soil crusting

vii) Low water and nutrient capacity excepting NH_4-N and phosphorus, and

viii) Dominance of 1:1 type of clay minerals

b. Lowland acid soils

i) Angular blocky or prismatic structure

ii) Medium to fine textured soil

iii) High organic matter content

iv) Low infiltration rate and moderate to low hydraulic conductivity

v) High water retention capacity and moderate permeability

vi) Sub-soil compaction, clodding and floppyness

vii) Low leaching loss of plant nutrients, and

viii) Drainage congestion

Therefore, constraints for both upland and lowland acid soils need to be ameliorated or well managed in bringing such problematic acid soils to a remunerative one. Crop cultivation during *kharif* season also suffer from severe limitation particularly under rainfed conditions where root growth got mechanical impedance and crust formation on the soil surface obstructing germination of small seeds. Hence, *rabi* or winter crops during dry season are only the option for enhancing crop productivity under acid soils regions on adoption of specific management practices.

2. Chemical constraints

a) Low soil pH

b) Low CEC due to dominance of 1:1 clay minerals

c) Low base saturation (16-67%)

d) Low organic matter status

e) Low water and nutrients availability

f) More susceptible to drought

g) Poor drainage of peat, marshy and acid sulphate soils

h) Crusting in red and lateritic soils

i) Direct acid toxicity on plants

j) Toxicity of Mn, Fe, Al and Mn with pH less than 5.4

k) Nutrient imbalances

l) Deficiency of Ca, Mg, Mo, B, Zn and Si

m) Low availability and high fixation capacity of P, and

n) Deficient in secondary nutrient like S

The acidity in soils of northern districts of West Bengal namely, Coochbehar, Jalpaiguri and Darjeeling and north-eastern regions are attributed to the presence of exchangeable Al^{3+}, its content varying from 0.12 to 4.5 in Coochbehar and Jalpaiguri, 0.1 to 4.1 in Sikkim; 0.13 to 0.63 in Nagaland; 0.15 to 0.55 in Tripura; 0.04 to 3.53 in Meghalaya and 0.75 to 3.0 cmol (p^+)/kg in the Brahmaputra valley of Assam.

3. Biological constraints

In acid soils the availability of Ca, Mg, P, Mo, B and Si is low, where as availability of Fe, Mn, Cu and Zn is high, which influence the microbial population, activity and survival. The bacteria and *actinomycetes* perform better in intermediate pH and higher pH range, whereas fungi perform better under acidic pH range. The activity of *Nitrosomonas* and *Nitrobacter* are significantly reduced with increased soil acidity. *Azotobacter* population is found to be almost nil in acid soils. The activity of non-symbiotic free living BGA and symbiotic rhizobia in acid soils is also less. Calcium and Mo deficiency in acid soils adversely affects nodulation as well as N fixation in legume while shortened root length due to high Al content leading to lesser site of nodule formation. Besides, increased attack by soil pathogens and accumulation of organic acids and other toxic compounds due to unfavourable redox conditions in acid soils is also noticed.

Diagnosis for evaluation of acid soils

For appropriate soil management practices vis-à-vis crop productivity, a proper diagnosis of acid soils in respect of identifying indicators is useful for soil fertility evaluation and management of acid soils. The following parameters need to be assed:

1. To evaluate the magnitude of soil acidity
2. To evaluate acidic soil quality: the capacity of the soil to function needs to be measured using physical (BD, MWHC and MWD etc), chemical (pH, Org. C, MBC, MBN, Avail. NPK and other nutrients), bio-chemical

(different enzyme activities) and microbiological (activity of bacteria, fungi, Al resistant and other acidophilic microorganisms) indicators.

3. To identify the magnitude of deficiency of plant nutrients, and
4. Measurement for the appropriate application of amendment like liming materials to neutralize soil acidity by determining the lime requirement (LR).

Reclamation and management of upland acid soils

The upland acid soils have diversified problems and need specific solution for each type of problems. The following physical methods are useful to solve the problems:

Conservation tillage: In order to arrest runoff and soil erosion, one summer ploughing with MB plough followed by twice disc harrowing is effective in controlling the weed growth.

Organic matter: Addition of organic matter, compost, paddy straw and green manure prior to adoption of conservation tillage help to increase the mean weight diameter of soil aggregates and soil water storage besides increasing the yield of crops.

Soil compaction: Soils with relatively low percentage of active fine fractions exhibit high percolation loss of water and nutrients which can be controlled by compaction with heavy rollers (4 to 6 passes of 800 kg roller) or heavy tractors. Soils that have very low clay, addition of tank silt and clay followed by compaction increases the water retentivity of soils. In case of sub-surface compaction, deep cultivation with chiseling/ploughing has been found to be beneficial and most economical.

Soil crusting: Application of gypsum @1 t/ha along with FYM @ 5 t/ha has been found to alleviate soil crusting and increasing the grain yield of upland rice in red-lateritic acid soil zone. Incorporation of straw mulch on seed lines of cotton, soybean, cowpea and powdered groundnut shells and paddy husk two weeks before setting is also effective in reducing the crust formation. Besides, application organic amendments such as FYM or decomposable green leaf manures (*Glyricidia*) is found to be quite effective in preventing crust formation and increasing moisture retentivity of the soil. Judicious application of lime, organic amendment and phosphate benefit the crop.

Permeability: Puddling twice by tractor drawn cage wheel or power tiller operated rotavator reduces the permeability of the soil and significantly increase the yield of rice.

Shallow depth: Construction of about 10 cm high ridge and furrow on shallow soils mitigate the depth related constraints. Further, addition of clay and organic matter improves the physical condition and crop growth. Under rolling topography, ridge plus live barriers has been found effective while for desurfaced soil, deep cultivation has been beneficial.

Soil amelioration

Liming has been found to be very effective soil ameliorant in improving the base saturation, inactivating Fe, Al and Mn toxicity and thus increasing P and Mo availability, stimulating microbial activity leading to mineralization of organic nitrogen and fixation of atmospheric nitrogen and thus increasing the yield of various upland crops. In liming programme, the first and foremost thing is to assess the lime requirement of the soil for optimum yield of the crop/crops. Lime requirement is estimated on the basis of exchange acidity and percentage base saturation of soil. Application of full dose of lime requirement (LR) has not been beneficial as the lime is lost by leaching from top soil of light textured soils because of their low exchange capacity. Split application is recommended to minimize the leaching loss. Lime should preferably be applied in smaller doses in every alternate year, rather than heavy dose one in three to four years. Application of full LR doses suppresses P availability and caused B deficiency. The excess application is thus harmful. Application of lime at the rate of 25% of LR in alternate year in Orissa and 10% of LR in Bihar has been found effective in raising the yield. In general, the lime requirement ranges from 3.5 to 15 tonnes/ha. The following practices are to be considered while initiating the liming programme to correct the soil abnormalities:

1. Lime application should be done in split doses
2. Smaller doses with frequent application
3. Application of lime in furrows at the time of sowing
4. Lime applied in every alternate year till the soil pH is brought to normal range, and
5. The programme must be conducted on area basis (block/panchayet)

A. Choice lime material

Source of lime material, its cost effectiveness and easy acceptability to the farmers is an important aspect of acid soil management. Amongst the naturally occurring lime sources, commercial limestone and dolomite limestone powder are the most used amendments, but these are not economically feasible as agricultural lime. Several industrial wastes such as basic slug and blast furnace slag from steel industry, lime sludge (65-84% $CaCO_3$) from paper mills, press mud from sugar mills using carbonation process, cement kiln wastes and precipitated $CaCO_3$from fertilizer factories have been successfully and economically used an amendment for acid soils which are eco-friendly and quite effective as ground limestone. Application of oyster shell powder (99.8% $CaCO_3$) and fly ash @ 10 tonnes/ha is rated equally effective with lime. The marketable lime materials should preferably have at least 25% CaO and ground to less than 80 mesh size. However, economic feasibility of liming has to be looked into before one recommends such a practice. Liming is much desirable when multi-cropping is followed in acid soils with legume as one of the crops in the sequence.

B. Crop response to lime application

Liming is widely accepted amelioration practice for efficient utilization of native soil nutrients and economic use of added fertilizer sources. It is also desirable for raising the productivity of several crops. However, crops differ in their response to lime application. Poor lime responsive crops like paddy, potato, minor millets and finger millets indicates their tolerance to acidity. The crops which show medium response to liming are maize, groundnut, gram, lentil and pea. However, crops like soybean, french bean, pigeonpea and cotton responded well to liming indicating their sensitivity to acidity (Table 6). The response to lime application in acid soils increased further when applied with recommended fertilizers or with FYM @ 5 tonnes/ha. Accordingly, crops are classified according to their relative response to liming. In red acidic soils under rainfed conditions, highly responsive crops (*calciphilic* crops) such as groundnut, cotton, soybean and pigeon pea should be grown for the first two years of liming, followed by medium response crops such as maize, black gram and green gram in the next two years. The low responsive crops like paddy and small millets should be grown in the last two years.

Table 6: Relative response of crops to liming

Crop	Relative yield units			
	Control	NPK	Lime	NPK + Lime
High response				
Cotton	100	2270	1083	3887
Arhar	100	394	1133	1927
Soybean	100	207	503	788
Medium response				
Maize	100	429	202	660
Lentil	100	301	266	627
Gram	100	256	223	606

Pea	100	313	295	596
Wheat	100	215	109	267
Sorghum	100	139	196	346
Groundnut	100	197	304	333
Low response				
Barley	100	265	102	327
Aus paddy	100	124	122	165

Source: Mandal *et al.* (1975)

Crop selection

Selection of alternative crops and cultivars tolerant to soil acidity is an effective tool to counter the soil problem and breeding of such varieties is of specific importance for attaining higher productivity, particularly in areas where liming is not an economic proposition. The crops can be grouped on the basis of their performance in different soil pH range (Table 7). Rice has certain amount of tolerance to acidity since flooding of the field creates favourable condition for growth of rice. The rainfall distribution pattern in the subhumid tropics enables the farmers to grow rice in the medium and low lands in the *kharif*season because major portion of the annual precipitation is received within July to September and fields remain ponded with standing water. In the post rainy season some pulses, oilseeds and vegetables that have adaptability with local environment are grown with the help of residual soil moisture.

Table 7: Soil reaction preference of different crops

Crops	Optimum pH range
Cereals	
Maize, wheat, barley	6.0-7.5
Millets	5.0-6.5
Rice	4.0-6.0
Oats	5.0-7.7
Legumes	
Field beans, soybean, pea, lentil	5.5-7.0
Berseem	6.0-7.5
Groundnut	5.3-6.6
Miscellaneous crops	
Sugarcane	6.0-7.5
Cotton	5.0-6.5
Potato	5.0-5.5
Tea	4.0-6.0

Source: Mishra, 2003

Crops and cropping system

In the acid soils region, rice is the predominant crop grown during the rainy season irrespective of type of land i.e. high, medium and lowland. With the availability of standing water during monsoon season, local farmers used to grow traditional paddy varieties in the medium and lowland. Crop diversification has been confined only to upland situation under rainfed conditions where maize, minor millets, pigeon pea, green gram, black gram and sesame are grown. In the post rainy season, some pulses, oilseeds and vegetables, those have established themselves by the process of natural selection; local peculiarity and needs in favour of survival rather than high productivity are being grown with the help of residual soil moisture. The important crops of eastern states are rice, wheat, maize, bean, horse gram, pigeon pea, pea, bengal gram, lentil, lathyrus, groundnut, mustard, sesame, niger, linseed, safflower, soybean, jute, sugarcane, cotton, potato, onion, vegetables, tea and coffee. Growing winter crops followed by rice is the common practices in eastern regions. The important crops and cropping systems in different land situations in rainfed as well as irrigated acid soils of West Bengal is represented in Table 8.

Table 8: Crops and cropping systems in rainfed and irrigated acid soils of West Bengal

Land type	Cropping system
Rainfed condition	
Upland	Rice/Maize/Jowar/Bajra/Ragi/Arhar/Niger/Mesta/Groundnut/ VegetablesIntercropping: Pigeonpea + Maize/Rice/Groundnut/ Soybean/BlackgramSequence: Maize/Groundnut – Linseed/Barley
Medium land	Rice – LentilJute/Aus paddy - Aman paddy - Khesari/Mustard/ Vegetable
lowland	Improved local rice – Lentil – GM – Aman paddy
Irrigated condition	
Upland(alluvial)	Rice – Wheat/Potato/Vegetables/Mustard-Jute/Sesamum/Moong/ Groundnut
Medium land (alluvial)	Rice – Wheat/Vegetables - Jute/Moong/Groundnut/Vegetables
Medium land (coastal)	Jute/Aus paddy/Vegetables - Aman paddy - Vegetables/Mustard/ Potato
Lowland (alluvial)	Rice – *Boro* rice - GM
Lowland (coastal)	Jute – Aman paddy-*Boro* paddy

Crop planning in dry season

The vast majority of the acid soils are mono-cropped where no other crops are possible to grow after harvesting *kharif* rice due to lack of water availability. Some site-specific devises are therefore necessary to make these lands under suitable crops either by utilization of residual soil moisture or by harvested rainwater management. Crop planning in dry season coupled with the improved technologies is of paramount importance for enhancing productivity of acid soils. The important components for efficient crop management are delineated below:

a) Rainwater harvesting, conservation and recycling

b) Management of soil profile moisture

c) Off season deep tillage and minimal tillage

d) Conservation tillage

e) Soil and crop residues mulching

f) Choice of crops, cropping pattern and plant geometry

g) Integrated nutrient management with the use of organic, chemical and bio-fertilizers

h) Integrated pest management through weeds, pests and diseases control, and

i) Organic farming

a) Choice of crops under rainfed ecosystem

1. Fodder maize and legumes to support livestock
2. Gourds and watermelon having less water requirement
3. Introduction of less water requiring crops like blackgram, greengram, lentil etc
4. Oilseed crops like mustard, sesame and sunflower
5. Winter vegetables
6. Flowers like marigold and tuberose.

b) Choice of crops under irrigated ecosystem

1. Major areas under irrigated rice in dry season
2. Crop diversification high value oilseed crops like groundnut, sunflower, sesame and mustard and pulses like soybean, arhar, blackgram, greengram, tuber crops like potato and sweet potato and winter vegetable crops

Soil management

The acid soils of dry land areas suffer from a variety of physical, chemical and biological constraints which could be managed through :–

1) Control of water erosion – agronomic and engineering measures

2) Deep, minimal, zero and conservation tillage as per requirement

3) Soil amendments
4) Incorporation of organic residues and green manuring and bio-fertilizers
5) Integrated plant nutrient supply and management
6) Balanced fertilizers including indigenous rock phosphate application and management

Conclusions

1. Acid soils in India have been formed due to drastic weathering associated with hot humid and high rainfall situations. Laterization, podzolization and accumulation of undecomposed organic matter in marshy conditions has caused soil acidity.
2. Poor growth of crops on acid soils is attributed to the presence of toxic amounts of Fe, Mn and Al and deficiency of B, Mo and Zn and less activity of soil micro organisms.
3. In eastern and north eastern regions of India soil acidity poses a great problem for crop production due to very poor fertility status and nutrient imbalances etc. and hence, it needs a special attention to address such acidity problems in relation to crop production.
4. The management of secondary and micronutrients in acid soils can be made either by the addition of cost-effective lime or liming materials to counteract the abnormalities of soils by raising pH as the first step and subsequently by modifying agricultural practices and applying balanced fertilizers including limiting nutrients viz. S, B, Mo and Zn at optimal levels based on soil test values for maximum crop production.
5. The integrated nutrient management system for the maintenance of soil fertility, sustainable agricultural productivity and improving profitability through the judicious and efficient use of mineral fertilizers especially secondary and micronutrients , organic matters, green manures, bio-fertilizers in acid soils.
6. Rainwater harvesting, its efficient storage and subsequent uses as supplemental irrigation especially in the areas of rainfed situations.
7. Selection of crops and cropping systems in different acid soils region need to be developed.

Present research needs for acid soils

1. Soil and crop related research programme are needed for enhancing crop productivity after removing production constraints of the acid soils.
2. Survey of soil related production constraints in the acid soils regions of the country.
3. Diagnosis of Al toxicity in soils and their amelioration.
4. Use of $MgCO_3$ and Si and also surface drainage for amelioration of acid sulphate soil.
5. Role of soil microbes in promoting soil aggregation.
6. Soil fertility management for various cropping systems.
7. Biological control of soil erosion by maintaining plant cover on the soil surface.
8. Soil management under shifting or jhum cultivation.
9. Soil test crop response and fertilizer recommendation for crops including plantation crops.
10. Selection of acid loving crops and varieties beyond what has already been prevailed.
11. Feasibility of fly ash for amending the acid soils through multi-location trials.
12. Crop diversification in rainfed uplands with emphasis on other crops on substitution of rice.
13. Rainwater harvesting and management for efficient use of irrigation water.

Future research strategy

1. More research on the characterization of acid soils considering the nature of acidity in a site-specific area for the efficient management of soils in relation to crop production.
2. More research on the efficacy of locally available low cost liming materials for amelioration of acid soils in different areas.
3. More acid tolerant crops and varieties adapted to acid soils of different intensities and locations needs to be developed through appropriate plant breeding techniques.
4. Al-resistant microorganisms where exchangeable Al is predominantly responsible for the development of acid soil and also other acidophilic

microorganisms are to be identified and maintained for overcoming constraints due to poor microbial activity in acid soils.

5. More research on acid soils regions of eastern India relating to microbiological and biochemical attributes should also be given top priority to evaluate acid soil quality.

6. An emphasis should also be given to integrate physical and chemical parameters with microbial parameters in developing a minimal set of acid soil quality indices.

7. Integrated farming system, involving agronomic and horticultural crops, livestock, dairy, fishery, agro-forestry etc with a view to increase crop productivity, employment opportunity and income to the resource poor farmers of acid soils region.

References

Das, D. K. 2004. In: Introductory Soil Science. Kalyani Publishers, pp170-193.

Moitramani, D.P. 1971. Soil and Water Research in India, Retrospect and Prospect. P 98 ICAR Publication.

Panda, N. 1987. Acid soils of Eastern India, their chemistry and management. Directorate of Agric. & Food Production, 5th J.N. Mukherjee Lecture, Pune.

Tamboli, P.M. 1976. Soil and Water Research in India, Retrospect and Prospect. P 132 ICAR Publication.

Velayutham, M., Mandai, D.K., Mandai, Champa and Sehgal, J. 1998. Agro·ecological sub regions of Indiafor planning and development.NBSS publication No, 35, NBSS & LUP, Nagpur. P 372.

Section 4

Sustainability in Agriculture

14

Need Based Sustainable Agricultural Production

Bishnu Prasad Mishra[1], *Banashri Rath and Aamlan Saswat Mishra*

[1]*SOET, CUTM, Paralakhemundi-761211, Odisha, India*
Email: dean.soet@cutm.ac.in

Abstract

Never ending additional need of farmers have forced them to grow more cash crop along with crops used for the staple food. The prime crops as per the food habit, social need has to be produced in the respective locality. The micro planning for agricultural and food production has to be planned for different blocks of a district to meet above needs. The production planning has to be simulated first and programmed for trial production. After successful trail, the same has to be replicated in larger area. Mathematical model development approach is one of the solutions to predict the need of the people and plan production depending on the availability of the input materials like natural resources and other inputs. In addition to this food and nutrient security for the people living in those different areas are very important in present context. Global population is growing fast with 1.2% per annum. Feeding these people increases the responsibility of farmers for growing more crops without considering the soil health. Improper crop sequencing has resulted in imbalance in soil nutrient which results in unsustainable production of forth coming crops. Only crop rotation cannot address sustainability issue. Considering the environmental factor, social and economic factor of the area the crop production model must be prepared and implemented. Multiple goals with many constraints are the situation of sustainable crop production simulation. Judicious use of natural resources for crop production as per the need for survival of people of that area with priority has been considered. GOAL program has been formulated taking the Odisha agricultural production data and solved. The nutrient and natural resources consumed to produce the agricultural produces has been computed and balance nutrient remaining for future crop has been obtained. The increase or decrease in crop production requirement will require new set of simulation. Prevailing facts has been used as boundary conditions to simulate the optimal production keeping natural reserve intact. WHO food recommendation for cereals, pulses and oil seeds has been considered to find the maximum food requirement for the district; Thirty districts of Odisha may have different priorities. Different GOAL programming and simulation will be needed while producing the crops for entire Odisha in a sustainable manner. Only one model has been developed for 30 district of Odisha. The separate computation of all 30 districts of Odisha or for different agro climatic combination will provide better understanding about sustainability production potential in varying food demand conditions. The average present market price for agricultural produces (26 crops) with 5 priorities has been considered for computing maximum production possible in a sustainable manner.

Keywords: Crop rotation; environmental factor; social and economic Judicious use of natural resources for crop production; GOAL programme

Introduction

World is passing through population explosion and cultivable land to produce food for these people is limited. Limits to growth has estimated the population growth to reach its peak by year-2050 and food production to reach its' peak by year-2030. This is a big challenge for the researchers to find out solution to this type of global issue. The increase in per capita income has created demand for niche foods. The targeted market may be rural, peri urban or urban but affordability, availability and accessibility are real issues for food security and finally sustainability. A researcher has to use suitable mathematical tool like GOAL programming to compute the possibility of optimal agricultural production using minimum natural resources as per the priority and requirement of the people. Use of natural resources in a regulated manner is the difficult part while implementing it. Agricultural policy has to incorporate such issues and implement it as an adoption initiative.

The secondary crop production and natural resource published data collected from Agriculture Department; Odisha has been taken for agricultural crops production simulation purpose. GOAL programming simulation tool has been used for simulation as it is the best tool for such simulation. Suitable priorities have been considered for formulation of the equations and solved for finding the best optimal solutions fulfilling different local needs. The crop production parameters for 26 crops grown in 30 districts of Odisha have been considered in this simulation. These 26 crops are most commonly grown in Odisha as per the food habit and social requirement of farmers. The crop production inputs for all these 26 crops has been collected from researchers of OUAT and used in writing the models equation. The priorities to be used in this GOAL programming have been obtained from the villagers and ratified by the KVK persons working in the region. The set of solutions have been obtained by using QSB package. This package has been developed by the researchers of IIT, Kharagpur. The results obtained are many, inter dependable nature of output makes it complicated for interpretation. Some results have been simplified by considering a few outputs like area to be cultivated under different crops to make the crop production sustainable. The sensitivity analysis has been carried out to cross check the possibility of variation of agriculture production in that region.

Review of Literatures

In a geographical area like Odisha, it is important to note the change in cropping pattern over years. This may be due to change in food habit, change in climate, knowledge base, policy and other unforeseen parameters. This has a stake on sustainability. Increase in population will certainly affect the sustainable development model to a large extent.

Crop Production Trend

The crop production data of Odisha for major nine crops have been collected for 50 years, analyzed [8,9] and plotted for interpretation. The crops like rice, wheat, course cereals, pulses, oil seeds, cotton, sugarcane, other crops and fiber crops are considered in this analysis. Change in production for above 9 crops in Odisha state for 50 years has been shown in Figure. 1. The Figure. 1, shows that there has been increase or decrease of production with respect to different crops during year-1950 to year-2000. to year-production with respect to n of different crops.crops are considered in this analysis. The increase in population, changes in food habit, environmental factors have influenced cropping practice to large extent. It is a matter of surprise to note that, production of rice, staple food for Odisha people is not varying much. It is about 23% to 24% of total crop production and stable over last 50 years. The production of other crops has increased from 8% to 17%, which may be treated as need based. The production of coarse cereal has decreased from 30% to 16% during last 50 years. The adoptions of modern farming, change in economic status have replaced this course cereal area with cash crops. Sustainable planning must include these clauses while framing the sustainable model and its production implementation policy [14]. All above crop uses the same natural resources but productivity increase using modern technique will make soil healthier. The crop sequencing using DSS will make it more realistic and SMART.

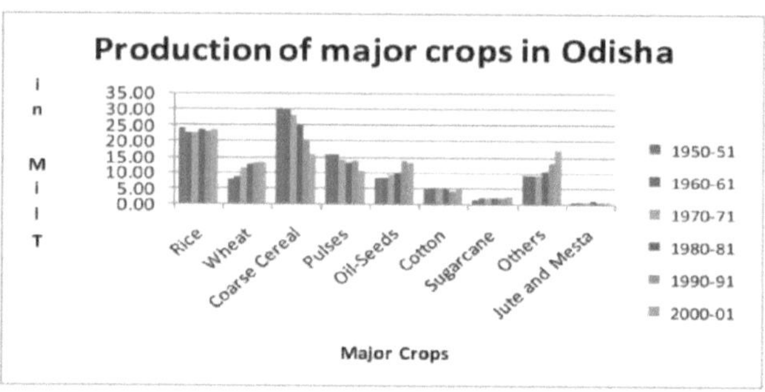

Fig. 1: Crop Production trend over last 50 years

The pulses production data in Odisha have been plotted on a bar graph in Figure. 1. Odisha people consume a lot of pulses. Still then pulses crop production is declining over time. The pulses crops are grown as pyre crop all over Odisha. Non-availability of residual moisture may be one of the reasons for decrease in pulses production. Further, to increase productivity of pyre crops no innovation has been reported in Odisha context. Since 1999 till date the production of pulses is increasing slowly. Presently rice fallow areas are being monitored using GIS technique and pulses crops are scientifically grown in those areas. Irrigation of pulses crop still remains untouched. Solar pumping systems are being promoted presently to address these issues. Production trend of millets “a rich man’s delicacy” is negative. Availability of cereals (PDS) at very low cost has kept small farmers away from growing it. Processed millets have huge national / international market potential. These are short duration crops which require less water and have short time for production [Figure. 2]. Production production oduction [Figure. .on of pulses crop still remains untouched. Solar pumping systems are being promoted presntly Policy interventions for promotion of millet crops are there but the efficiency of implementation is very less. The spread effect of millet cultivation has not been observed although the program is implemented for many years.

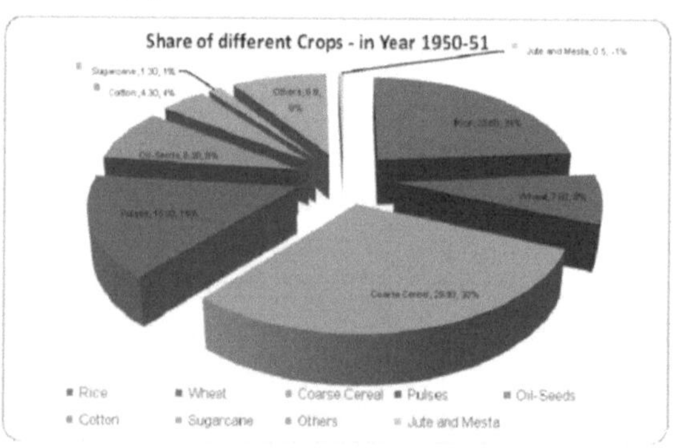

Fig. 2: Share of different Crops – Odisha 1950

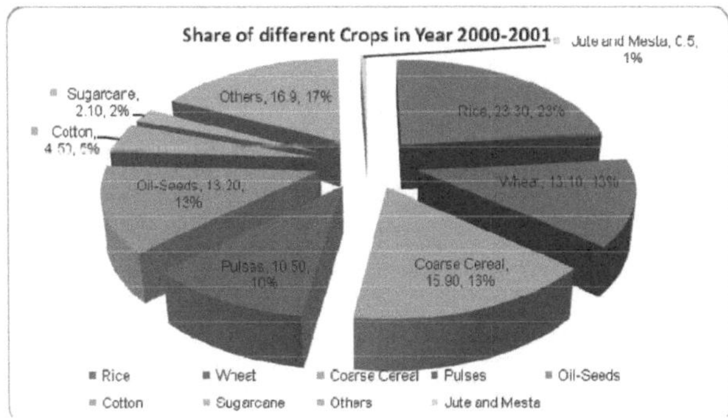

Fig. 3: Share of different Crops, Odisha 2000

Further, the data collected are plotted using pie chart and compared to observe change in area under above nine crops. The charts show that the share of paddy production has been steady over 50 years (23%). It has been reported by agricultural scientists of Odisha that production share of pulses has decreased as the storage and processing has not been sufficient to preserve it. Similarly, due to less profitability in paddy growing, farmers are diversifying to other crops.

Input Management for crop production

The major inputs data like land, water, seed, fertilizer, finance and machinery for production of all above nine crops in all the 30 districts of Odisha have been collected. The collected data have been used for formulation of GOAL programming model.

Table1: Inputs for Agricultural Production, Odisha

District	Production	Net Area Sown	Rain fall	Loan	Seed	Fertilizer	Machinery
Jharsuguda	102.45	107.61	1653.6	23.15	1653	9408	64418
Deogarh	58.21	108.08	1346.6	12.07	1080.3	3321	18693
Boudh	93.16	139.13	1525.3	24.64	3307.8	7152	17124
Gajapati	43.61	141.61	1279.3	17.67	4986	5187	22805
Jagatsingpur	179.12	186.36	1688.5	53.03	6980	11710	77414
Phulbani	73.97	190.25	1686.2	24.64	1262	1425	8384
Sonepur	246.74	198.4	1488.7	37.87	9642	9342	112225
Malkangiri	112.35	221.29	1349.8	20.42	6533	6417	25424
Nayagarh	133.22	222.75	1497.6	94.98	2081	7334	33198
Khordha	163.11	232.65	1447.4	196.97	5752	13917	60238
Bhadrak	344.45	237.96	1285.3	124.84	6844	34584	43973
Rayagada	98.82	251.56	1285.7	42.19	7862	8513	45445
Dhenkanal	185.35	262.79	1428.7	131.45	5921	8122	35245

Kendrapara	200.87	268.45	1658.9	75.45	5302	9217	94971
Nawapara	150.77	268.89	1312.4	42.82	9832	10931	32319
Jajpur	212.26	272.72	1469.2	81.11	7127	32609	50631
Sambalpur	337.13	281.59	1879	80.28	12320	30595	121203
Nabarangpur	195.69	286.1	1838.2	60.47	13210	27875	17910
Puri	206.1	286.7	1419.5	80.21	11013	18808	72288
Cuttack	189.22	309.39	1853	201.76	8457	23012	81525
Angul	134.08	321.34	1422.2	78.24	5055	8833	23120
Balasore	351.58	332.61	1701	227	16273	43045	105723
Sundargarh	300.56	377.67	1496.2	101.39	9658	12764	87298
Koraput	212.66	394.29	1647.8	77.32	16886	14066	39248
Keonjhar	278	439.95	1298.8	75.51	11133	11098	58566
Bargarh	633.71	447.36	1279.1	211.11	29464	47760	111950
Bolangir	311.67	472.05	1641	57.36	19666	19710	54649
Mayurbhanj	542.59	518.84	1763	132.86	10777	21844	100605
Kalahandi	349.96	606.14	1885.4	72.39	24630	36134	132726
Ganjam	474.55	686.09	1161.9	140.92	24871	40136	94349

Cropping pattern practiced in Odisha for soil nutrient management is an attempt to arrive at sustainability in crop production. Most popular eight numbers of rice based cropping patterns being followed in the state of Odisha. They are found to be not enough to arrive at sustainable crop production.

Problem Formulation

The crops generally grown by the farmers of Odisha have been noted and the inputs /resources used for production of all these crops have been given below Table. 2. The inputs like water, labour, fertiliser, animal power, plant protection, finance, revenue have been considered in the GOAL model development. Twenty-six crops commonly produced in Odisha for so many years has been considered as the sustainable production model.

Crops	Water	Human (Days)	Fertilizer (Kg)	Animal power (days)	PP (Days)	Cost	Yield Q/Ac	Price
Paddy	100	82	48	26	6	20666	20	1100
Wheat	40	45	42	18	2	24470	11	1050
Barley	50	45	32	18	2	24320	12	900
Green Gram	15	30	32	8	1	6549	4	3700
Cowpea	15	40	32	11	2	19860	7	1200
Arhar	40	40	32	10	3	19450	8	2400
Groundnut	48	70	40	14	2	19768	6	3700
Sesame	25	36	24	10	2	18980	4	4000
Castor	58	30	70	11	0	17730	4.5	1500
Cotton	62	109	64	15	8	24608	6	5000
Tobacco	55	114	56	13	3	37962	5	9000
Sugarcane	170	262	144	16	6	63587	400	800
Jute	48	86	48	14	3	29750	11	1200

Mesta	45	62	26	13	2	24090	7	1100
Niger	25	30	32	11	1	16560	4	7000
Sunflower	40	30	30	11	2	16790	3	1200
Mustard	50	40	48	11	3	28220	6	2100
Lentil	40	30	32	8	1	15330	5.5	1500
Field pea	15	10	40	0	1	7870	8	1400
Horse gram	20	27	15	7	0	8164	6	4500
Black gram	20	30	32	8	1	8514	4	4200
Bengal Gram	15	27	24	8	1	8050	7	4000
Ragi	40	57	32	10	1	21510	10	9500
Maize	55	84	64	16	2	17534	14	950
Jawar	55	55	48	11	2	21975	9	1150

Formulation of the Sustainable Goal Program

Considering the available natural resources, resources/inputs used in crop production, crops to be produced for meeting the social, economic and environmental load, the GOAL Programming model has been formulated. The goals like water use, labour use, production enhancement, revenue generation and food security has been considered as the conflicting goals to be achieved with priorities. The priorities have been changed to know the matching results. It has been observed that the priority 1 and one or two more priorities have been achieved in the proposed model.

Goal 1

Annual water required to produce 5 crops such as paddy, maize, green gram, black gram and groundnut has been considered for simulation. Water has been assigned priority level 1 as it is very important for crop production. Water available for cultivation has to be used for crop production. The sufficiency/deficiencies will be evaluated.

Goal 2

Maximum value of the crop produced from Odisha has been the second level goal. The value of the crops assigned was about 3 trillion INR. It is based on the population and per capita income from agriculture.

Goal 3

The agricultural labourer available has been obtained from the statistical facts. The solution obtained requires about 8.05 million-man days for all above crops. Non-availability of agricultural labourers can be well addressed by using farm machinery where ever it can substitute human labour. The alternative mode of power input for sustainable crop production has to be developed.

Goal 4

The fourth goal was to use all arable land for crop production. The use of land relates to available of land and other inputs for cultivation. Simulation output will be interpreted for recommendation.

Goal 5

Production of different food grains in right quantity as per the population growth rate must be computed and incorporated in to the production policy. Presently, WTO has given maximum importance for food security. As per ICMR recommendation, considering 400 grams of cereals, 150 grams of pulses and 50 grams of oil as the basic per capita per day food requirement, problem formulation has been developed to address food security.

All the solutions will not optimal but are fair in nature. The formulation of Goal programming has been given below.

	Goal/Constraint/Bound
Min:G1	1X27+1X30
Min:G2	1X32
Min:G3	1X34
Min:G4	1X36
Min:G5	1X38+1X40+1X42+1X44+1X46+1X48+1X50+1X52+1X54+1X56+1X58+1X60+1X62+1X64+1X66+1X68+1X70+1X72+1X74+1X76+1X78+1X80+1X82+1X84+1X86+1X88+1X90+1X92+1X94
C1	100X1+40X2+50X3+15X4+15X5+40X6+48X7+25X8+58X9+62X10+55X11+170X12+48X13+45X14+25X15+40X16+50X17+40X18+15X19+20X20+20X21+15X22+40X23+55X24+55X25+55X26-1X27+1X28=3834457
C2	100X1+40X2+50X3+15X4+15X5+40X6+48X7+25X8+58X9+62X10+55X11+170X12+48X13+45X14+25X15+40X16+50X17+40X18+15X19+20X20+20X21+15X22+40X23+55X24+55X25+55X26-1X29+1X30=3417229
C3	82X1+45X2+45X3+30X4+40X5+48X6+70X7+38X8+30X9+109X10+114X11+262X12+86X13+62X14+30X15+30X16+40X17+30X18+10X19+27X20+30X21+27X22+57X23+84X24+84X25+55X26-1X31+1X32=2.35E+07
C4	48X1+42X2+32X3+32X4+32X5+32X6+40X7+24X8+70X9+64X10+56X11+144X12+48X13+26X14+32X15+30X16+48X17+32X18+40X19+15X20+32X21+24X22+32X23+64X24+64X25+48X26-1X33+1X34=873709
C5	26X1+10X2+10X3+8X4+11X5+10X6+14X7+10X8+11X9+15X10+13X11+16X12+14X13+13X14+11X15+11X16+11X17+8X18+7X20+8X21+8X22+10X23+16X24+16X25+11X26-1X35+1X36=1.15205E+09
C6	6X1+2X2+2X3+1X4+2X5+3X6+2X7+2X8+8X10+3X11+6X12+3X13+2X14+1X15+2X16+3X17+1X18+1X19+1X21+1X22+1X23+2X24+2X25+2X26-1X37+1X38=1.02E+07
C7	20666X1+24470X2+24320X3+6549X4+19860X5+19450X6+19760X7+18980X8+17730X9+24600X10+37962X11+63587X12+29750X13+24090X14+16560X15+16790X16+20220X17+15330X18+7870X19+8164X20+8514X21+
C8	20X1+11X2+12X3+4X4+7X5+8X6+6X7+4X8+4.5X9+6X10+5X11+400X12+11X13+7X14+4X15+3X16+6X17+5.5X18+8X19+6X20+4X21+7X22+10X23+14X24+14X25+9X26-1X41+1X42=2E+07
C9	1X1-1X43+1X44=4226
C10	1X2-1X45+1X46=18
C11	1X3-1X47+1X48=15
C12	1X4-1X49+1X50=836
C13	1X5-1X51+1X52=51
C14	1X6-1X53+1X54=135
C15	1X7-1X55+1X56=248
C16	1X8-1X57+1X58=261
C17	1X9-1X59+1X60=15
C18	1X10-1X61+1X62=74
C19	1X11-1X63+1X64=8
C20	1X12-1X65+1X66=41
C21	1X13-1X67+1X68=9
C22	1X14-1X69+1X70=16
C23	1X15-1X71+1X72=93
C24	1X16-1X73+1X74=21
C25	1X17-1X75+1X76=93
C26	1X18-1X77+1X78=112
C27	1X19-1X79+1X80=10
C28	1X20-1X81+1X82=35
C29	1X21-1X83+1X84=244
C30	1X22-1X85+1X86=617
C31	1X23-1X87+1X88=42
C32	1X24-1X89+1X90=179
C33	1X25-1X91+1X92=253
C34	1X26-1X93+1X94=9
Integer:	
Binary:	
Unrestricted:	

The detailed solution with the changing priority has been given below. The solution, constraint summary details are also given below in a tabular form.

	G1	Goal	Value	(Min.) =	0	(Alternate	Solution	Exists!!)
	G2	Goal	Value	(Min.) =	20,350,000.00			
	G3	Goal	Value	(Min.) =	0			
	G4	Goal	Value	(Min.) =	1,151,854,080.00			
	G5	Goal	Value	(Min.) =	199,235,125,248.00			

	Constraint	Left Hand Side	Direction	Right Hand Side	Slack or Surplus	Allowable Min. RHS	Allowable Max. RHS	ShadowPrice Goal 1	ShadowPrice Goal 2	ShadowPrice Goal 3	ShadowPrice Goal 4	ShadowPrice Goal 5
1	C1	3,834,457.25	=	3,834,457.00	0	3,417,320.00	28,851,628.00	0	-0.82	0	-0.26	-206.92
2	C2	3,417,229.25	=	3,417,229.00	0	-M	3,834,457.00	0	0	0	0	0
3	C3	23,500,000.00	=	23,500,000.00	0	3,149,920.00	M	0	1.00	0	0	0
4	C4	873,709.06	=	873,709.00	0	-M	1,843,732.75	0	0	0	0	0
5	C5	1,152,850,048.00	=	1,152,850,048.00	0	995,968.00	M	0	0	0	1.00	0
6	C6	10,200,000.00	=	10,200,000.00	0	229,889.00	M	0	0	0	0	1.00
7	C7	199,999,995,904.00	=	199,999,995,904.00	0	794,066,944.00	M	0	0	0	0	1.00
8	C8	20,000,000.00	=	20,000,000.00	0	781,964.00	M	0	0	0	0	1.00
9	C9	4,226.00	=	4,226.00	0	-M	38,263.35	0	0	0	0	0
10	C10	18.00	=	18.00	0	0	85,111.38	0	-12.20	0	-7.60	-16,206.20
11	C11	15.00	=	15.00	0	0	M	0	0	0	0	1.00
12	C12	836.00	=	836.00	0	0	M	0	0	0	0	1.00
13	C13	51.00	=	51.00	0	0	M	0	0	0	0	1.00
14	C14	135.00	=	135.00	0	0	M	0	0	0	0	1.00
15	C15	248.00	=	248.00	0	0	M	0	0	0	0	1.00
16	C16	261.00	=	261.00	0	0	M	0	0	0	0	1.00
17	C17	15.00	=	15.00	0	0	M	0	0	0	0	1.00
18	C18	74.00	=	74.00	0	0	M	0	0	0	0	1.00
19	C19	8.00	=	8.00	0	0	M	0	0	0	0	1.00
20	C20	41.00	=	41.00	0	0	20,062.97	0	-122.60	0	28.20	-28,816.60
21	C21	9.00	=	9.00	0	0	70,920.15	0	-46.64	0	-1.52	-19,831.84
22	C22	16.00	=	16.00	0	0	M	0	0	0	0	1.00
23	C23	93.00	=	93.00	0	0	M	0	0	0	0	1.00
24	C24	21.00	=	21.00	0	0	M	0	0	0	0	1.00
25	C25	93.00	=	93.00	0	0	M	0	0	0	0	1.00
26	C26	112.00	=	112.00	0	0	M	0	0	0	0	1.00
27	C27	10.00	=	10.00	0	0	M	0	0	0	0	1.00
28	C28	35.00	=	35.00	0	0	M	0	0	0	0	1.00
29	C29	244.00	=	244.00	0	0	M	0	0	0	0	1.00
30	C30	617.00	=	617.00	0	0	M	0	0	0	0	1.00
31	C31	42.00	=	42.00	0	0	M	0	0.00	0	0	1.00
32	C32	179.00	=	179.00	0	0	M	0	0	0	0	1.00
33	C33	253.00	=	253.00	0	0	M	0	0	0	0	1.00
34	C34	9.00	=	9.00	0	0	M	0	0.00	0	0	1.00

Problem 1: Solution with water as priority 1

12-23-2013 11:30:34	Decision Variable	Solution Value	Basis Status	Reduced Cost Goal 1	Reduced Cost Goal 2	Reduced Cost Goal 3	Reduced Cost Goal 4	Reduced Cost Goal 5
1	X1	38,263.35	basic	0	0	0	0	0
2	X2	18.00	basic	0	0	0	0	0
3	X3	0	at bound	0	-4.00	0	-5.00	-13,989.00
4	X4	0	at bound	0	-17.70	0	-4.10	-3,451.20
5	X5	0	at bound	0	-27.70	0	-7.10	-16,766.20
6	X6	0	at bound	0	-7.20	0	0.40	-11,185.20
7	X7	0	at bound	0	-30.64	0	-1.52	-9,844.84
8	X8	0	at bound	0	-15.50	0	-3.50	-13,814.00
9	X9	0	at bound	0	17.56	0	4.08	-5,734.14
10	X10	0	at bound	0	-58.16	0	1.12	-11,793.96
11	X11	0	at bound	0	-68.90	0	1.30	-26,590.40
12	X12	41.00	basic	0	0	0	0	0
13	X13	9.00	basic	0	0	0	0	0
14	X14	0	at bound	0	-25.10	0	-1.30	-14,788.60
15	X15	0	at bound	0	-9.50	0	-4.50	-11,393.00
16	X16	0	at bound	0	2.80	0	-0.60	-8,519.20
17	X17	0	at bound	0	1.00	0	2.00	-17,884.00

Problem 1: Solution summary with water as priority 1

12-19-2013 19:50:02	Decision Variable	Solution Value	Basis Status	Reduced Cost Goal 1	Reduced Cost Goal 2	Reduced Cost Goal 3	Reduced Cost Goal 4	Reduced Cost Goal 5
18	X18	0	at bound	0	2.80	0	2.40	-7,060.70
19	X19	0	at bound	0	2.30	0	3.90	-4,776.20
20	X20	0	at bound	0	-10.60	0	-1.80	-4,032.60
21	X21	0	at bound	0	-13.60	0	-2.80	-4,381.60
22	X22	0	at bound	0	-14.70	0	-4.10	-4,955.20
23	X23	0	at bound	0	-24.20	0	0.40	-13,245.20
24	X24	0	at bound	0	-38.90	0	-1.70	-6,170.40
25	X25	0	at bound	0	-38.90	0	-1.70	-6,170.40
26	X26	0	at bound	0	-9.90	0	3.30	-10,606.40
27	X27	0	at bound	1.00	-0.82	0	-0.26	-206.92
28	X28	0	at bound	0	0.82	0	0.26	206.92
29	X29	417,228.00	basic	0	0	0	0	0
30	X30	0	at bound	1.00	0	0	0	0
31	X31	0	at bound	0	1.00	0	0	0
32	X32	20,350,080.00	basic	0	0	0	0	0
33	X33	970,023.81	basic	0	0	0	0	0
34	X34	0	at bound	0	0	1.00	0	0
35	X35	0	at bound	0	0	0	1.00	0
36	X36	1,151,854,080.00	basic	0	0	0	0	0
37	X37	0	at bound	0	0	0	0	1.00
38	X38	9,970,111.00	basic	0	0	0	0	0
39	X39	0	at bound	0	0	0	0	1.00
40	X40	199,205,928,960.00	basic	0	0	0	0	0
41	X41	0	at bound	0	0	0	0	1.00
42	X42	19,218,036.00	basic	0	0	0	0	0
43	X43	34,037.35	basic	0	0	0	0	0
44	X44	0	at bound	0	0	0	0	1.00
45	X45	0	at bound	0	-12.20	0	-7.60	-16,206.20
46	X46	0	at bound	0	0	0	0	0
47	X47	0	at bound	0	0	0	0	1.00
48	X48	15.00	basic	0	0	0	0	0
49	X49	0	at bound	0	0	0	0	1.00
50	X50	836.00	basic	0	0	0	0	0
51	X51	0	at bound	0	0	0	0	1.00
52	X52	51.00	basic	0	0	0	0	0
53	X53	0	at bound	0	0	0	0	1.00
54	X54	135.00	basic	0	0	0	0	0
55	X55	0	at bound	0	0	0	0	1.00
56	X56	248.00	basic	0	0	0	0	0
57	X57	0	at bound	0	0	0	0	1.00
58	X58	261.00	basic	0	0	0	0	0

Problem 1: Solution summary with water as priority 1

12-19-2013 19:50:02	Decision Variable	Solution Value	Basis Status	Reduced Cost Goal 1	Reduced Cost Goal 2	Reduced Cost Goal 3	Reduced Cost Goal 4	Reduced Cost Goal 5
59	X59	0	at bound	0	0	0	0	1.00
60	X60	15.00	basic	0	0	0	0	0
61	X61	0	at bound	0	0	0	0	1.00
62	X62	74.00	basic	0	0	0	0	0
63	X63	0	at bound	0	0	0	0	1.00
64	X64	8.00	basic	0	0	0	0	0
65	X65	0	at bound	0	-122.60	0	28.20	-28,816.60
66	X66	0	at bound	0	0	0	0	0
67	X67	0	at bound	0	-46.64	0	-1.52	-19,831.84
68	X68	0	at bound	0	0	0	0	0
69	X69	0	at bound	0	0	0	0	1.00
70	X70	16.00	basic	0	0	0	0	0
71	X71	0	at bound	0	0	0	0	1.00
72	X72	93.00	basic	0	0	0	0	0
73	X73	0	at bound	0	0	0	0	1.00
74	X74	21.00	basic	0	0	0	0	0
75	X75	0	at bound	0	0	0	0	1.00
76	X76	93.00	basic	0	0	0	0	0
77	X77	0	at bound	0	0	0	0	1.00
78	X78	112.00	basic	0	0	0	0	0
79	X79	0	at bound	0	0	0	0	1.00
80	X80	10.00	basic	0	0	0	0	0
81	X81	0	at bound	0	0	0	0	1.00
82	X82	35.00	basic	0	0	0	0	0
83	X83	0	at bound	0	0	0	0	1.00
84	X84	244.00	basic	0	0	0	0	0
85	X85	0	at bound	0	0	0	0	1.00
86	X86	617.00	basic	0	0	0	0	0
87	X87	0	at bound	0	0	0	0	1.00
88	X88	42.00	basic	0	0	0	0	0
89	X89	0	at bound	0	0	0	0	1.00
90	X90	179.00	basic	0	0	0	0	0
91	X91	0	at bound	0	0	0	0	1.00
92	X92	253.00	basic	0	0	0	0	0
93	X93	0	at bound	0	0	0	0	1.00
94	X94	9.00	basic	0	0	0	0	0
	Goal 1:	Minimize	G1 =	0				
	Goal 2:	Minimize	G2 =	20,350,080.00				
	Goal 3:	Minimize	G3 =	0				
	Goal 4:	Minimize	G4 =	1,151,854,080.00				
	Goal 5:	Minimize	G5 =	199,235,125,248.00				

Problem 1: Solution summary with water as priority 1

	20:03:13		Thursday	December	19	2013						
	G1	Goal	Value	(Min.) =	0	(Alternate	Solution	Exists!!)				
	G2	Goal	Value	(Min.) =	963,902,208.00							
	G3	Goal	Value	(Min.) =	0							
	G4	Goal	Value	(Min.) =	901,236,416.00							
	G5	Goal	Value	(Min.) =	2,294.93							
	Constraint	Left Hand Side	Direction	Right Hand Side	Slack or Surplus	Allowable Min. RHS	Allowable Max. RHS	ShadowPrice Goal 1	ShadowPrice Goal 2	ShadowPrice Goal 3	ShadowPrice Goal 4	ShadowPrice Goal 5
1	C1	3,834,496.00	=	3,834,457.00	0	-M	967,736,640.00	0	-1.00	0	0	0
2	C2	3,417,216.00	=	3,417,228.50	0	-M	967,736,704.00	0	0	0	0	0
3	C3	23,500,032.00	=	23,500,000.00	0	-M	793,576,064.00	0	0	0	0	0
4	C4	873,728.00	=	873,709.19	0	-M	464,543,488.00	0	0	0	0	0
5	C5	1,152,850,048.00	=	1,152,850,048.00	0	251,613,632.00	M	0	0	0	1.00	0
6	C6	10,199,996.00	=	10,200,000.00	0	-M	58,063,460.00	0	0	0	0	0
7	C7	199,999,995,904.00	=	199,999,995,904.00	0	35,142,287,360.00	916,344,274,944.00	0	0.00	0	0.00	0
8	C8	20,000,000.00	=	20,000,000.00	0	-M	193,566,112.00	0	0	0	0	0
9	C9	4,226.00	=	4,225.69	0	-M	9,676,946.00	0	0	0	0	0
10	C10	17.55	=	17.55	0	0	M	0	0	0	0	1.00
11	C11	15.20	=	15.20	0	0	M	0	0	0	0	1.00
12	C12	836.03	=	836.03	0	0	M	0	0	0	0	1.00
13	C13	51.31	=	51.31	0	0	M	0	0	0	0	1.00
14	C14	135.39	=	135.39	0	0	M	0	0	0	0	1.00
15	C15	247.69	=	247.69	0	0	M	0	0	0	0	1.00
16	C16	260.62	=	260.62	0	0	M	0	0	0	0	1.00
17	C17	14.54	=	14.54	0	0	M	0	0	0	0	1.00
18	C18	74.37	=	74.37	0	0	M	0	0	0	0	1.00
19	C19	8.00	=	8.00	0	0	M	0	0	0	0	1.00
20	C20	40.84	=	40.84	0	0	3,143,709.00	0	-137.69	0	64.00	0
21	C21	9.46	=	9.46	0	0	M	0	0	0	0	1.00
22	C22	16.04	=	16.04	0	0	M	0	0	0	0	1.00
23	C23	93.35	=	93.35	0	0	M	0	0	0	0	1.00
24	C24	20.63	=	20.63	0	0	M	0	0	0	0	1.00
25	C25	93.35	=	93.35	0	0	M	0	0	0	0	1.00
26	C26	112.45	=	112.45	0	0	M	0	0	0	0	1.00
27	C27	10.35	=	10.35	0	0	M	0	0	0	0	1.00
28	C28	34.59	=	34.59	0	0	M	0	0	0	0	1.00
29	C29	244.01	=	244.01	0	0	M	0	0	0	0	1.00
30	C30	617.18	=	617.18	0	0	24,832,474.00	0	-23.95	0	2.13	0
31	C31	41.91	=	41.91	0	0	9,125,513.00	0	-64.08	0	17.06	0
32	C32	179.48	=	179.48	0	0	11,400,683.00	0	-29.84	0	6.06	0
33	C33	252.65	=	252.65	0	0	11,400,757.00	0	-29.84	0	6.06	0
34	C34	8.90	=	8.90	0	0	9,096,547.00	0	-51.33	0	16.65	0

Problem 2: Constraint Summary with labour as priority 1

12-19-2013 20:05:38	Constraint	Left Hand Side	Direction	Right Hand Side	Slack or Surplus	Shadow Price Goal 1	Shadow Price Goal 2	Shadow Price Goal 3	Shadow Price Goal 4	Shadow Price Goal 5
1	C1	3,834,496.00	=	3,834,457.00	0	0	-1.00	0	0	0
2	C2	3,417,216.00	=	3,417,228.50	0	0	0	0	0	0
3	C3	23,500,032.00	=	23,500,000.00	0	0	0	0	0	0
4	C4	873,728.00	=	873,709.19	0	0	0	0	0	0
5	C5	1,152,850,048.00	=	1,152,850,048.00	0	0	0	0	1.00	0
6	C6	10,199,996.00	=	10,200,000.00	0	0	0	0	0	0
7	C7	199,999,995,904.00	=	199,999,995,904.00	0	0	0.00	0	0.00	0
8	C8	20,000,000.00	=	20,000,000.00	0	0	0	0	0	0
9	C9	4,226.00	=	4,225.69	0	0	0	0	0	0
10	C10	17.55	=	17.55	0	0	0	0	0	1.00
11	C11	15.20	=	15.20	0	0	0	0	0	1.00
12	C12	836.03	=	836.03	0	0	0	0	0	1.00
13	C13	51.31	=	51.31	0	0	0	0	0	1.00
14	C14	135.39	=	135.39	0	0	0	0	0	1.00
15	C15	247.69	=	247.69	0	0	0	0	0	1.00
16	C16	260.62	=	260.62	0	0	0	0	0	1.00
17	C17	14.54	=	14.54	0	0	0	0	0	1.00
18	C18	74.37	=	74.37	0	0	0	0	0	1.00
19	C19	8.00	=	8.00	0	0	0	0	0	1.00
20	C20	40.84	=	40.84	0	0	-137.69	0	64.00	0
21	C21	9.46	=	9.46	0	0	0	0	0	1.00
22	C22	16.04	=	16.04	0	0	0	0	0	1.00
23	C23	93.35	=	93.35	0	0	0	0	0	1.00
24	C24	20.63	=	20.63	0	0	0	0	0	1.00
25	C25	93.35	=	93.35	0	0	0	0	0	1.00
26	C26	112.45	=	112.45	0	0	0	0	0	1.00
27	C27	10.35	=	10.35	0	0	0	0	0	1.00
28	C28	34.59	=	34.59	0	0	0	0	0	1.00
29	C29	244.01	=	244.01	0	0	0	0	0	1.00
30	C30	617.18	=	617.18	0	0	-23.95	0	2.13	0
31	C31	41.91	=	41.91	0	0	-64.08	0	17.06	0
32	C32	179.48	=	179.48	0	0	-29.84	0	6.06	0
33	C33	252.65	=	252.65	0	0	-29.84	0	6.06	0
34	C34	8.90	=	8.90	0	0	-51.33	0	16.65	0

Problem 2: Solution summary with labour as priority 1

12-19-2013 20:07:13	Decision Variable	Solution Value	Basis Status	Reduced Cost Goal 1	Reduced Cost Goal 2	Reduced Cost Goal 3	Reduced Cost Goal 4	Reduced Cost Goal 5
1	X1	9,676,946.00	basic	0	0	0	0	0
2	X2	0	at bound	0	-78.41	0	12.79	-1.00
3	X3	0	at bound	0	-67.68	0	12.60	-1.00
4	X4	0	at bound	0	-16.69	0	0.24	-1.00
5	X5	0	at bound	0	-81.10	0	13.99	-1.00
6	X6	0	at bound	0	-54.12	0	14.47	-1.00
7	X7	0	at bound	0	-47.65	0	10.87	-1.00
8	X8	0	at bound	0	-66.84	0	13.88	-1.00
9	X9	0	at bound	0	-27.79	0	11.31	-1.00
10	X10	0	at bound	0	-57.07	0	15.96	-1.00
11	X11	0	at bound	0	-128.69	0	34.76	-1.00
12	X12	40.84	basic	0	0	0	0	0
13	X13	0	at bound	0	-95.96	0	23.43	-1.00
14	X14	0	at bound	0	-71.57	0	17.31	-1.00
15	X15	0	at bound	0	-55.13	0	9.83	-1.00
16	X16	0	at bound	0	-41.24	0	10.12	-1.00
17	X17	0	at bound	0	-86.55	0	24.50	-1.00

Problem 2: Solution summary with labour as priority 1

12-19-2013 20:07:13	Decision Variable	Solution Value	Basis Status	Reduced Cost Goal 1	Reduced Cost Goal 2	Reduced Cost Goal 3	Reduced Cost Goal 4	Reduced Cost Goal 5
18	X18	0	at bound	0	-34.18	0	11.29	-1.00
19	X19	0	at bound	0	-23.08	0	9.90	-1.00
20	X20	0	at bound	0	-19.50	0	3.27	-1.00
21	X21	0	at bound	0	-21.20	0	2.71	-1.00
22	X22	617.18	basic	0	0	0	0	0
23	X23	41.91	basic	0	0	0	0	0
24	X24	179.48	basic	0	0	0	· 0	0
25	X25	252.65	basic	0	0	0	0	0
26	X26	8.90	basic	0	0	0	0	0
27	X27	963,902,208.00	basic	0	0	0	0	0
28	X28	0	at bound	0	1.00	0	0	0
29	X29	964,319,488.00	basic	0	0	0	0	0
30	X30	0	at bound	0	1.00	0	0	0
31	X31	770,076,096.00	basic	0	0	0	0	0
32	X32	0	at bound	1.00	0	0	0	0
33	X33	463,669,792.00	basic	0	0	0	0	0
34	X34	0	at bound	0	0	1.00	0	0
35	X35	0	at bound	0	0	0	1.00	0
36	X36	901,236,416.00	basic	0	0	0	0	0
37	X37	47,863,460.00	basic	0	0	0	0	0
38	X38	0	at bound	0	0	0	0	1.00
39	X39	0	at bound	0	0.00	0	0.00	0
40	X40	0	at bound	0	0.00	0	0.00	1.00
41	X41	173,566,112.00	basic	0	0	0	0	0
42	X42	0	at bound	0	0	0	0	1.00
43	X43	9,672,720.00	basic	0	0	0	0	0
44	X44	0	at bound	0	0	0	0	1.00
45	X45	0	at bound	0	0	0	0	1.00
46	X46	17.55	basic	0	0	0	0	0
47	X47	0	at bound	0	0	0	0	1.00
48	X48	15.20	basic	0	0	0	0	0
49	X49	0	at bound	0	0	0	0	1.00
50	X50	836.03	basic	0	0	0	0	0
51	X51	0	at bound	0	0	0	0	1.00
52	X52	51.31	basic	0	0	0	0	0
53	X53	0	at bound	0	0	0	0	1.00
54	X54	135.39	basic	0	0	0	0	0
55	X55	0	at bound	0	0	0	0	1.00
56	X56	247.69	basic	0	0	0	0	0
57	X57	0	at bound	0	0	0	0	1.00
58	X58	260.62	basic	0	0	0	0	0

Problem 2: Solution summary with labour as priority 1

In problem 1; water has been taken as priority 1. For solving this problem, a 94 X 39 matrix has been solved with 77 numbers of iterations. In this solution G1, G3 are zero. This implies optimal solution have been obtained with respect to water and production. The solutions with respect to labour, revenue and food security have been partially achieved. The result thus obtained shows that water requirement as proposed under the simulation ranging from 2.36 million ha-cm to 11.81 million ha-cm is sufficient to grow above five crops. Hence the goal is achieved.

Similarly, in problem 2, labour has been assigned to priority number one. In the solution, it is found that the optimal solution has been obtained with respect to labour and production. The solutions with respect to water, revenue and food security have been partially achieved. Presently farm machinery is available for ploughing, irrigations, weeding, harvesting, threshing, plant protection, fertigation and transportation operation. This can be well achieved through maximum use of the farm machinery.

Solving the modified GOAL program for problem no 3, production has been considered as priority no 1. The optimal solutions with respect to production and labour have been achieved. Satisfactory results have been obtained for other three goals like water, revenue and food security. In case of sustainable use of land for crops production, crop like maize can be cultivated over more area to achieve the target.

In problem 4; the revenue has been taken as priority 1. Here optimal have been obtained with respect to revenue, production and labour. The satisfactory goal has been achieved with respect to water and food security. The value of all crops produced in a sustainable environment amounts to 2.95 trillion INR.

In Problem 5, Solving the modified formulation with food security as first priority, the solution obtained used all the water and labour resources available in Odisha. For sustainable development precision irrigation system and mechanization can resolve these issues. The food security for people of Odisha is also achieved.

Sensitivity Analysis Summary

The above table slows that in three (water, labour, production) cases two objectives have been achieved. In other two (revenue and food security) cases three objectives have been fulfilled. The optimal solution in other cases has not been achieved. The realistic solution has been presented in the sensitive analysis table.

Description	Priority					Optimal Solution					
	Highest	High	Medium	Medium low	Low	$x_1 .. X_{26}$	Goal $_1$	Goal $_2$	Goal $_3$	Goal $_4$	Goal $_5$
Water	S_1+, S_2-	s_3-	s_4-	s_5-	$s_6-..s_{26}-$		0	2.00E+07	0	1.15E+09	1.91E+11
Labour	s_3-	S_1+, S_2-	s_4-	s_5-	$s_6-..s_{26}-$		0	9.63E+08	0	9.01E+08	2.20E+03
Production	s_4-	S_1+, S_2-	s_3-	s_5-	$s_6-..s_{26}-$		0	9.63E+08	0	9.01E+08	2.20E+03
Revenue	s_5-	S_1+, S_2-	s_4-	s_3-	$s_6-..s_{26}-$		0	4.43E+09	0	0	3.40E+03
Food Security	$s_6-..s_{26}-$	S_1+, S_2-	s_4-	s_5-	s_3-		0	4.01E+08	0	1.00E+10	0

Conclusion

GOAL programming approach has re-invented the requirement of inputs for the targeted production for sustainable development for Odisha state. Considering the soil health, food security, nutritional security, long term sustainability the inputs, area to be put to cultivation for production of cereals, pulses, oilseeds etc. has been computed considering multiple goals. The soft constraints like seed supply, mechanization, finance, and number of crops in a year (cropping intensity) have been determined and specified under present condition. When the condition changes in a particular area, the simulation can be modified and the corresponding value of the soft constraints can be determined. It will help the planners to make the inputs available for targeted production. This exercise will justify the present notification – "Agriculture as Industry". The area under different crops can be computed. The *rabi* and *kharif* program can be drafted as a SMART production goal. This simulation has used 150% cropping intensity for paddy and realistic cropping intensity for other crops. The total area under production has been considered as 73 lakh hectares. The production from five crops like paddy, green gram, black gram, maize and ground nut has been simulated. The revenue from these five crops has been simulated to be 2950 crores with present input parameters.

References

Alston, J. M., and D. M. Larson. 1993. "Hicksian vs. Marshallian Welfare Measures: Why Do We Do What We Do?" *Amer. J. Agr. Econ.*75:764-769.

Angehrn, A.A. and Dutta, S.1992. Integrating case-based reasoning in multi-criteria decision support systems: *INSEAD Working Paper* 92/54TM.

Angehrn, A.A.1991. Designing humanized systems for multiple criteria decision making. Human Systems Manage. 10:221-231.

Anonymous, 2000. "Agricultural Modernization in Rural Odisha: First Impact on the Weaker Sections", *IASSI Quarterly*, 14 (3&4): 85-93.

Anonymous, 1991.Census of India.

Anonymous, 2001. Census of India.

Anonymous, 2004. Economic Survey, 2003-2004, Bhubaneswar: Directorate of Economics and Statistics, Planning and Co-ordination Department.

Anonymous, Odisha Reference Annual 2004, Bhubaneswar: Information and Public Relations Department, 2004.

Anonymous, 1979.Statistical abstract of Odisha, Department of Statistics, Government of Odisha.

Anonymous, 1991.Statistical abstract of Odisha, Department of Statistics, Government of Odisha.

Anonymous,2010.Statistical abstract of Odisha, Department of Statistics, Government of Odisha.

Ata, A, Sennaroglu, B., 2008. The usage of GOAL techniques in determining the constants of criteria's weight of warship design, *Journal of Naval Science and Engineering*; 4(1):1-16

Banashri, R, Mangaraj, BK, Mishra, BP, 2012. Fuzzy logic based simulation for modelling of sustainable marketing policy for modern rice mill in Odisha, *International Journal of Supply Chain Management*, 1(3):34-42.

Banashri, R. 2012. Decision support system for drafting of sustainable agriculture policy for Odisha, *International Journal of Supply Chain Management*, **2**(2):92-98.

De, P.K. and Yadav, B. 2011.An algorithm to solve Multi-Objective Assignment Problem Using Interactive Fuzzy Goal Programming approach, *Int.J. Contemp. Math. Sciences*, 6, (34):1651-1662.

Donella, H. Meadous *et.al.* 1972., Limits to Growth; Universe Books, New York.

Government of India, *Odisha Development Report*, New Delhi: Planning Commission, 2002.

Laxminarayan, V. and Rajgopanan, S.P.1977. Optimal cropping pattern for basins in India, *J. Irrigation and drainage Engineering*, ASCE, 103(IRI): 53-59.

Rath, B, Mohanty, R.P. and Mishra, B.P. 2003. GOAL approach rural common resource planning for sustainable rural development, *Siddhant, A journal of decision making*, 3(1):135.

Sinha, S B, Rao KA, Mangaraj BK, 1988, Fuzzy goal programming in multi-criteria decision systems: a case study in agricultural planning.*Journal of Information and Optimization Science,* 9(1):257-274.

Sinha, S B. 1989.Fuzzy technique for Agricultural Planning, *Journal of Information and Optimization Science,* 10(1): 257-274.

15

Combatting Climate Change by Organic Agriculture: Mitigation and Adaptation Options for Sustaining Food Security

Sagar Maitra, Tanuj Kumar Mandal, Tanmoy Shankar and A. Zaman

M.S. Swaminathan School of Agriculture, CUTM, Paralakhemundi-761 211 Gajapati, Odisha, India
Email: sagar.maitra@cutm.ac.in

Abstract

The climate is one of the main determinants of agricultural production. At present, climate change has been appeared as a tremendous global crisis and modern agriculture is responsible for anthropogenic emissions of Greenhouse Gases (GHGs) which are important causing factors. The ill effects of global warming and climate change have been reflected adversely in agriculture causing threat to food productivity and security. Thus there is an urgent need for adoption of ecologically sound and economically viable farming practices to mitigate the adverse effects of global warming as well as climate change. Organic agriculture is one of the most prominent and widely practiced ecologically sound farming systems and it might be a suitable option as it is based on the principles of 'care, health, ecology and fairness' with the capability to reduce emission of GHGs. Moreover, organic agriculture provides several gains like contribution to the conservation of biological diversity and animal welfare, supply of local products of natural origin, reduction of fossils fuel energy and protection of the environment from pollution with chemicals, carbon sequestration and restoration of soil fertility, assurance to agricultural sustainability, and thus ensuring food security to ever growing population of the world.

***Keywords*:** Climate change; GHGs emission, organic agriculture; mitigation; adaptation; sustainability; food security

Introduction

The climate is one of the main determinants of agricultural production and climate alteration might cause variability in agricultural production system. As climate pattern shifts, changes in the distribution of plant diseases and pests might also have adverse effects on agriculture and the ingredients of climate might affect negatively on growth and productivity of the crops. At the same time, agriculture proved to be one of the most adaptable human activities for livelihood to varied climate conditions (Mendelson *et al.*, 2001). Agriculture is not only affected by climate change but also contributes to it by emitting greenhouse gases (GHGs). According to the Inter-governmental Panel on Climate Change (IPCC, 2007) agriculture currently accounts for 10-12% or 5.1–6.1 Gt of the total anthropogenic annual GHGs emissions of CO_2-equivalent year^{-1} in 2005. That is 10-12% of the *total* global *anthropogenic* emission of greenhouse gases (Smith *et al.*, 2007)due to human food and fibre production. GHGs attributed to agriculture include emissions from soils, enteric fermentation (GHGs emis-sions from the digestion process of ruminant animals), rice production, biomass burning and ma-nure management. There are other indirect sources of GHGs emissions like changes in land-use, use of fossil fuels for mechanization, irrigation, transportation and agro-chemicals and fertilizers production. In addition, intensive agriculture has led to deforestation and overgrazing that result in soil degradation. These changes in land use also contribute considerably to global CO_2 emissions. Sustainable agriculture and food supply systems are thus more urgently needed than ever before.

To combat with the negative impacts of climate change, the importance of enhanced action on adaptation has been realized (IPCC, 2007; UNFCCC 2007). Organic farming has an inherent potential to both reduce GHG emissions and to enhance carbon sequestration in the soil. The main organic strategies are diversifications and an increase of soil organic matters, which both could enhance resilience against extreme weather conditions. Organic farming avoids nutrient exploitation and increases soil organic matter content, hence soils under organic farming capture and store more water than soils under conventional cultivation. Organic farming accordingly addresses key consequences of climate change, namely increased occurrence of extreme weather conditions and aberrations and problems related to soil quality (IFOAM, 2006; IPCC 2007a; MacRae, R. 2009). The Food and Agriculture Organization of the United Nations (FAO) regards organic agricultural as an effective strategy for mitigating climate change and building robust soils that are better adapted to extreme weather conditions associated with climate change (Niggli *et al.* 2009; FAO, 2011). Organic agriculture enhances biodiversity, protects fragile soils, improves the nutritional quality of food, ensures high standards of animal welfare, maintains equilibrium of ecosystem and provides increased employment in rural areas. At the same time, organic agriculture reduces GHGs emissions and fossil fuel energy use, cuts chemical nutrient and pesticide pollution and stops potentially harmful chemical residues entering our food chain. Thus organic agriculture builds resilient and sustainable farming systems capable of combating climate change and securing local food supplies and is effective in carbon sequestration (Muller and Aubert, 2013). In true sense, organic agriculture is accomplished by using agronomic, biological and mechanical methods, as opposed to using synthetic materials, to fulfill any specific function within the farming system. Organic farming is a systemic and encompassing approach to sustainable livelihoods in general, where due importance is given to relevant factors of influence for sustainable development in terms of physical, economic, or socio-cultural levels (Eyhorn, 2007; Ching, 2011). In India, organic farming has a long tradition as a farming approach and it has been adapted for many climatic zones and local conditions.

Organic farming is practiced worldwide by 1.2 million producers in 141 countries with production of or-ganically grown food continuing to steadily increase by at least about 15 per cent per year (The World of Organic Agriculture, IFOAM, 2009). There are now nationally adapted organic agriculture standards and quality control in most of the countries including India. Organic agriculture has a wealth of production, certification and marketing experience for last few decades and practices devel-oped over many years in different climatically, topographically and culturally diverse and even fragile and vulnerable regions. While most of the organic markets are in developed countries, developing countries are becoming important suppliers and exporters as organic practices are particularly suited for the conditions of developing nations comprised of small and marginal farmers and subsistence farming.

Policy makers across the globe are now referring to restoration of soil organic matter, carbon sequestration, ecosystem services and holistic approaches for ecological development to combat the ill effects of climate change and all of which are long established as core pillars of organic agriculture. International Federation on Organic Agriculture Movement (IFOAM) clearly mentioned the four basic principles of organic farming in the beginning of the formation of the concept which are: 'care, health, ecology and fairness' that can be remembered in this aspect (IFOAM, 2009). Besides, Organic agriculture is nothing but the well-established practices that can mitigate climate change, build resilient farming systems, poverty alleviation and improve food security. This farming practice emits much lower levels of GHGs and quickly, affordably and effectively sequesters carbon in the soil. In addition, or-ganic agriculture makes farms and farmers more resilient to climate change and thus assures sustainable livelihood through adoption of proper mitigation and adaptation strategies.

Mitigation of climate change by organic agriculture

Mitigation of climate change is achieved through organic farming practices. These prac-tices include enhancing soil biological processes, soil fertility and structure, creating organic matter in forms that are more effective at producing soil carbon, integrating crop and livestock based farming systems, increasing the propor-tion of vegetation cover which promotes the soil's micro-organisms that stabilize soil carbon (Klein *et al.*, 2007).

Conventional agriculture relies on fossil fuel based chemical nitrogen (N) fertilizers and agrochemicals manufac-tured in energy intensive factories transported to farms. Chemical fertilizers release both CO_2 and N_2O during their manufacturing process. Organic agriculture replaces these synthetic chemical nutrients with naturally occurring nutrients like manure, compost, and incorporation of leguminous plants. Organic crops tend to have deeper and denser roots and scavenge nutrients more efficient-ly than non-organic crops. In organic agriculture, soil fertility is managed by a dynamic activity of abundant micro-organisms present in soil to mineralize organic manures and inputs.

Chemical inputs inhibit the natural biological activity of the soil. Organic agriculture also incorporates sufficient amounts of organic matter, avoids bare fallows and generally has a higher proportion of grasslands and cover crops within farming systems (Khan *et al.,* 2007). Ensuring the soil is always covered with vegetation prevents the soil from being exposed to processes that accelerate GHG emissions from stored soil carbon and the inclusion of cover crops and green manures within organic farming systems retains nutrients for future utilization. Cover vegetation is also important in providing a greater and more continuous supply of the root exudates that support the soil's micro-organisms to build the soil carbon store. Crop residues can also be left on the field as a protective layer for the soil in organic agriculture to avoid CH_4 and N_2O emissions caused by an unscientific practice, the burning of crop residues in conventional agriculture. Besides, due to the lower N input in organic agriculture, N_2O emissions are lower than in conventional agriculture. Organic N inputs such as N fixed by legumes and applied through manure considerably decreases the concentration of easily available mineral N in soils. As a consequence organic N is released more slowly than chemical N, thus minimizing N_2O emissions.

Organic agriculture assures soil carbon build-up which is mainly stimulated by the additional biomass return to the soil from organic manures. (Pimentel *et al.,* 1995; Drinkwater and Wagoner, 1998; Leifeld and Fuhrer, 2010). Organic agriculture utilizes cover and catch crops, green and animal manures, crop residues and compost to build soil organic matter, minimize nutrient loss and maximize productivity. These practices also contribute to the build-up of a healthy soil structure which facilitates the sequestration of carbon and minimizes soil erosion and carbon loss. Organic agriculture improves the biological conditions for soil carbon accumulation. Biological factors significantly affect the proportion of carbon that is converted to stable soil carbon (humus) thus contributing to increased carbon sequestration in soils. The application of different forms of compost and the avoidance of chemical inputs stimulates soil microbial processes. Many studies showed enhanced levels of microbial biomass, microbial enzymatic activities, earthworms and soildwelling insects in organic agriculture (Gerhardt, 1997; Mader *et al.,* 2002; Pfiffner and Luka, 2003; Fliessbach *et al*., 2007). Soil micro-organisms including fungi and earthworms play a major role in securing carbon by producing polysaccharide gums that form soil aggregates which contribute to the stabilization and protection of humus.

Organic agriculture utilizes a range of tillage to build soil fertility and struc-ture, enhance biological functions and optimize the build-up of carbon in the soil. Tillage manipulates the soil aggregates that contain soil carbon and therefore exposes the carbon to the processes of mineralization, oxidation and erosion. No-till or zero-till avoids such damage to the soil carbon structures and therefore contribute to the conserva-tion of soil carbon in the top layers of the soil (Kreismane, 2015). No-till is a passive approach to building soil carbon whereas organic agriculture systematically incorporates carbon rich organic matter into the soil. Besides, zero and minimum till may concentrate carbon sequestration in the top soil (Barton *et al.,* 2008). Such systems may be particularly beneficial in drought prone regions where disturbed soils are supposed to soil carbon loss.

Organic agriculture systems combine different perennial plant species such as trees, shrubs, and grasses with different annual crops including native perennial trees and hedgerows from the surrounding landscape. Perennial plants develop their roots and branches over many years to store carbon in the vegetation and soil. There is great potential to improve mitigation by increasing the use of perennials and reducing annual crops, especially to provide animal feed for livestock production. The agroforestry systems are having a great potential for sequestrating carbon. Combining permanent and annual crop species also en-hances the ecofunctionality and productivity of the farming system.

There is debate on the issue of direct GHG emissions from livestock (enteric fermentation, respiration, decomposition of organic wastes into manure) considered much higher than estimated by the IPCC. If the whole life cycle of the meat production is considered such as the land-use changes linked to animal feed production, slaughtering, processing, refrigeration and transportation the contribution of meat production to climate change is found very significant to global warming and climate change. About 33% of arable land worldwide is estimated to be used for producing feed for animals rather than food for humans or for producing plant-based alternatives to fossil fuels. The demand for meat is enhancing deforestation resulting in CO_2 being released into the atmosphere, the loss of habitats, biodiversity and disturbance in ecosystems as well as the loss of vegetation capable of carbon sequestration. Organic agriculture livestock production emits significantly less GHGs than conventional livestock produc-tion as in later animals are primarily raised in pastures reducing stall feed requirement and their manure contributes to the build-ing of soil carbon. Organic agriculture integrates crops and animals in a single farming system (integrated or diversified farm. This allows efficient feed and manure use, thus reducing GHGs emissions and rather in organic agriculture livestock is as an integral part of the integrated organic farming system. Improved grassland management, a common practice in integrated organic agriculture system, is another option for carbon sequestrationin the soil especially by building up carbon in their root systems. Grasses also have fine root hairs and high mycorrhizal organisms which enhance soil aggregation. Many of the arable soils with permanent grasslands were found with very high carbon contents. Legumes are strongly promoted on organic grasslands and pastures, as they increase ni-trogen uptake in the soil, raise protein levels in animal feedstuff and avoid the need to purchase off-farm animal feed concentrates.

Organic agriculture encourages reliance on on-farm inputs by maximizing efficiency and synergy amongst the components of farming system, reduces emissions from transportation and increase access to local food production and therefore enhances food security. The promotion of local food production in food insecure regions is critical for increasing food accessibility and food security and thus reduces emissions.

Adaptation strategies

Agriculture is highly vulnerable to climate change and our food supply relies on successful adaptation strategies in mitigating adverse effects on the issues. Adaptation actions include restoration of the resilience on ecosystems and sustainable economic development. Organic agriculture increases the ability of the farming system to continue functioning even under the adverse effects of climate change by enhancing resilience within the agro-eco-system (Barron, 2006; Ensor, 2009; Muller, 2009).The financial implications for adaptation of organic system of agriculture are comparatively cheaper. This is particularly important for the empowerment of vulnerable and poor people in rural populations that rely on agriculture for their livelihoods. This section narrates main organic principles and how organic agriculture fulfills many of the requirements for suc-cessful adaptation strategies?

Rice production is a key agricultural emitter of greenhouse gases, particularly methane. Innovative and alternative rice production methodology like system of rice intensification (SRI) based on organic principles through decreased flooding offers potential to significantly reduce CH_4 emissions. Besides, aerobic rice cultivation by adoption of organic agriculture, another methodology of direct seeded rice (DSR) farming, does not advocate water stagnation of during a major period crop growth (Zaman, *et al.,* 2016) Satyanarayana and Uphoff 2007) which could play role to reduce CH_4 emission.

Worldwide about 38% of agricultural land is significantly degraded, resulting in vulnerability of agroecological functions and poor crop yields (Pretty *et al.,* 2006). Agricultural erosion through overuse, undernourishment and chemical inputs damaged the biological activity in the soil. Biologically alive soils are crucial in the adaptation of agriculture to climate change, providing sufficient nutrients for plant pro-duction, achieving high yields in times of water scarcity and preventing and reversing soil erosion (Pimental *et al.*, 1995). Organic practices ensure use of organic instead of chemical fertilizer; cover crops, catch crops, green manure; composting; appropriate tillage; and the integration of perennials and trees into the farming system which ultimately lead to a sound ecology including soil conservation and restoration of degraded land. By increasing the soil organic matter content, organic ag-riculture is able to build up soil carbon and soil fertility from low levels and bring severely degraded land back into production (FAO, 2008).

The majority of the world's farmers are still depended on rainfed agriculture wherein rainfall is predicted to become more erratic and unpredictable due to climate change. Organic agriculture systems along with on-farm water harvesting assure use water more efficiently to give better soil structure and higher levels of humus. Organic matter also enhances drainage in soils, significantly reducing the risks from water-logging and surface-water flooding. Water capture in organic fields can also be higher than in conventional fields during torrential rains (Rodale, 2008). Under organic agriculture the soil organic matter captures and retains more water in the crop root zone. Other practices such as crop residue reten-tion, mulching and agro-forestry, conserve soil moisture and protect crops against microclimate extremes.

Building healthy soils is the basis for growing climate resilient plants that are better able to withstand environmental aberrations. Organic agriculture strengthens the inherent immune systems of plants and thus the defense and self-healing capacities of crops against biotic and abiotic stress. Plants that obtain their nutrients through natural biological processes are more resilient to environ-mental stress than crops that obtain their primary nutrition artificially through highly soluble chemical fertilizers. In addition organic crops tend to have longer and denser roots that are able to seek out water reserves deeper in the soil profile and which are also more resilient to desiccation.

Organic agriculture has a big role to play in protecting the world's agricultural genetic resources as well as biodiversity and it encourages the use of locally adapted varieties and decentralized participatory breeding programs especially in-situ (on-farm) based conservation, breeding and production (Hole *et al.*, 2005). In-situ approaches of maintaining varieties for future needs while allowing them to continuously adapt to environmental pressures such as climate change. The agro-genetic diversity is a critical resource in the effec-tive ongoing adaptation to continuous changes in climate.

Diversification is a fundamental aspect of organic agriculture. Practices that enhance biodiversity allow farms to mimic natural ecological processes, enabling them to better respond to change and reduce risk. Crop diversity (both temporal and spatial) provides a variety of rooting depths that enhance soil stability and structure, improves use of nutrient and water, and con-tributes to a stabilized microclimate. Farmers who increase inter-specific diversity via organic agriculture suf-fer less damage compared to conventional farmers planting monocultures (Bolwig and Odeke, 2007; Altieri and Koohafkan, 2008). The robust and resilient nature of organic agriculture systems also helps to protect sequestrated carbon from climatic disturbances increasing the permanence of sequestration.

Most of the world's farmers are smallholders and organic agriculture encourages the use of local and indigenous traditional knowledge and observation techniques. However, smallholder farmers throughout the world have developed a multitude of practices and innovations that should be seen for what they are; the basis for any realistic development. Indigenous and traditional knowledge are key sources of information on adaptive capacity, centered on the selective, experimental and resilient capabilities of farmers (Meredith, 2008 and Niggli *et al.,* 2009). Organic agriculture strengthens the rich ecological knowledge of the world's small and marginal farmers resulting in enhanced productivity, greater resil-ience and resource efficiency, and improved access to food and economic stability.

Conclusion

There appears to be a large potential for synergies between mitigation and adaptation of organic system in agriculture. In order to realize the full potential for present-day agriculture in a climate change context, new agricultural organic production systems need to be developed that integrate bioenergy and food and feed production domain to ever increasing population.

References

Altieri, M.A. and Koohafkan, P. 2008. Enduring farms: Climate change, smallholders and traditional farming communities. TWN Environment and Development Series 6, Third World Network, Penang.

Barton, L., R. Kiese, D. Gatter, K. Butterbach-Bahl, R. Buck, C. Hinz, and D. Murphy. 2008. Nitrous Oxide Emissions from a Cropped Soil in a Semi-arid Climate, *Global Change Biology,* 14: 177-92.

Borron, S. 2006. Building resilience for an unpredictable future: How organic agriculture can help farmers adapt to climate change. FAO, Rome.

Bolwig, S. and Odeke, M. 2007. Household food security effects of certified organic export production in tropical Africa - a gendered analysis.http:// www.grolink.se/epopa/ Publications/EPOPA Report on Food Security impact of organic production.pdf , Accessed on 15 December, 2017.

Ching, Lim Li. 2011. Mitigating and Adapting to Climate Change through Ecological Agriculture. ISBN: 978-967-5412-46-2, Third World Network, 131 Jalan Macalister, 10400 Penang, Malaysia, pp. 18.

David Pimentel, C. Harvey, P. Resosudarmo, K. Sinclair, D. Kurz, M. Mcnair, S. Crist, L. Shpritz, L. Fitton, R. Saffouri, and R. Blair. 1995. Environmen-tal and economic costs of soil erosion and conservation benefits. *Science*, 267:1117–1123.

Drinkwater, L.E. and Wagoner, P., 1998: Legume-based cropping systems have reduced carbon and nitrogen losses. *Nature: International Weekly Journal of Science,* 396: 262-265.

Ensor, J. 2009. Biodiverse agriculture for a changing climate, Practical Action, UK.

Eyhorn, F., R. Ramakrishnan, and P. Mäder. 2007. "The Viability of Cotton-Based Organic Farming Systems in India," *International Journal of Agricultural Sustainability,* 5 (1): 25–38.

FAO 2008.Feeding the World - Sustainable Management of Natural Resources. Food and FAO Environmental Assessment and Management Unit (NRCE), Rome, ftp://ftp.fao.org/ docrep/fao/010/ai549e/ai549e00.pdf , Accessed on 10 May, 2017.

FAO, 2011. Organic agriculture and climate change mitigation. A report of the Round Table on Organic Agriculture and Climate Change, December 2011, Rome, Italy, pp.68

Fliessbach, A., Oberholzer, H. R., Gunst, L., Mader, P. (2007): Soil organic matter and biological soil quality indicators after 21 years of organic and conventional farming. Agriculture, *Ecosystems and Environment*, 118:273-284.

Gerhardt, G.A. 1997. A comparative analysis of the effects of organic and conventional farming systems on soil structure, *Biol. Agric. Hort.*, 14:139-157.

Hole, D.G., Perkins, A.J., Wilson, J.D., Alexander, I.H., Grice, P.V. & Evans, A.D. 2005. Does organic farming benefit biodiversity? *Biological Conservation*, 122: 113–130

IFOAM (International Federation for Organic Agriculture Movements). 2006. "Mitigation and Adaptation to Climate Change: The Role of Organic Agriculture", Bonn, Germany.

IFOAM, 2009. Organic Agriculture, a Guide to Climate Change and Food Security, pp.27.

Intergovernmental Panel on Climate Change (IPCC). 2007. Synthesis report. *In* O.R.D. Metz, P.R. Bosch, R. Dave, and L.A. Meyer (eds). Fourth Assessment Report: Climate Change, Cambridge University Press, Cambridge, UK.

Intergovernmental Panel on Climate Change (IPCC).2007a. Summary for Policy Makers. In IPCC Fourth Assessment Report, "Working Group II Report: Impacts, Adaptation, and Vulnerability," adaptation, chapter 17: adaptation and mitigation, chapter 18: vulnerability.

Khan, S. A:., Mulvaney, R. L., Ellsworth, T.R. and Boast, C.W. 2007. The Myth of Nitrogen Fertilization for Soil Carbon Sequestration, *Journal of Environmental Quality*, 36(6): 1821-1832.)

Klein, R.J.T., S. Huq, F. Denton, T.E. Downing, R.G. Richels, J.B. Robinson, F.L. Toth, 2007: Inter-relationships between adaptation and mitigation. Climate Change 2007: Impacts, Adaptation and Vulnerability. Contribution of Working Group II to the Fourth Assessment Report of the Intergovern-mental Panel on Climate Change, M.L. Parry, O.F. Canziani, J.P. Palutikof, P.J. van der Linden and C.E. Hanson, Eds., Cambridge University Press, UK, 745-777.

Kreismane, D. 2015. Effect of organic farming on climate change. NJF 25th Science Congress, 292-295.

Leifeld, J. and Fuhrer, J. 2010. Organic farming and soil carbon sequestration: what do we really know about the benefits? *Ambio*, 39(8): 585–599.

MacRae, R. 2009. Comparing energy use and GHG mitigation potentials in organic vs. conventional farming systems, A discussion paper for the Organic Agriculture Centre of Canada,.33 p, (available at organicagcentre.ca/Docs/DiscussionPapers09/Energy_Use_rmc.pdf), Accessed on 12 December, 2017.

Mader, P., Fliessbach, A., Dubois, D., Gunst, L., Fried, P., Niggli, U. (2002): Soil Fertility and Biodiversity in Organic Farming. *Science*, 296: 1694-1697.

Mendelson, Robert., Dinar, Ariel. And Sanghi, Apurva. 2001. The Effect of Development on the climate sensitivity Of Agriculture. *Environment and Development Economics*, 6: 85-101.

Meredith Niles. 2008. Sustainable Soils: Reducing, Mitigating, and Adapting to Climate Change with Organic Agriculture. Sustainable Development Law & Policy, pp.68-69.

Muller, A. 2009. Benefits of Organic agriculture as a Climate Change Adaptation and Mitigation Strategy in Developing Countries, EFD Discussion Paper 09-09, Environment for Development Initiative and Resources for the Future,http://www.efdinitiative.org /research/ publications/publications-repository/benefits-of-organic-agriculture-as-a-climate-change-adaptation-and-mitigation-strategy-for-developing-countries, Accessed on 16 October, 2017.

Muller A., Aubert C. 2013. The Potential of Organic Agriculture to Mitigate the Influence of Agriculture on Global Warming – A Review, *In:* Bellon S., Penvern S. (Eds.). *Organic Farming, Prototype for Sustainable Agricultures*, Springer Science Business Media, Dordrecht, chapter 13, p. 239-259.

Niggli, U., A. Fliessbach, P. Hepperly and N. Scialabba. 2009. Low Greenhouse Gas Agriculture: Mitigation and Adaptation Potential of Sustainable Farming Systems. FAO, April 2009, Rev. 2.

Pfiffner, L. and Luka, H. (2003): Effects of low-input farming systems on carabids and epigeal spiders – a paired farm approach. *Basic and Applied Ecology* 4:117-127.

Pretty, J.N., A.D. Noble, D. Bossio, J. Dixon, R.E. Hine, F. Penning de Vries, and J. Morison. 2006. "Resource-Conserving Agriculture Increases Yields in Developing Countries," *Environmental Science and Technology* 40(4): 1114–19.

Rodale. 2008. Farm Systems Trial, The Rodale Institute, 611 Siegfriedale Road, Kutztown, PA 19530-9320.

Satyanarayana, T. and Uphoff, N. 2007: Opportunities for water saving with higher yield from the system of rice intensification. *Irrigation Science*, 25(2): 99-115).

UNFCCC (United Nations Framework Convention on Climate Change)., 2009. Online document. The Bali Action Plan, United Nations Framework Convention on Climate Change. http:/ /unfccc.int/files/meetings/cop_13/application/pdf/cp_bali_action.pdf. Accessed February 12, 2017.

Zaman, A, Gangarani Devi, T. and Zaman, P. 2016. Growing Direct Seeded Rice.Lambart Academic Publishing, Germany: pp 1- 185.

16

Role of ITKs in Sustainable Development of Innovative Agriculture

A. Zaman and Sagar Maitra

Centurion University of Technology and Management
Paralakhemundi-761 211, Odisha, India
Email: aftab@cutm.ac.in

Abstract

More than 73% of its total population lives in rural areas and depends primarily on agriculture for livelihood. The performance of agriculture determines food and nutrition security which is important for reducing poverty and achieving sustainable growth. Production and productivity has increased manifold and yields of major crops like paddy, pulses, oilseeds and vegetables have more than trebled in the last four decades. The impressive long term growth in agriculture helped in taking the state out of famines and serious food shortages into one of the food surplus states and ensuring food, nutrition and livelihood security. The state is characterized by wide diversity and considerable spatio-temporal variations in growth and productivity. In fact, agriculture isincreasing with public investment in infrastructure including irrigation, establishing appropriate farming systems, developing suitable and affordable technologies and crop varieties, augmenting credit delivery and designing region and crop-specific plans. In this backdrop, a study was conducted to document the indigenous and organic farming technologies in selected communities of the state in tribal dominant Gajapati district of Odisha intended to help the farmers to be aware of their own innovative technologies for weakening balances on the environment due to intensive use of fertilizers and agrochemicals. Thestudy encompasses the methodology that includes physical observation and questionnaire surveys that consisted of open and close-ended questions addressing the farm management skills on indigenous and organic farming practices. There are practices/technologies representing successful ways by which farmers have dealt with poor soil quality and pest problems and their valuable experiences are shared to other farmers' communities, that showed the ingenuity of the farmers in improving indigenous practices into modern technologies considering its effect on soil, plant, human and ecosystem.

Keywords: Organic farming, indigenous knowledge, technological learning, nature farming

Introduction

Agriculture is a vital sector of the economy of Odisha and a good deal has been achieved in this sector during the plan period. Odisha, situated in eastern part of the country, is basically an agrarian state and wherein agriculture plays a vital role in state's economy. Farm production has increased manifold and yields of major crops such as paddy, pulses, oilseeds and vegetables have more than trebled in the last four and half decades. The impressive long term growth in agriculture has helped in taking the state out of famines and serious food shortages into one of the food surplus states in the country and ensuring food, nutrition and livelihood security. But although the need of making available an adequate food supply of reasonable quality for the population has been accomplished, the corresponding need of sustainable and equitable agricultural growth still remains a problem in the state. In fact, agriculture in Odisha is characterized by wide diversity and considerable spatio-temporal variations in growth and productivity. Such disparities are a cause of concern for obvious reasons. In such view of the facts, a study on spatio-temporal variations in agricultural development of Odisha is worth pursuing.Variations in the Indian agro-climate from tropical to temperate conditions have led to the development of indigenous skills based on innovations at farmers' fields. Unfortunately, these skills do not find much appreciation, as these are not tested in the field by researchers. Although many of these traditional practices are supported by scientific findings, yet these have not been converted into technological recommendations to enable farmers to reduce their costs and make agriculture more sustainable. Such skills are termed as indigenous technological knowledge or (ITK).

Review of literatures

Most of the studies focused on the disparities in agricultural performance at the state level in the country. However, their findings do not reveal a clear trend over time and lack a definite pattern across states. The studies by Nayak (1998), Birthal *et al.* (2011) and Chand and Parappurathu (2012) reveal the existence of wide variations in productivity and overall agricultural performance among the states. Somasekharan *et al.* (2011) studied the regional development in agriculture in 15 major states of India for the period 1971-2007. Their findings indicated that regional disparities in agricultural performance, which increased during 1971-88, moderated during 1988-2007 suggesting a kind of convergence. Introduction high yielding irrigation-responsive and fertilizers responsive HYVs varieties of rice and intensive cropping system three to four crops in a year in the same piece of land required huge amount of water. The water was supplied either surface or from ground water resources. As results, the depletion of water table occurred to be severe (Zaman *et al.*, 2017). Swain (2002) and Swain *et al.* (2009) conducted three benchmark surveys in recent years (1980-81, 1990-91 and 1998-99) in Odisha. A notable finding of the studies is that regional disparities have manifested and moderated in the post-reform period 1991-1999 because of implementation of backward area development programmes by the government. Tripathy *et al.* (2011) studied covering the period 1980-81 to 1992-93 that also pointed out the uneven performance across districts with the coastal districts and as well as Sambalpur district of western region that exhibited better performance than others. Pattanayak and Nayak (2004) have studied the regional disparities in agricultural development in different districts and zones of Odisha during the benchmark years 1980 and 2000. The findings indicated high inter-district variations in agricultural productivity with higher productivity in the coastal and plain land areas than others .In a recent study concerning Odisha agriculture covering the pre-liberalization (1971-90) and post-liberalization (1991-2008) periods, Reddy (2013) observed that the perceptible regional disparities in agricultural growth with Coastal Plains and Central Table Land zones showed commendable performance compared to Eastern Ghat and Northern Plateau zones in both the periods. In contrast the study by Chand *et al.* (2009) which did not show large variations in productivity levels across various districts in Odisha. The survey of literatures in the state pointed out to a number of deficiencies of existing research works on the subject. First, the findings appear to be mixed and diverse. Second, with agricultural performance of an area influenced significantly by its natural – institutional-technological characteristics (which differ widely amongst and within the states), existing research focusing largely at the state level and the evolution of bottom up and micro-level planning emphasizing the importance of district level approach to agricultural development where in need for studies at the lower level becomes apparent. Third, there is a paucity of studies on agricultural performance at the disaggregate level in Odisha. Whatever little studies do exist are confined to the 13-district set up and are based on a very limited number of variables covering periods up to the early 2000s. Meanwhile 17 new districts have been carved out by reorganizing most of the existing districts raising their number to 30. Besides a lot of changes in cropping pattern, irrigation coverage, cropping intensity, fertilizer consumption, average size of operational holdings and percentage of agricultural workers have taken place in state agriculture during the recent years. These changes might have affected the agricultural performance of districts differently and needless to say they could not have been grasped in the existing studies. There is thus a need for overcoming these research gaps and enriching the existing literature. The present study is a humble attempt in this direction. Average productivity of agri-inputs is a summary measure of agricultural performance of an area/district. Since land and labour are the two principal factor inputs in agriculture, information relating to output per unit of both land and labour can be used to assess the level of agricultural development. But measuring labour productivity is a difficult task and whatever assessment can be made in this regard is bound to be inadequate and unreliable for obvious reasons. Furthermore, the state is considered to be the poor agrarian states and characterized overcrowding in agriculture, land is the scarcest factor and it is relatively easy to compute output per unit of land. The land productivity as a key indicator of agricultural development and such productivity are measured in terms of yields of food grains and paddy per hectare of cropped area. The food grains and paddy have been selected because they constitute 74% and 45.5%, respectively of GCA of the state. District-wise data in respect of per hectare yield of food grains and paddy for the three benchmark years was given in Table 1.

Table 1: Inter-District Variations in Agricultural Productivity in Odisha

Districts	Productivity of Food Grains (kg/ha)			Productivity of Paddy (kg/ha)		
	2001-02	2006-07	2011-12	2001-02	2006-07	2011-12
1	2	3	4	5	6	7
Sundargarh	880	1073	1482	1481	1841	2678
Debagarh	1095	998	1157	1976	1895	2449
Mayurbhanj	1403	1396	1651	2326	2357	2812
Kendujhar	1064	1087	1363	2015	2060	2557
Balasore	1624	1376	1999	2661	2163	3207
Bhadrak	1754	1628	1555	2919	2648	2487
Jajpur	1150	1051	862	2416	2037	1667
Cuttack	1163	1063	1239	2529	2256	2860
Jagatsinghpur	1203	1110	1627	2583	2231	3633
Kendrapara	1086	890	1106	2170	1677	2279
Puri	1283	1103	1124	2525	2046	2258
Khordha	1132	1275	1298	2258	2443	2339
Nayagarh	1060	981	520	2454	2241	967
Ganjam	1375	1322	616	3270	3148	887
Gajapati	1293	1193	957	2934	2673	1539
Rayagada	1032	1182	1176	2294	2567	1876
Kandhamal	1181	1105	943	2288	2089	1709

Koraput	1056	1217	1211	2266	2537	2196
Nabarangpur	1172	1488	1249	1925	1987	1051
Malkanagiri	906	1098	729	1643	2101	852
Kalahandi	1085	930	894	2085	1670	1442
Nuapada	836	795	649	1912	1650	1182
Bolangir	1101	1049	486	2314	2228	642
Subarnapur	1621	1768	1772	2855	3132	3287
Boudh	993	1226	951	1875	2391	1782
Sambalpur	1508	1588	1458	2763	3036	2710
Bargarh	1713	1718	1787	2971	3055	3219
Jharsuguda	1295	1512	987	2277	2789	1822
Dhenkanal	1309	1051	1392	2810	2161	3139
Angul	1048	918	725	2484	2042	1476
C.V.	19.68	20.93	33.91	17.47	18.47	39.34

Source: Patra, 2014

Research Methodology

The importance of district level approach is appreciated for agricultural developmental indices that required the need to study at the lower level as the priority to identify a number of deficiencies of existing research works on the subject area. First, the findings appear to be mixed and diverse. Second, with agricultural performance of an area influenced significantly by its natural – institutional- technological characteristics (which differ widely among and within states), existing research focusing largely at the state level and the evolution of bottom up and micro-level planning emphasizing the issue. Third, there is a paucity of studies on agricultural performance at the disaggregate level in the state excepting a few that does exist are confined to the 13-district set up and are based on a very limited number of variables covering periods up to the early 2000s. Meanwhile 17 new districts have been carved out by reorganizing most of the existing districts raising their number to 30. Besides a lot of changes in cropping pattern, irrigation coverage, cropping intensity, fertilizer consumption, average size of operational holdings and percentage of agricultural workers have taken place in Odisha's agriculture during the recent years. These changes might have affected the agricultural performance of districts differently and needless to say they could not have been grasped in the existing studies. There is thus a need for overcoming these research gaps and enriching the existing literature. The present study is a humble attempt in this direction to know the ITKs available and its role in changing over the agricultural scenario of the state. In order to obtain the feedback of the farmers regarding the ITKs uses in agriculture, the survey schedule developed with a proforma that initiated in treated and untreated villages in different agro-ecological regions through the project centers. Farmers' awareness and adoption of ITKs practices were attempted in the study and recorded for necessary analyzing the data to quantify the ITKs role in present day agriculture.

The process followed during documentation was as follows:

1. The ITK proformae consisting of general information of village, views and observations on soil and water conservation measures by applying homemade organic manures, in-situ moisture conservation practices, technical information on run-off management practices and Indigenous technical knowledge was developed.
2. The proformae for documenting ITKs in agriculture as a whole in its management was finalized duly after getting the feedback from participants. The documentation and analysis of ITKs, a proforma was developed. This proforma contains the following (i) Title (ii) Purpose of use (iii) Location/Zone, (iv) Description (v) Advantages (vi) Constraints with feasibility of adoption and (vii) Researchable issues. The researchable issues open up new vistas for furthering the technologies towards development of farmer-friendly doable technologies. This is just an honest beginning of a validation process wherein a few ITKs have already been documented in the limited time frame.

Documented ITKs

The ITKs were documented under following specific categories

- Agronomic Measures
- Tillage
- Bunding and Terracing (Mechanical and Vegetative barrier)
- Land Configuration
- Soil Amendment and Mulching
- Erosion Control and Runoff Diversion Structures
- Water Harvesting, Seepage Control and Ground Water Recharge

Some popular ITKs identified in the process

Some potential ITKs identified for further study as well as adaption, research and development of new projects.In view to study in depth scientifically that could change these ITKs adding to modern technical knowledge (MTK) keeping identities of prevailing ITKs.All the ongoing projects on resource conservation and management should focus on the viable and appropriate ITKs relating to soil and water for sustainable development and dissemination of the local technology. As an initiative, the ITKs for other agro climatic regions may be documented and validated later and refined at local level. Exposure visit and farmer-to-farmer interaction on the subject may be encouraged for better adoption.

Possible stakeholders of the ITKs

The stakeholders who can be partners for promoting the ITKs in agriculture are:

1. Farmers
2. NGOs
3. Government Extension Agencies
4. Research Institutes / Scientists
5. Administrators
6. Policy makers / people's representatives

Some Identified ITKs

Although many of these traditional practices are supported by scientific findings, yet these have not been converted into technological recommendations to enable farmers to reduce their costs and make agriculture more sustainable. Farmers constantly learn and unlearn, choosing appropriate knowledge in their struggle for a sustainable livelihood. Let's take a look at a few important ITKs still prevalent in the state:

1. **Milk treatment:** Reduction of viral diseases has been found to occur mostly in infected plants of *Solanaceae*, *Piperaceae* and *Malvaceae* families by milk spray. It was tested that fresh milk spray reduces nearly 73 per cent of TMV infections in tobacco. Fermented buttermilk (11 in 20 l of water) sprayed on tobacco plants cures leaf curl disease. Fresh goat milk sprayed over plants like chilli (*Capsicum annum*), brinjal and species like black pepper also helps control fruit and leaf curl. The neutral to near-neutral pH of milk not only washes off the fungal spores but the adhesive property of milk fat also prevents germination of spores.\
2. **Neem and ash**: Before transplanting paddy, seedlings are kept in small plots of pounded water mixed with ash and pulverisedneem seeds. For a plot of 15 sq. feet, half a kilogram of neem seeds and a kilo of ash is sufficient for mixing with water to accommodate 50 bundles of seedling at a time for a period of one hour. Treated seedling produce a crop free of pest and diseases.
3. **Lantana leaves:** Leaves of lantana, an obnoxious weed, can be utilised to preserve and process sweet potatoes. Insect attack can be restricted to a bare minimum with no use of poisonous insecticides.

4. **Inducing conception:** An indigenously developed concept of giving tonic to dairy animals is gaining popularity. The cows, if fails to conceive, a tonic made of the bark of *Cassia fistula*, tender leaves of jackfruit tree, root of *Plumbagozeylanica* and leaves of *Clerodendruminerme* are powered and the mixture is fed to animals along with jaggery.
5. **Deworming:** A reddish-brow powder collected from the fruit of the widely grown kaamal tree (*Mallotusphilippinenis*) is used for the treatment of worm infection and or constipation in cattle and buffaloes. Depending on the type of ailment, 50-60 g of kamila powder mixed in lukewarm water/butter milk is administered orally.
6. **Udder swelling**: A few days after calving, dairy cows suddenly stop milking due to swelling of the udder. To treat this, pulverized quartz stones are passed through a muslin cloth, mixed with butter and smeared over the udder. It is applied twice a day till the animal is cured.
7. **Treating fractures:** A paste for treating a fractured bone of an animal is prepared by mixing 500 g bark of khakhra (*Buteamonosprema*) and 500 g seed of guvar (*Delonixelata*).

Conclusion

Many ITKs on in-situ in agriculture are not adopted everywhere throughout the country because of constraints in adoption and unawareness of the effectiveness of such practices. The present documentation process has definite bearing on the future course of action in framing new projects. This short-term documentation project may lead to the following future activities:

(i) Similar exercise can be undertaken to document the ITKs from all the Agroecological regions of the country.

(ii) The potential ITKs may be tested for their suitability and adoption in other Agro-ecological regions as a dissemination strategy.

(iii) The documented ITKs may be published/translated in all regional languages for the benefit of the farming communities.

(iv) Validation of the ITKs is a logical step to qualify and quantify the effectiveness of these practices. Suitable modifications of the traditional practices through on-farm research would help in developing appropriate and acceptable technologies for different local environments.

(v) The effect of conservation measures on resource losses can be studied in detail through experimentation and use of stimulation model.

(vi) As a policy matter the local ITKs should be in built in the resource conservation programme.

Hence, the process of documentation, refinement and promotion of ITKs should form the basis for implementing natural resources in advancement of agriculture removing the regional disparities in the state.

References

Birthal, P.S., Singh, H. and Kumar, S. 2011. Agriculture, economic growth and regional disparities in India. *Journal of International Developmen*, 23: 119-131.

Chand,R. and Parappurathu, S. 2012. Temporal and spatial variations in agricultural growth and its determinants. *Economic and Political Weekly*, 47(26 & 27): 55-63.

Chand, R., Garg, S. and Pandey, L. 2009.Regional variations in agricultural productivity- A district level study. *Report of the National Professor Project, NCAEPR,* ICAR, New Delhi.

Nayak, P. 1998. Unbalanced agricultural development in North-East India. *Journal of Assam University,* 3(1): 24-32.

Pattnayak, M. and Nayak, B.P. 2004.Performance of agriculture in the changing structure of the Orissa Economy: Issues Revisited. *Paper presented at the 40th Annual Conference of the Indian Econometric Society,* http//ssrn.com.

Patra, R. 2014. Agricultural development in Odisha: Are the disparities growing? *International Journal of Food and Agricultural Economic.*2(3): 129-44

Reddy, A.A. 2013. Agricultural productivity growth in Orissa, India: Crop diversification to pulses, oilseeds and other high value crops. *African Journal of Agricultural Research*, 8 (19): 2272-2284.

Somasekharan, J., Prasad, S. and Roy, V.P.N. 2011. Convergence hypothesis: Some dynamics and explanations of agricultural growth across Indian states. *Agricultural Economics Research Review,* 24(2): 211-216.

Swain, M. 2002. Performance of agriculture in Orissa : Intertemporal and Spatial *Variations in Orissa.Development Report, Planning Commission, New Delhi.*

Swain, M., Swain, M. and Das, D.K. 2009.Regional disparity in agricultural development in Orissa in the pre and post-reform period.*Social Change*, 39(2): 195-215.

Tripathy, U., Das, M. and Padhi, P.L. 2011. Regional variation in development profile of Orissa: An econometric analysis of indicators of development. *International Journal of Multi-Disciplinary Research,* 1(6):107-130.

Zaman, A., Zaman, P. and Maitra, S. 2017.Water resource development and management for agricultural sustainability.*Journal of Applied and Advance Research*, 2(2):73-77.

17

Role of Geomorphological Features to Determine Sustainable Flood Management Strategies in Bhagirathi-Padma River Bank Areas

A. Gayen[1] and A. Zaman[2]

[1]Senior Hydrogeologist, Rajiv Gandhi National Ground Water Training & Research Institute (RGNGWT &RI), Central Ground Water Board (CGWB) Raipur, Chhattisgarh, India
Email: anadigayen1968@gmail.com
[2]Emeritus Professor, Centurion University of Technology and Management Paralakhemundi-761211, Odisha, India
Email: profazaman@gmail.com

Abstract

The district Murshidabad lies within the Younger Deltaic Plain and the Bhagirathi Recent Surface has been considered as the severe flood affected district of West Bengal state, India. Murshidabad is highlighted one in terms of occurrence of flood due to its typical geographic location and diverse geomorphological characteristics. The district witnessed frequent floods because of high intensity of rain fall in the basin area of Bhairavi-Jalangi-sealmari rivers. Inflow of flood water from Ganga-Padma in high spate condition further aggravates the congestion of drainage. Lack of proper drainage is also responsible for flood in the upstream side in the western part as well as beyond the Berhampur town. Flood management may be undertaken through physical and socio-economic measures, which includes the scientific control of reservoirs; construction of new barrages, improvement of river channels, removal of human encroachment along river side, timely flood warning and forecasting, preparedness of essential services along with proper flood plain planning are indispensable. Addressing of socio-economic problems may help to combat the flood situation in an effective way. Careful study on surficial geomorphic features would suggest flood mitigation measures to be adopted in various geomorphological set ups.

Keywords: Geomorphology, rain fall, socio-economic problems, flood analysis, mitigation

Introduction

In recent times, flood is one of the most disastrous natural hazards in India and especially in West Bengal state. The success of flood control measures of the state can be achieved through an analysis of flood types, which includes the essential component like geomorphology of the river basin. The United Nations Development Programme (UNDP) flood policy has emphasized the detailed study of river geomorphology. Historical data reveal that even after continuous efforts to control the floods, flood damages and flood-affected areas in India have increased with time (Agarwal and Narain, 1996). Statistically streams will equal or exceed the mean annual flood once every 2.33 years (Leopold, et at., 1964). Flooding is a natural and recurring event for river or stream and is the consequential effect of heavy and continuous rainfall exceeding the absorptive capacity of soil and flow capacity of the rivers, streams and coastal areas. Finally it causes a watercourse to overflow its banks and is spreaded over the adjacent lands. Flood is defined as a relatively high flow, which overtakes the natural channel provided for run-off. Another definition of flood is "a body of water which rises to overflow land which is not normally submerged" (Ward, 1978).

The flood in the West Bengal state may be categorised into two types: i) River flooding ii) Coastal flooding. Out of 8.88 m. ha area in the state, 2.65 m. ha area are liable to flood almost every year. Monsoon rainfalls can cause disastrous flooding in any tropical countries due to their extended periods of rainfall. When tropical cyclones move from sea to land may create waves of up to 8 meters high, which may cause huge flood to the areas in the river side areas. This type of flooding occurs regularly along the coastal part of West Bengal.

Murshidabad district (Fig.1), situated in the eastern part of Bhagirathi is low-lying and alluvial. The climate is humid and the land is liable to flood by the spill of the Bhagirathi and other rivers. The soil is amazingly fertile. On the western side, the relief is higher than the eastern part and the surface is undulating. The soil is hard clay. The climate is drier than in the eastern tract. Although the western tract or the 'rarh' is slightly high and undulating than the eastern tract 'Bagri', the land is interspersed with numerous swamps and paleo-channels.

The radiating channel pattern with natural levees on each bank is well preserved. Interestingly large filled valley-cuts are found near the junction of Older Deltaic Plain and the Bhagirathi recent surface. The sedimentary structures are common in Young Deltaic Plain. The fluviatile land forms like ox-bow lakes, levees and meanders are also preserved.

The Bhagirathi recent surface is confined into the present day channels and their immediate vicinity. The belt of terrace is 3-5 kilometres in width with slope 0.2 meters per kilometres. The land is very low in some places and there is a sharp break present between Young Deltaic Plain and the Bhagirathi recent surface (e.g. flood plain between Bhagirathi and Dwarka River). The surface is composed of loose and completely unconsolidated sediments structures. The river Bhagirathi meanders strongly between Jangipur and Berhampore police station. Geologist assumes that the river is a misfit and likely to be an ancient course of the Ganges. Big meanders scars, cut off, abandoned channels and levees are some of the examples of preservation of landforms. Presently the river courses are oscillating frequently (blocks like Suti-I, Suti-II, Raghunathganj-I, Lalgola and Bhagawangola-I are mainly affected).

The older landforms are exposed to erosion; therefore, the land forms of the younger ones are better preserved. The land form near rivers are cut to produce geo-morphologically younger plain which is filled by sediments of the younger plain, through 'tongue' gradually close the older plain. This is how an oscillating river maintains the sequence from oldest to youngest as Lateritic Upland - Older Deltaic Plain-Younger Deltaic Plain- Bhagirathi recent surface.

Methodology

The present study has taken into consideration of the flood data of 40 years of the Murshidabad district. The toposheets on 1:50,000 scale (Survey of India), the Satellite Imageries, field studies as well as analysis of annual rainfall data for the period of last 10 years have been used to draw the conclusion of the present study. Consultation of literatures and various reports of the Department of Agriculture, Department of science and Technology, National Bureau of Soil Survey and land Use Mapping (NBSSLUPM) followed by compilation and correlation to geomorphology to understand the nature, cause, extension and character of floods occurred in the study area in previous years. Future management strategies deciphered based on the careful scrutiny of the nature of flood history and thereby the root causes responsible for occurrence of flood in and around the Murshidabad.

Study area

Murshidabad district is situated between 23Ú432 -24Ú502 N latitude and 89Ú492 -88Ú462 E Longitude in the middle part of West Bengal state. The district covers an area of 5,324 sq. km and consists of 26 Community Development Blocks. The district has typical geomorphological characteristics and occupies the interfluve region between the Padma, which roughly constitutes its eastern boundary and the Bhagirathi, the northernmost distributary of the Padma. Kandi Sub-Division is frequently affected by flood disaster. Location map of Murshidabad district is shown in fig.1.

Fig. 1: Location map of Murshidabad district, West Bengal

Flood in West Bengal

West Bengal forms a part of the Ganga delta has a long history of record flood since 1985 as because the major part of the state has been formed by the Ganga- Padma river system through the process of delta building. At present 42.3% of the total area of the state is susceptible to flood spread over 110 blocks in 18 districts. The highest flood affected area as recorded in 1978, is about 30,607 sq km. About 23,970 sq km of area was devastated by flood in 2000 (Choudhary *et al.* 2010).

West Bengal state is located in the eastern part of India having International border with Bangladesh, Nepal and Bhutan. The state covers an area of 88,752 sq. km. of which 22% is very prone to flood in each year.

Genrally floods in west Bengal is caused by precipitation from Bay of Bengal depressions. Comparative studies of the historical flood records indicate that the frequency of high-magnitude floods has increased significantly in recent years. Flood study also reveals that high flood flows are responsible for bedrock erosion as well as transportation of coarse material. The basic reason of flood in this zone is the shape of the catchment area, its steep slope from a high level plateau area sloping sharply down to a flat terrain near the outfall and also adverse outfall condition because of its limited intake capacity. In the state except only five years, in the last last fifty years (1959 to 2009), after last 1959 devastating flood, the state suffered consecutively in 1978, 1985, 1998, 1999 and 2000.

Flood status in Murshidabad

Amongst six flood prone districts of West Bengal state, the Murshidabad district has the highest land area (5,34,100 ha), which is being flooded every year. The most flood affected districts of West Bengal, i.e., Murshidabad is most prominent because of its vulnerability and geographical position. Almost each and every year this district suffers from flood and creates landforms common to flood prone regions. Recently in the year 2015, vast area of cropping fields and households in areas under Kandi and Jangipur sub-divisions of Murshidabad district have been immersed due to sudden outburst of flood in the rain-fed rivers. Heavy rains followed by flows of upland streams from Jharkhand have inundated a number of lowland villages under Bahutali and Haaru Gram Panchayats of Suti area, while the overflowing rivers like Kana Mayurakshi and Dwarka have caused floods in the villages of Bharatpur, Barwan, Khargram blocks in Kandi Sub-division. Before 2015, there were severe floods in Murshidabd during 2003 and 2006, wherein main reason for flood was continuous monsoonal downpour. Record reveals that seventy two hours of continuous and concentrated rainfall over the western river basin areas of Bhagirathi viz. from the Pagla-Bansloi to the Ajoy, generated so huge water volume that all embankments on the eastern side of the Bhagirathi were almost washed away and the whole of Nadia and larger part of Murshidabad district were flooded and remained under water for a long period. Bhagirathi-Hooghly basin accumulates enormous quantity of water during peak monsoon. Flood prone areas of Murshidabad district is shown in fig.2. Probable reasons of flood in the district are as below:

1. The drainage congestion at its outfall because of the high stage of Bhagirathi.
2. High intensity rainfall causes water logging in the basin area of Bhairabi-Jalangi-Sealmari.
3. Siltation in the Bhagirathi –Padma River causing reduction of capacity of water drainage.

4. Inflow of flood water from Ganga-Padma at its high spate.
5. Sudden release of large quantity of water from Upper River dams under Damodar Valley Corporation (DVC) in the River Mayurakshi.
6. Habitations on the natural flood plain causing hindrance to drain out water during heavy rainfall.

Murshidabad has a long flood history. In Bhagirathi River and its tributaries has the average flood duration of continues more than a month and it is noteworthy that the maximum duration was of 59 days in 1971. Recently it has been reported that the Ganga River at Farakka has attained the danger flood level of 22.76 during July, 2016.

Discussion

Geomorphological study correlating flood occurrence

The Indo-Gangetic Plains are drained by some of the largest river systems in the world. These plains are also severely affected by frequently occurring disastrous floods and are presently regarded as the worst flood affected region in the world (Agarwal and Narain, 1996). The district Murshidabad lies generally within the Younger Deltaic Plain or Kandi Plain and the Bhagirathi Recent Surface. The river Bhagirathi, flowing from north to south through the district had divided the area into almost two equal halves, namely 'Rarh' and 'Bagri'. The tract to the west of the river is locally known as 'Rarh' and the tract to the east as 'Bagri'. The Ajay- Bhagirathi Valley was divided into four surficial geomorphic plains a) Lateritic upland, b) Older deltaic Plain, c) Younger Deltaic Plain and d) Bhagirathi Recent Surface (A. Bhattacharya & S.N. Banerjee (1979). The Quaternary (Recent Pleistocene) sediments have been classified with Geomorphic subdivision and its characteristics (Bhattacharya & Banerjee (1979).

Geomorphic sub-division	Morphological characteristics
1. Younger Deltaic Plain formation (YDP):	River terrace sediments
	Alluvial sediments with soil containing soft ferruginous nodules (ODP): Older Deltaic Plain Formation
	Alluvial sediments with soil
2. Lateritic Formation:	Hard lateritic and mottled clay with underlying alluvial sediments containing calcareous nodules

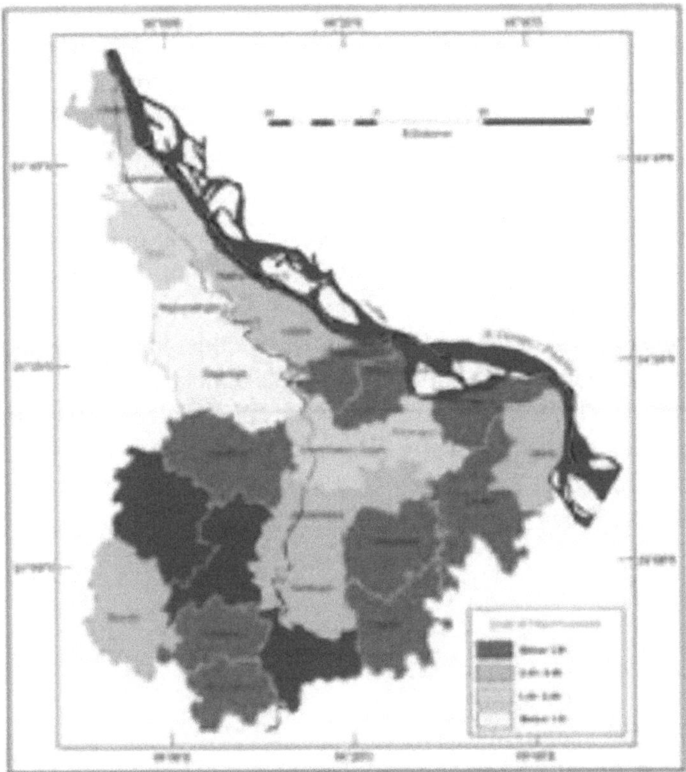

Source: Irrigation & Waterways Department, Govt. of West Bengal
Fig 2: Flood prone areas of Murshidabad district

Hydrological character of a river basin is mainly controlled by the geomorphology and therefore, an organised scientific approach to flood management studies should include the thorough understanding of the geomorphological set up of that river basin. Floods have long been considered as purely hydrological phenomenon, and therefore, flood management programmes have essentially focused on hydrological variations with diversified geomorphological signatures. Flood plain is defined as "a strip of relatively smooth land bordering a stream and overflowed at a time of high water" (Leopold, 1964). Topographically, it is very flat and lies adjacent to stream. Geomorphologically, flood plain is categorised as a landform composed primarily of unconsolidated depositional material derived from sediments being transported by the related stream. But, from hydrological point of view, it is to be defined as a landform subject to periodic flooding by a parent stream. A combination of these characteristics perhaps comprises the essential criteria for defining the floodplains" (Schumdde, 1968). Very simply flood plain can be described as "flood-prone" and hazardous to the development activities subjected to the exceeding of tolerance level. The abundance of ox bow lakes and misût river channels are also characteristic of this part of Gangetic West Bengal; stagnant water bodies and marshy land dot the landscape, some of which are spill channels of the Damodar River that had lost their headwater to silting or to a shift in its course (Spate *et al.,* 1967).

The river Bhagirathi meanders strongly between Jangipur and Berhampore area. Big meanders scars, cut off, abandoned channels and levees are some of the examples of preservation of landforms. Presently the river courses are oscillating frequently and the blocks like Suti-I, Suti-II, Raghunathganj-I, Lalgola and Bhagawangola-I are highly affected. As the older land forms are subjected to erosion; the younger land forms are protected in better way. The land form is cut to produce younger plain near rivers and is filled by sediments of the younger plain through 'tongue' gradually closes the older plain. In this manner, the sequence from oldest to youngest as Lateritic Upland -Older Deltaic Plain- Younger Deltaic Plain- Bhagirathi recent surface is maintained by an oscillating river.

Soil

The 'Bagri' surface is characterised by loose and completely unconsolidated sediments having diversified sedimentary structures. Brown mottled soils of 1-2 m. thickness with soft and incipient ferruginous concretion have been developed on Younger Deltaic Plain (YDP). The 'Rarh' region is mainly a continuation of the Sub-Vindhyan region of lateritic clay and nodular Kankar. The soil is greyish or reddish mixed with lime and iron oxide and scattered nodular lime Kankar. Based on soil characteristics, the 'Bagri' region may be classified into four goups:a) Ganga Riverine lands, b) Ganga Flat land, c) Ganga Lowland and d) Ganga Uplands. Soil map of Murshidabad district is shown in Fig 3.

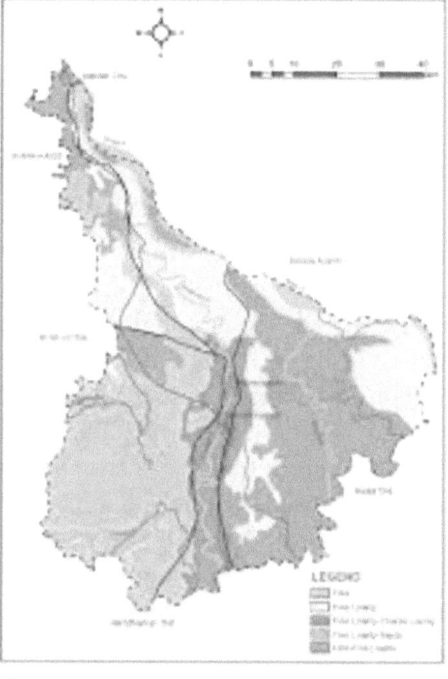

Source: NBSS&LUP, Government of India

Fig 3: Soil map of Murshidabad district, West Bengal

River Network

Two rivers channels distributes throughout the district – a) Ganga-Padma River and their distributaries and b)Mayurashki river and distributaries. The main rivers of Ganga-Padma River system are Bhagirathi, Jalangi and their distributaries like Sialmari and Bhairab. These rivers are flowing through the 'Bagri' region of Murshidabad. These distributaries are seasonal and remain dry in remaining months of the year. Bhairab River has become saucer shaped because it receives silt from Padma River in every year. At present the Bhairab river bed is used for '*boro*' paddy cultivation.

The Mayurakshi,-Dwarka- Brahmani River comprises of: a) river Basloi, river Pagla, and river Gumani from the north western Rajmahal hills and b) end part of Mayurakshi, - Dwarka- Brahmani River in the south. These palaeo-channels fed the 'Rarh' blocks of Murshidabad during monsoon accumulates sand and shield and form a deltaic plain (after Willocks).

The rivers are in their decaying stages and land building processes have been stopped. The elongated depression area between the Bhagirathi and Jalangi rivers is demarcated as flood prone zone (Sanyal and Lu, 2006). The interûuves of the numerous distributaries are ill drained (Spate. *et al*, 1967) and causes frequent waterlogging during June–September month with the southwest monsoon and finally leading to stagnation of water followed by the development of palaeo-channels. The overall geomorphology of the study area depicts a degenerating ûuvial system. These rivers and adjoining tract of the land is flushed by flood spills. The Bhairab was one of the most active delta builders and had quite a number of spill channels. Following the diversion into the Padma resulting in reduction and almost cutting off head waters supplies, the lower reaches of these rivers silted up by tidal action and these river courses began to deteriorate. The mouth of the Jalangi has also silted up. The Chhoto Bhairab River has been originated from the river Bhairab, but its mouth has been silted up. Basloi River and other hill streams from the west Rajmahal hill flood the area every year (Gazetteer by A. Mitra, 1979; pp: 10, 15).

The local landlords with the power vested on them by the Permanent Settlement started to encroach into the detention basins for revenue maximization by constructing circuit embankments to reclaim the low lying areas of *kharif* cultivation. The large portion of land that used to absorb the flashy rivers flood got fragmented with unscientifically aligned and designed circuit embankments. The progressive deterioration of the carrying capacity of the detention basins locally known as bils and Bhagirathi over long years aggravates the flood situation in this part. Since historical period, encroachment of these low bils areas for livelihood has come in practise. Floods in the western part of the Bhagirathi are caused by a number of flashy rivers having their origins in the Chhotonagpur Plateau. Earlier a large number of detention basins on the courses of the rivers used to absorb the flash floods and gradually released the same when flow of Bhagirathi reduced. The natural sequence of floods of different rivers in this region was that the floods of the Pagla-Bansloi and Ajay drain first, followed by the floods of the Mayurakshi-Babla system and the floods of the Brahmani-Dwarka drain last of all.

As far as the source of flood is concerned, the nature of flood problem of the Suti-I, Farakka blocks are different from other blocks. The rivers like Bagmari, Gumani, Trimohini, Kanloi, Madhabjani plays an important role to create severe flood situation in these blocks. All the rivers outfall into the Bhagirathi except Bagmari. Since the outfall channels of these rivers are not in a good condition, it again aggravates the flood situation in this region. Huge rainfall during monsoon has increased the river levels in river catchments in the Murshidabad district.

Physiographical set up

The river Bhagirathi, flowing from north to south through the district, broadly divides the district into two almost equal portions, which form a striking contrast in their geology, their physical characteristics and agriculture form among each other. Traditionally the district was divided into four tracts:

a) "Rarh" consisting areas to the south of the Ganga and west of Bhagirathi, b) "Barendra" lying north of the Ganga and between the Mahananda and Karatoya rivers, c) "Bagri" is the south Bengal and d) "Vanga" is the eastern Bengal.

The whole of the "rarh" region lies at an elevation as the range of Rajmahal hills slopes gently down in this tract. The north of the district is an area of about 150 sq. miles is a strip of low-lying region which becomes a vast lake in the monsoon. The villages are like islands during this season. South of this narrow strip of land is a plain almost tree-less at the confluence of the Mor and the Dwaraka popularly known as 'Hijal'. The "Bagri" is low area, exposed to inundations during monsoon and is a tract of fertile soil. Here in the south eastern corner of the district about 50 sq. miles of area is swampy tract of dark clay and is popularly known as 'Kalantar'. It receives the drainage of the swampy rivers in some part of the district.

Sustainable flood management strategy

Flood is such a natural phenomenon that is uncontrolled and to some extent unpredictable. Man has been toiling for ages to control flood but all in vain. Geologically southern part of West Bengal, is a part and parcel of Ganga-Brahmaputra delta, prone to flood almost each and every year. Flood damages all the economic aspects of the state and led to get poverty, illiteracy, starvation death and migration etc. The flood situation in the severe flood affected Murshidabad district can sustainably be managed by different physical and socio-economic measures which are indispensable:

i) Scientific control on reservoirs.

ii) Construction of new barrages, lock gates.

iii) Embankment maintenance.

iv) River channel improvement.

v) Removal of human encroachment from the area, where typical geomorphological features like Younger Deltaic Plain and Bhagirathi Recent Surface are present along the banks of Padma and Bhagirathi rivers.

vi) Flood warning and forecasting.

vii) Enhancement of essential services along with proper flood plain planning.

viii) Adoption of Controlling measures of Messanjore Dam.

ix) Sheet Erosion Control and Soil Conservation by River Bank Protection and construction of dykes etc.

x) Flood planning considering flood history and rainfall pattern over the past many years.

Socio-Economic Problems

As per Human Development Index (HDI) report of Govt. of West Bengal and Sachar Committee Report, Murshidabad is ranked as one of the backward districts of West Bengal in view of literacy, economic status and agricultural development. The flood related socio-economic problems in Murshidabad includes poverty, poor agricultural growth despite having fertile land, unemployment, child labour, illiteracy, insufficient educational services, dense rural settlements along the riverside which are highly affected by flood, loss of livestock, submerged clay-made walls of the houses during flood, lack of desired level of industrial growth and transport facilities.

Conclusion

West Bengal floodplain is one of the most destructive floodplain in India. Owing to be the part of Ganga-Brahmaputra rivers delta, the southern part of West Bengal is frequently affected by flood with unmanageable discharge. The main flood situation in this area arises because of inflow from Ganges-Padma in their high stage, rainfall in the own catchment area and also tide lockage. A Ganga Riverine land adjacent to the banks of the Ganga-Padma River and Jalangi River is susceptible to frequent floods. Proper and scientific management of flood and floodplains of Murshidabad can reduce the damage of the state and can give the relief for human civilization. Adoption of community participation through involvement of local people to identify flood prone areas followed by mitigation measures could help to yield expected result. Efficiency and capacity building measure of the flood management team may be enhanced to expedite the activities like emergency warnings, evacuation, rescue, relief shelter jobs. Three blocks in the western part of the study area and four blocks in the eastern part of the study area are very high flood prone areas. Therefore, special attention is needed for the planners for flood management in these areas. Based on flood history, a flood hazard map can be prepared, which will help to develop a suitable area specific flood management strategy for Murshidabad district.

Acknowledgement

The corresponding author is thankful to the Regional Directors for according permission and encouragement to write this paper. The author is indebted to the Chairman, Central Ground Water Board for according kind permission and accepted for oral presentation in the 6th International Conference on Water and Flood Management (ICWFM, 2017) organized by Bangladesh University of Engineering and Technology (BUET), Dhaka, Bangladesh during 4-6 March, 2017.

References

Aziz, F., N.K. Tripathi, O. Mark, and M. Kusanagi, 2003. Flood warning and evacuation system using MIKE 11 and GIS, *Asian Journal of Geoinformatics* 3(3): 31-39.

Bagchi, K. 1945: Ganges Delta, University of Calcutta Press, PP. 50-70 Curray.

Boyle, S.J., I.K. Tsanis and P.S. Kanaroglou, .1998.: Developing geographic information system for land use impact assessment in flooding condition, *Journal of Water Resources Planning and Management* 124: 89-98.

Chorley R.J., Schume S.A., Sugden.D.E. 1984, Geomorphology, Methuen, London, pp.32-131'.

Dong, Y., B. Forster, and C. Ticehurst,.1997. Radar backscatter analysis for urban environments, *International Journal of Remote Sensing* 18(6): 1351-1364.

Goodbred, S. L. and Kuehl, S. A. 1989: *Geology*, 27, pp. 559– 562.

Government of West Bengal. 2009. Flood Preparedness and Management Plan, Office of the District Magistrate, Murshidabad, pp.12-19.

Government of West Bengal.2007. Report of the Technical Committee on Floods in the District of Murshidabad and its Adjoining Districts in West Bengal, Irrigation and Waterways Department, Volume I, pp. 69- 83.

Government of West Bengal1999. Report on Hydrological investigation in Moribund Deltaic plain of Murshidabad district, West Bengal State Water Investigation Directorate, Geological Circle III, pp- 6-89.

Government of India.2006. Report on Water Resources for XI Five Year Plan (2007-2012).

Government of West Bengal. 2009. Flood Preparedness and Management Plan, Office of the District Magistrate Murshidabad, Kolkata.

Government of West Bengal. 2008. District Statistical Handbook, Bureau of Applied Economics and Statistics, Kolkata. pp.75-87.

Hirst, F.C.1915: Report on Nadia Rivers, Reprinted by West Bengal Districts Gazatteers, Government of West Bengal, Kolkata.

Irrigation and Waterways Directorate. 1992. Kandi Area, Flood Protection Scheme, Report, Estimate and Drawing, Investigation and Planning Circle –1, Government of West Bengal, Kolkata.

Imhoff, M.L., C. Vermillion, M.H. Story, A.M. Choudhury, A. Gafoor and F. Polcyn, .1987. Monsoon Flood Boundary Delineation and Damage Assessment Using Space borne Imaging Radar and Landsat Data, *Photogrammetric Engineering & Remote Sensing,* 53(4): 405-413.

Kron W. 2005. Flood Risk, Hazard Values Vulnerability, Water International, Volume 30, Issue 1, Routledge publication, DOI: 10.1080/ 025080605086918 37, pp.68. http://portal. worldwaterforum5.rg/wwf5/en-us/Panels/High-Le.

Leopold, Wolman, Miller. 1964: Fluvial processes in Geomorphology, Eurasia Publishing House (Pvt). Ltd.pp.317-322, pp.337-357.

Majumdar, S.C. 1942. Rivers of Bengal, Reprinted by West Bengal State Gazetteers Department, Government of West Bengal, Kolkata, P.19.

O' Mally L.S.S. 1997. Rivers of Bengal, Bengal District Gazetteers, pp.1-25, pp.97-154 and Pp.177-204, 1914, Reprint, 1997.

Parua P.K. 2000, Reason of Ganga River Bank Erosion and Management: A Discussion, Sechpatra, Monthly News Paper, Irrigation and Water Ways Directorate, Kolkata, pp.3.

Parker, D.J. (ed). 2000. Floods, Volume I., Routledge, London, UK.

Ray Chandan. 2004. Floods and Role of the People Perspective of West Bengal, Chief Engineer Irrigation and Waterways Dept. Govt of West Bengal; www.unescap.org/.../ Flood%20and%20role%20of%20people.doc.

Rudra, K. 2008: Banglar Nodikatha (in Bengali), Shishu Sahitya Samsad Pvt. Ltd, Kolkata, pp.1-36.

State Inter Agency Group. 2009. South Bengal Flood: A Rapid Multisectoral Assessment, United For Humanitorian Response, Directorate of disaster Management, Govt of West Bengal, Kolkata, 2009. www.iagwestbengal.org.in.

Spate, O. H. K., Learmonth, A. T. A. and Learmonth, A. M. 1967. India and Pakistan: a general and regional geography, Methuen & Co Ltd, Bungay, Suffolk, pp- 588.

Sanyal, J. and Lu, X. X. 2006. GIS-based flood hazard mapping at different administration scales: A case study in Gangetic West Bengal, India. Singapore *J. of Tropical Geography*, 27: 207-220.

Sanyal, J. and Lu, X. X. 2006. GIS-based flood hazard mapping at different administration scales: A case study in Gangetic West Bengal, India. *Singapore J. of Tropical Geography*, 27: 207-220.

Wasson, R. J. 2003. A sediment budget for the Ganga– Brahmaputra catchment, *Current Science*, 84, (8):1041-1045.

18

Agricultural Sustainability for Crop Production

A. Zaman[1] and Md. Hedayetullah[2]

[1]Emeritus Professor, Centurion University of Technology and Management M.S. Swaminathan School of Agriculture, Paralakhemundi-761 211 Odisha, India
Email: profazaman@gmail.com
[2]Assistant Professor, Department of Agronomy, Directorate of Research Bidhan Chandra KrishiViswavidyalaya, Kalyani-741235, Nadia, West Bengal, India
E-mail: hedaye.bckv@gmail.com

Abstract

The word "sustain" is derived from the Latin verb '*sustinere*'(to keep in existence or maintain) and implies long-term support or permanence. The accepted definition of sustainable agriculture has proved to be elusive, given the extraordinary diversity and complexity of agricultural land use, and the perspective taken including producers, consumers, policy makers and workforces. Sustainable agriculture is defined as 'the ability of farmland to produce food grain and other agricultural products to satisfy human needs indefinitely as well as having sustainable impacts on the broader environment and maintained equilibrium in productivity level'. That in other words, requires agriculture to avoid severe or irreversible damage to the endogenous or external ecosystem services upon which it depends, notably soil fertility, irrigation water, genetic variability and pollinating agencies and have acceptable impacts on the broader environment under the environmental stewardship.

***Keywords*:** The objectives; producers; consumers; policy makers; workforces

Introduction

The principle of sustainability implies the use of resources at rates that do not exceed the capacity of ecosystems to replace them. By definition, dependency on non-renewable inputs is unsustainable, even if in the short term it is necessary as part of a trajectory toward sustainability. There are many difficulties in making sustainability operational. Over what spatial scale should food production be sustainable? Clearly an overarching goal is global sustainability, but should this goal also apply at lower levels, such as regions, nations, or farms? Could high levels of consumption or negative externalities in some regions be mitigated by improvements in other areas, or could some unsustainable activities in the food system be offset by actions in the non-food sector (through carbon-trading, for example)? Though simple definitions

In a broad sense, sustainability is the capacity to maintain equilibrium in productivity levels what we achieved. The concept is applied to ecosystems and human development efforts and its meaning is to a considerable extent contextual. Ecosystems are sustainable when they maintain ecological processes, functions, biodiversity and productivity into the future. Sustainability is the potential for long-term maintenance of wellbeing, which will in turn depend on the responsible use of natural resources.

1. Sustainability is often defined as resting on three pillars or having three dimensions: (i) environmental, (ii) social and (iii) economic sustainability. Some have argued the need to integrate these dimensions or redress the balance between them, others have pointed out the vagueness of the concept. Adams (2006) contrasts the hugely expanded awareness for sustainable development in recent years with the mounting evidence for the "global human enterprise becoming rapidly less sustainable" putting this down, in part, on the looseness of the concept and that it means different things to different people.

2. The conventional understanding of sustainable development, based on the 'three pillars' implies that trade-offs can be made between environmental, social and economic dimensions of sustainability. A distinction is often drawn between 'strong' sustainability (where such trade-offs are not allowed or are restricted) and 'weak' sustainability (where they are permissible) (Adams 2006). The concept of 'critical natural capital' is also used to describe elements of the biosphere that cannot be traded off (e.g. critical ecosystems). However, in practice, development decisions by governments, industries and other actors do allow trade-offs and have traditionally put greatest emphasis on the economy above other dimensions of sustainability. This is a major reason

why the environment continues to be degraded and development does not achieve desirable equity goals (Adams 2006).

Environmental sustainability is the ability to maintain rates of renewable resource harvest, pollution creation, and non-renewable resource depletion that can be continued indefinitely.

Economic sustainability is the ability to support a defined level of economic production indefinitely.

Social sustainability is the ability of a social system, such as a country, to function at a defined level of social well-being indefinitely.

A more complete definition of *sustainability* is thus environmental, economic, and social sustainability. This forms the goal of The Three Pillars of Sustainability

The concept of sustainable development

Sustainable development is development that meets the needs of the present without compromising the ability of future generations to meet their own needs. That contains within it two key concepts:

The concept of 'needs', in particular the essential needs of the world's poor, to which overriding priority should be given; and

The idea of limitations imposed by the state of technology and social organization are on the environment's ability to meet present and future needs.

"Agro" refers to agricultural production. High Ag production relies heavily on the quality of a soil ecosystem

The quality of a soil eco-system can directly affect crop production and livestock production

Sustainable agriculture is the act of farming using principles of ecology. Ecology is defined as the study of relationships between organisms and their environment. An ecosystem exemplifies the study of ecology.

Importance of sustainable agriculture

Sustainable agriculture is an agricultural production and distribution system that that can achieve the integration of natural biological cycles, can controls, protect and renews soil fertility and natural resource base, optimizes the management and use of non-farm resources, reduces the use of non-renewable resources and purchased production inputs, provide an adequate and dependable form of income and promotes opportunity in family farming and farming communities

The keywords of sustainable agriculture

1. Understanding the three pillars of sustainable agriculture and the indicators of sustainable farming;
2. Understanding how ecosystems function and the services they provide where good soil health is the foundation of a healthy ecosystem;
3. How sustainable farming practices enhance and protect soil health, water quality and water quantity, biodiversity, manage insect pests, disease, and weeds; and
4. The importance of local and regional foods systems to sustainable agriculture.

Progress towards food security

Much of the progress towards food security that was made between the 1970s and the 1990s has been reversed in recent years. At the start of the 1970s, 920 million people were chronically undernourished, with insufficient food for even low levels of activity. During the 1990s, despite continuing population growth, the figure had been reduced to just over 800 million – about 20% of the total population of the South. However, this has since risen to 963 million (2008), the vast majority of whom live in developing countries. Of these, 65% live in only seven countries: India, China, the Democratic Republic of Congo, Bangladesh, Indonesia, Pakistan and Ethiopia.

Unfortunately, the food supplies in some countries and regions are poor and in decline. Agricultural production has not kept pace with population growth in sub-Saharan Africa where many countries are in a worse position nutritionally than they were 40 years ago. Also the dramatic rises in the price of food over the last couple of years (most significantly during 2007) has impacted the ability of people in many countries to buy even their basic food needs.

Agriculture has changed dramatically, especially since World War II. In mid 1970s in our country, the production of food and fibre crops and animal products has increased markedly due to new farming technologies, including new fertilizer and irrigation responsive high-yield varieties of crops, partial modernization of agriculture, and the increased application of chemical pesticides. Government policies have also encouraged farmers to maximise production. These changes have allowed fewer farmers to produce the majority of the world's agricultural crops.

Although these changes have had many positive effects and reduced many risks in farming, there have unfortunately also been some significant costs. These include: the clearing of native vegetation, soil erosion, and the decline in soil fertility, groundwater contamination, increasing costs of production and declining returns on family farms, and the closing of services in rural communities. In many cases, these changes have forced small farmers to leave the land and move to urban areas.

The sustainable agriculture movement has grown in the past three decades as a solution to the environmental and social problems caused by conventional agricultural systems and practices. Sustainable agriculture is getting increasing support within mainstream agriculture also. This is because sustainable agriculture also offers innovative and economically viable opportunities for farmers, consumers, policymakers and many others in the food system.

The definition for sustainable agriculture suggested and recommended an holistic approach to farming that means 'comprehensive' and 'integrated' taking all possible elements and systems into account and valuing their interdependent relationships.

A holistic approach to agriculture would recognise the linkages between the soil, vegetation, air and water and the way that these both influence, and are influenced by, the farmer's beliefs, perceptions, ambitions, skills and knowledge, and the social, economic, cultural, and political systems in which the farm operates.

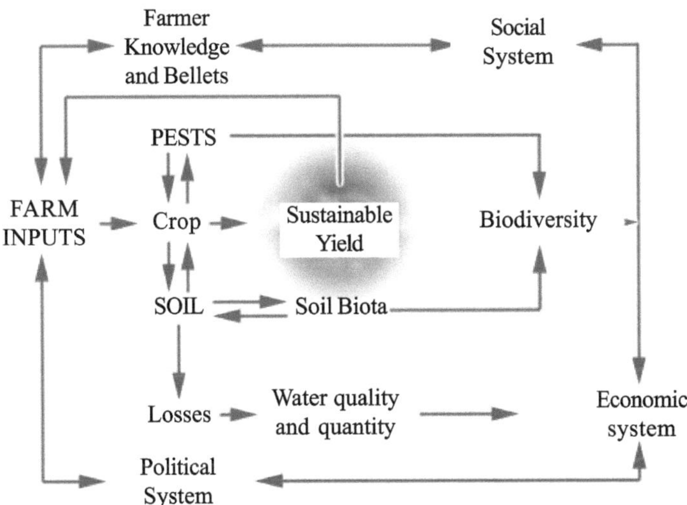

Fig 1: The holistic nature of sustainable agriculture in a diagram

An important connection between the natural and socio-economic systems that influence agriculture is between these systems and the farming methods that are used. There are many ways to improve the sustainability of a farming system with location-specific variations. Farmers trying to take a more sustainable approach share some common practices. All these practices contribute to long-term farm profitability, environmental stewardship and rural quality of life.

Integrated Pest Management (IPM)

IPM manages pests by combining biological, cultural, physical and chemical tools in a way that minimises economic, health and environmental risks.

Rotational Grazing

Management-intensive grazing systems take animals out of barns to pastures to provide high-quality forage and reduced feed costs while avoiding manure buildup.

Soil Conservation

Many soil conservation methods, including strip cropping, reduce tillage and "no-till," help prevent loss of soil due to wind and water erosion.

Water Quality/Wetlands

Water conservation and protection are important parts of sustainable agriculture. Many practices improve quality of drinking and surface water, as well as to protect wetlands. Wetlands play a key role in filtering nutrients and pesticides, in addition to providing wildlife habitat.

Cover Crops

Growing plants such as rye or clover in the off season after harvesting a grain or vegetable crop provides many benefits, including weed control, erosion control, and improved soil nutrients and soil quality.

Crop/Landscape Diversity

Growing a greater variety of crops on a farm reduces risks from extremes in weather, market conditions or crop pests. It also contributes to soil conservation, wildlife habitat and increased populations of beneficial insects.

Nutrient Management

Central management of nitrogen and other plant nutrients improves the soil and protects the environment. Increased use of on-farm nutrient sources, such as manure and leguminous cover crops, reduces the need to buy fertilizer.

Agroforestry

Agroforestry covers a range of tree uses on farms, including interplanting trees with crops or pasture, better managing woodlots, and using trees and shrubs along streams as riparian buffer strips.

Marketing

Farmers often find that improved marketing provides a key way to enhance profitability. Direct marketing of agricultural goods to consumers, such as farmers' markets, roadside stands and community-supported agriculture, is becoming much more common.

The cost of modernization of agriculture

The process of agricultural modernisation has had an important influence on farm productivity and improved living standards for many farmers. However, farmers need access to: modern seeds, water, labour, capital or credit, fertilisers and pesticides. Many poorer farming households simply cannot adopt the whole package. If one element is missing, the seed delivery system fails or the fertiliser arrives late, or there is insufficient irrigation water, then yields may not be much better that those for traditional varieties. Even if farmers want to use external resources, very often delivery systems are unable to supply them on time.

Where production has been improved through theses modern technologies, all too often there have been adverse ecological and social impacts.

Ecological costs

Many ecological problems have increased dramatically in recent years. These include:

- contamination of water by pesticides, nitrates, soil and livestock wastes, causing harm to wildlife, disruptions of ecosystems and possible health problems in drinking water;
- contamination of food and fodder by residues of pesticides, nitrates and antibiotics;
- damage to farm and natural resources by pesticides, causing harm to farmworkers and public, disruption of ecosystems and harm to wildlife;
- contamination of the atmosphere by ammonia, nitrous oxide, methane and the products of burning, which play a role in ozone depletion, global warming and atmospheric pollution;

- overuse of natural resources, causing depletion of groundwater, and loss of wild foods and habitats, and of their capacity to absorb wastes, causing waterlogging and increased salinity;
- the tendency in agriculture to standardise and specialise by focusing on modern varieties, causing displacement of traditional varieties and breeds;
- new health hazards for workers in the agrochemical and food-processing industries.

Social costs

Agricultural modernisation has also helped to transform many rural communities, both in the South and the North. The process has had many social impacts. These include the loss of jobs, the further disadvantaging of women economically if they do not have access to the use and benefits of the new technology, the increasing specialisation of livelihoods, the growing gap between the well-off and the poor, and the cooption of village institutions by the state.

The success of India's agriculture is attributed to a series of steps that led to availability of farm technologies which brought about dramatic increases in productivity in 70s and 80s often described as the so called green revolution. The major sources of agricultural growth during this period were the spread of modern crop varieties, intensification of input use and investments leading to expansion in the irrigated area. In areas where 'Green Revolution' technologies had major impact, growth has now slowed. New technologies are needed to push out yield frontiers, utilize inputs more efficiently and diversify to more sustainable and higher value cropping patterns. At the same time there is urgency to better exploit potential of rainfed and other less endowed areas if we are to meet targets of agricultural growth and poverty alleviation. Given the wide range of agroecological setting and producers, Indian agriculture is faced with a great diversity of needs, opportunities and prospects. Future growth needs to be more rapid, more widely distributed and better targeted. These challenges have profound implications for the way farmers' problems are conceived, researched and transferred to the farmers. On the one hand agricultural research will increasingly be required to address location specific problems facing the communities on the other the systems will have to position themselves in an increasingly competitive environment to generate and adopt cutting edge technologies to bear upon the solutions facing a vast majority of resource poor farmers.

Sustainability concerns

Several indicators highlight increasing concerns of sustainability in areas which have largely contributed to increased production in the 'Green Revolution' era.

Adoption of high yielding cultivators is virtually complete. Almost entire wheat and rice crops in the states of Punjab, Haryana and Western Uttar Pradesh are irrigated. In the higher production regions yields are plateauing and most traditional sources of productivity growth having been exhausted future gains in production have to come from elsewhere.

At farmers' level concerns are being expressed in several ways. Many farmers believe that the input levels have to be continuously increased in order to maintain high yields. In sixties and seventies most farmers used only nitrogenous and phosphate fertilizers to achieve high yields. Due to widespread deficiencies of several secondary and micronutrients , most farmers now have to apply higher doses and a greater variety of fertilizers to maintain crop yields. Results from many long term studies on rice-wheat cropping system showed a declining yield trend when input levels were kept constant – thus the growth rate of system productivity has been declining relative to growth rate of nutrients use. Lowering of groundwater tables due to intensive rice-wheat system in many areas is resulting in increased costs of lifting water in the intensively cultivated high production areas, diseases and pest problems are turning more serious than ever before and pose both short and long large problems. It is reported that some weeds have developed resistance to the commonly used herbicides. What this implies is that the farmers are applying increasing amount of herbicide incurring increasing cost without the benefit of effective control. Pesticide residues entering the food chain and overall safety in use of pesticides continue to be serious problems.

Other emerging problems threatening sustainability of intensive cropping system e.g. rice-wheat include loss in biodiversity related issues. Large areas planted to single/few varieties of a crop is a potential cause of concern. As the diversity is reduced natural processes that control and affect habitat quality and genetic expression weaken and for this reason internal and natural control mechanisms must be replaced by more externally applied artificial controls in the form of management and inputs which in due course lead the system towards unsustainability.

Groundwater is the major source of meeting the irrigation needs of irrigated agriculture. Currently about half the area under irrigation in the country is irrigated from groundwater sources. Large-scale groundwater development has led to fall in the water table in many areas. Over pumping is leading to declining water table levels and failure of tube-wells. Pumping costs are increasing, as is the energy consumption. In the coastal areas this has led to ingress of sea water, with serious environmental implications.

Changes in water quality are adversely affecting agriculture and vice-versa. Inefficient and/or over use of fertilizers and pesticides in agriculture and untreated disposal of industrial and urban wastes are leading to increasing contamination by such elements as lead, zinc, copper, chromium, cadmium particularly in areas having high industrial activity e.g. in districts of Ludhiana, Faridabad, Kanpur, Varanasi etc.

An increase in the content of arsenic has been reported in several of the districts of West Bengal. This is attributed amongst other causes to the lowering of groundwater table due to excessive groundwater withdrawal and is leading to serious and widespread toxicity problems adversely affect the health of hundreds of thousands people of the region.

Changing land-use and future of agriculture

One of the most important consequences of growing pressure on land is the declining trend in the average farm size and the pattern of holdings (Tables 10 and 11). According to the latest Agricultural Census in 1970-71 there were 70 million holdings operating 162 million ha. By 1990-91 there were 105 million holdings operating 165 million ha. The average farm size decreased from 2.30 ha in 1970-71 to 1.57 ha in 1990-91. As of 1990-91 about 78 percent of holdings were small (1.0 to 2.0 ha) and marginal (<1.0 ha). A little more than 20 percent of the farmers were semi-medium (2.0 to 4.0 ha) and medium (4.0 to 10.0 ha). Large farmers (>10.0 ha) constituted only 1.6 percent of the total holdings. Over the twenty-year period since 1970 the proportion of marginal farmer has increased from 50 to 59 percent and that of large farmer has declined from about 4 to 1.6 percent. The proportion of total area operated by marginal farmers increased from nine percent in 1970-71 to nearly 15 percent in 1990-91 while the proportion of large farmers declined from about 31 percent to 17 percent in the same period. The size of average holding is very unevenly distributed among the states. States with relatively large average size of operational holding are a mixed lot – they include states with large tracts of barren lands e.g. Rajasthan, Maharashtra and Madhya Pradesh on the one hand and agriculturally advanced state like Punjab on the other. These trends in farm size changes will have a profound effect on the future agricultural development strategies.

Although India's population growth rate has slowed from 2.1 percent in 1980s to 1.8 percent in the 1990s and is expected to slow further in the coming decades, yet the population is projected to reach 1.33 billion by 2020 from the current one billion. The urban share of total population is projected to increase from 26 percent to 35 percent of the total population. Although incidence of poverty is falling, it is estimated that in 93-94 (upto which data is available) 320 million people constituting 36 percent of the population were below the officially defined poverty line.

The nature of the poverty line has been shifting. About 30 years ago 48.4 percent of those living in rural areas were poor and 20 percent of those living in the urban areas were classed as poor. Recent studies show that the number of poor in urban areas has been increasing at relatively higher rate compared to the rural areas. At present those below the poverty line in rural sector constitute 37 percent of the population while in the urban sector the percentage is 32 percent. In the context of poverty alleviation, therefore, emphasis will be required to be placed both on production of food by the poor as well as on the availability of food for the urban poor. It needs to be recognized that a large proportion of the rural poor are located in regions of low potential for food production e.g. arid and semi-arid areas, hilly regions, degraded land and forest areas. Widespread hunger and malnutrition are the direct manifestation of poverty and will call for increasing efforts to produce more food at affordable price.

Increasing population and economic growth are changing patterns of land use making potentially unsustainable demands on the country's natural resources.

- Since early fifties the net area sown was expanded rapidly at first but at a diminishing rate since 1970 to reach approximately 142 million ha at present. During 1950s and 1960s areas under agriculture expanded substantially as the fallows were reduced and cultivable wastes were put under the plough. The net area sown increased from 119 million ha in 1950-51 to 133 million ha by 1960-61 and further to 140 million ha by 1970-71. Fallow lands declined from 28 million ha in 1950-51 to 20 million ha by 1970-71. Cultivable wastelands declined from 23 to 17.5 million ha.
- Land use intensity i.e. fraction of net sown area to total geographical area increase from 36 percent in 1950-51 to 40.5 percent in 1960-61 and 43 percent by 1970-71 where it has since stabilized.
- Cropping intensity i.e. gross sown area as percent of net sown area increased from 111 percent in 1950-51 to 115 percent in 1960-61, 118 percent in 1970-71 and 130 percent by mid 1990s.
- While the contribution of increased area in the growth of agriculture has declined over time, that of productivity has increased. The yield of all crops grew at 1.5 percent per annum between early 1950s and mid-1960s, the pace accelerated to 1.7 percent in the 1970s and then to 3 percent per annum between early 1980s and mid-90s. Unlike the gains in area, which benefited non-foodgrains, the gains in productivity accrued mostly to foodgrains.
- India's forest resources have been dwindling. According to the 'State of Forest Report' (1997) the total forest cover of the country is estimated at

63.34 million ha i.e. 19.27 percent of the geographic area of the country. Of these the dense forest (crown density more than forty percent) and open forest (crown density 10 to 40 percent) occupying about 11 and 8 percent of the geographic area respectively and mangroves occupy 0.15 percent of the geographic area. The country has lost about 5482 sq. km. of forest cover since the 1995 assessment. By any estimate the area under forest is far below the national policy goals and many areas nominally under forest are being used for non-forest purposes. Similarly 'uncultivated lands' such as permanent pastures, miscellaneous tree crops, cultivable wastes and fallow is subject to increasing competition from uses other than feeding livestock.

- The growth of livestock population is an important source of competition for land. The increases in number of major classes of livestock are shown in Table.
- The area sown to fodder crops is not recorded. Information available from other sources provide an estimate ranging from 4 to 5.5 percent of the net sown area and suggest that the area under fodder crops will have to increase to 10 percent or more to support increasing livestock based activity.

Table 1: Livestock and poultry population (million)

Category	1982	1987	1992
Cattle	192.45	199.69	204.53
Buffaloes	69.78	75.96	83.49
Sheep	48.76	45.70	50.79
Goats	95.25	110.20	115.28
Total (Livestock)	419.59	445.28	470.15
Total Poultry	207.74	275.32	307.10

Source: I P Ibrol, 2012

Table 2: Per capita net availability of cereals and pulses per day (g)

Year	Cereals	Pulses	Total
1951	334	61	395
1961	400	69	469
1971	418	51	469
1981	417	38	455
1991	468	42	510
1998	451	33	484

Table 3: Production of milk, eggs and fish

Year	Milk production/availability million tonnes/annum	g./day	Fish (000 tonnes)	Eggs (million)
1950-51	17.0	124	752	1832
1960-61	20.0	124	1160	2881
1970-71	22.0	112	1756	6172
1980-81	31.6	128	2442	10060
1990-91	53.9	176	3836	21101
1998-99	75.0	210	5388	28400

Table 4: Number of holdings (million)

Category of Farmers	1970-71	1980-81	1990-91
Marginal	35.7 (50.6)*	50.1 (56.4)	62.1 (59.0)
Small	13.4 (19.0)	16.1 (18.1)	20.0 (19.0)
Semi-medium	10.7 (15.2)	12.5 (14.0)	13.9 (13.2)
Medium	7.9 (11.2)	8.1 (9.1)	7.6 (7.3)
Large	2.8 (3.9)	2.2 (2.4)	1.7 (1.6)
All groups	70.5 (100)	89.0 (100)	105.3 (100)

**Figures in parenthesis represent the percentage of holdings*

Table 5. Area operated - % of total area

Category of Farmers	1970-71	1980-81	1990-91
Marginal	9.0	12.1	14.9
Small	11.9	14.2	17.3
Semi-medium	18.5	21.1	23.1
Medium	29.7	29.6	27.2
Large	30.9	23.0	17.4
All groups (Million ha)	162.1 (100)	163.8 (100)	165.6 (100)

Table 6: India – Land Use Categories (million ha)

Classification	1960-61	1970-71	1980-81	1990-91	1993-94
Reporting area	298.5	305.0	305.0	305.0	305.0
Area under non-agricultural uses	14.8	16.5	19.7	21.1	22.0
Barren and un-cultivable	35.9	28.2	20.0	19.4	19.0
Cultivable waste	19.2	17.5	16.7	15.0	14.5
Old fallow	11.2	8.8	9.9	9.7	9.7
Forests	54.1	63.9	67.5	67.8	68.4
Permanent pastures	14.0	13.3	12.0	11.8	11.2
Land under misc.trees etc.	4.5	4.3	3.6	3.8	3.7
Net sown area	133.2	140.3	140.0	143.0	142.1
Current fallow	11.6	11.1	14.8	13.7	14.3

The pressure on India's land and water resources is seriously threatening native plant and animal diversity. India has uniquely rich and diverse genetic base. With increasing agriculture and economic development the genetic pool is declining. This decline, if unchecked and poorly managed can have unforeseen and adverse consequences for the sustainability of agriculture of the region.

Agriculture in the changing global scenario

Steady globalization of trade has profound implications for future agricultural development. The diversity of India's agro-ecological setting, high bio-diversity and relatively low cost of labour provide potential for agricultural competitiveness in a globalized economy. It is expected that with increasing globalization of markets over the years there will be demands for agricultural intensification. This will also be favoured because of greater backward and forward linkages between agriculture and food industry. Therefore, increase in production and productivity are bound to be strategically important to economy. Intensification will not only favour alleviation of rural poverty but will also improve resource conservation particularly in the small farming sector where farmers can be encouraged to take up organized production of high value crops such as fruits, specialty vegetables, flowers medicinal and aromatic herbs etc. Stronger demands for crops of the small farmers' will not only improve incomes and welfare but will also make investments in technology and resource conservation more attractive.

The General Agreement on Tariff and Trade (GATT) and liberalization of global trade is bound to have impact on future land use and production pattern. Understanding the local, national and international environment under which agricultural production is taking shape will be crucial in developing our own strategies.

Extension strategies

Since early fifties a number of public by funded agricultural development programmes have been sponsored. These have included programmes like the National Extension Service (NES) Blocks in 1953, the Intensive Agricultural District Programme (IADP) in 1961-62, the Intensive Agricultural Area Programme (IAAP) 1964-65, the High Yielding Variety (HYV) programme 1966-67 and the Small and Marginal Farmers' Development Programmes (SMFDP) in 1969-70. Though these programmes had a perceptible impact the efforts did not get replicated over different areas and categories of farmers. In mid-seventies based on pilot level project in Rajasthan Canal and Chambal command area a 'Training and Visit' (T&V) system of extension was promoted in different states. Extension efforts of the Indian Council of Agricultural

Research through its research Institutes and the State Agricultural University were largely limited to demonstration of new technologies through such programmes as National Demonstration Project, Operational Research Project, the Lab to Land Programme and the KrishiVigyanKendras. However, there appears much to be desired in the way that extension programmes are conceived and implemented.

At present extension programmes are implemented in largely a top-down fashion leaving little scope for localized planning and action. Farmers are almost passive receivers and their involvement in the process of technology generation and adoption is almost absent. Extension services, at present, are almost exclusively in the public sector domain and there is no effort or institutional support for other operators e.g. the NGOs, the corporate bodies etc.

Extension programmes sponsored by the government operate largely in isolation and there appears a strong need to view the extension programmes as an integral part of the research and development process.

The challenges facing agricultural development call for fundamental changes in our approach to technology transfer/extension programmes. Changes are necessary in the context of changing economic environment following policy adjustments in relation to privatization, deregulation and globalization calling for greater efficiency and effectiveness of the extension system. More importantly there is need for

- Greater emphasis on providing producers with knowledge and understanding needed to overcome the problems or to exploit opportunities of their own specific production systems. Correspondingly there will be a need to de-emphasize 'package of practices' or the blanket recommendations, top down approach followed thus far.
- Shift in the focus of public extension systems from promoting inputs use to one on sustainable management of resources and improvements in the production system as a whole.
- Closer interaction between farmers, extension scientists and production system researchers in diagnosing problems and identifying location specific recommendations emphasizing participation and education rather than being prescriptive.
- Widening the range of extension delivering agencies. While the publicly operated extension systems will continue to be important, there will appear a greater role for NGOs, farmers' associations and corporate sectors in particular situations. Role of commercial suppliers of seeds, agrochemicals, machinery, vaccines and medicines in providing advisories, as is already

being done in a limited way, will need to be encouraged and factored into public system's own priorities.

- Wider and more creative use of mass media in tune with current developments in information technology to get information across to the farming community whose ability to overcome constrains at farm level will increasingly depend on access to reliable and up-to-date information.

Technological needs and future agriculture

It is apparent that the tasks of meeting the consumption needs of the projected population are going to be more difficult given the higher productivity base than in 1960s. There is also a growing realization that previous strategies of generating and promoting technologies have contributed to serious and widespread problems of environmental and natural resource degradation. This implies that in future the technologies that are developed and promoted must result not only in increased productivity level but also ensure that the quality of natural resource base is preserved and enhanced. In short, they lead to sustainable improvements in agricultural production.

Productivity gains during the 'Green Revolution' era were largely confined to relatively well endowed areas. Given the wide range of agroecological setting and producers, Indian agriculture is faced with a great diversity of needs, opportunities and prospects. Future growth needs to be more rapid, more widely distributed and better targeted. Responding to these challenges will call for more efficient and sustainable use of increasingly scarce land water and germplasm resources.

Technical solutions required to solve problems will be increasingly location-specific and matched to the huge agro-ecological/climatic diversity. Detailed indigenous knowledge and greater skills in blending modern and traditional technologies to enhance productive efficiency will be more than ever before, key to the farming success and sectoral growth. Most technological solutions will have to be generated and adapted locally to make them compatible with socio-economic conditions of farming community.

New technologies are needed to push the yield frontiers further, utilize inputs more efficiently and diversify to more sustainable and higher value cropping patterns. These are all knowledge intensive technologies that require both a strong research and extension system and skilled farmers but also a reinvigorated interface where the emphasis is on mutual exchange of information bringing advantages to all. At the same time potential of less favoured areas must be better exploited to meet the targets of growth and poverty alleviation.

These challenges have profound implications for products of agricultural research. The way they are transferred to the farmers and indeed the way research is organized and conducted. One thing is, however, clear – the new generation of technologies will have to be much more site specific, based on high quality science and a heightened opportunity for end user participation in the identification of targets. These must be not only aimed at increasing farmers' technical knowledge and understanding of science based agriculture but also taking advantage of opportunities for full integration with indigenous knowledge. It will also need to take on the challenges of incorporating the socio-economic context and role of markets.

With the passage of time and accelerated by macro-economic reforms undertaken in recent years, the Institutional arrangements as well as the mode of functions of bodies responsible for providing technical underpinning to agricultural growth are proving increasingly inadequate. Changes are needed urgently to respond to new demands for agricultural technologies from several directions. Increasing pressure to maintain and enhance the integrity of degrading natural resources, changes in demands and opportunities arising from economic liberalization, unprecedented opportunities arising from advances in biotechnology, information revolution and most importantly the need and urgency to reach the poor and disadvantaged who have been by passed by the green revolution technologies.

Another important implication of increasing globalization relates to the need for greater attention to the quality of produce and products both for the domestic and the foreign markets. This would imply that production must be tuned to actual rapidly changing product demand. Such adaptation to global markets would require state of the art research, which can be achieved only by setting global standards of research, focus on well-defined priorities and mechanisms which permit close interaction of farmers with researchers, the private sector and markets.

References

Adams WM. 2006. The Future of Sustainability: Re-thinking Environment and Development in the Twenty-first Century. The World Conservation Union, 19 p. http://cmsdata.iucn.org/downloads/iucn_future_of_sustanability.pdf (accessed 20 July 2009)

CMIE. 1997: Directory of Indian Agriculture, pp402 Centre for Monitoring Indian Economy Pvt. Ltd., Mumbai.

Indian Council of Agricultural Research, 1998: National Agricultural Technology Project, pp 193, ICAR, New Delhi.

Indian Agriculture in Brief 2000: Directorate of Economics and Statistics, Department of Agriculture and Cooperation, Ministry of Agriculture, Government of India, New Delhi.

Pretty, J. 1998.Regenerating Agriculture, Earthscan, London, p. 4.

Rowley, J. 1993.Bhorletar: The sustainable village, People and the Planet, 2 (4), pp. 14-19.

19

Women Empowerment, SHGs and Social Organization's Credo: Looking Lights from the Tribal Villages of East Singhbhum

Atanu Deb[1] *and Arunava Sengupta*[2]

[1]*CUTM, Paralakhemundi-761211, India*
[2]*IRTDM Faculty Centre, RKMVU, Ranchi- 834008, India*
Email: atanu.literacyindia@gmail.com

Abstract

'Women Empowerment' has been now an important phrase of discussion for the social scientists. The concept of 'Empowerment' was firstly highlighted by the Political Scientists. Undoubtedly, effective utilization of human potentials and realization of knowledge about him/herself are believed to be the most critical capabilities of individuals. Probably this is the root of 'Empowerment'. The term 'Women Empowerment' symbolizes women's realization about their potentials. On the other hand, SHG approach intends to highlight towards unfolding the potentials of women through group approach, enabling them to help themselves by their own capacity. In spite of this, the condition of women SHGs is in doldrums. This might be quite difficult to generalize the whole scenario. The study is a modest endeavor to revisit women SHGs (developed by NGOs) of East Singhbhum districts of Jharkhand. Of the various factors attributed to tribal socio-economic lives of family based farmers, gender role is considered as the important factor. Tribal women in almost all walks of life are sharing responsibilities in domestic as well as economic sector.The study is based on Ex- post facto research design. The methods of data collection were quantitative and qualitative in nature and were aimed to seek answers to certain research questions. Participatory Rural Appraisal and FGD are being the dominant methods. Keeping a foreside to the main objectives, the study emphasizes on socio-economic, political, entrepreneurial aspects of empowerment. Finally, the study suggests some measure against the problems and constraints of the SHGs.

Keywords: SHG; FGD; women empowerment; participatory rural appraisal; Jharkhand; gender role

Introduction

Women Empowerment

'Women Empowerment' has been now an important phrase of discussion for the social scientists. The concept of 'Empowerment' was firstly highlighted by the Political Scientists. Undoubtedly, effective utilization of human potentials and realization of knowledge about him/herself are believed to be the most critical capabilities of individuals. Probably this is the root of 'Empowerment'. The term 'Women Empowerment' symbolizes women's realization about their potentials.

After independence all round efforts have been made to promote the welfare of the women. Article 15 of Constitution confers the equal rights of both sexes, though the state could make any special provisions for women (Kabeer, 1998). The introduction of English education and the contact with the western nations kindled the rationalistic spirit of the people (Mishra, 2006). Biju, (2006) opined that social differentiation & inequality are universal phenomenon. Today gender has become a common problem all over the world. However the form & extent of gender problem varies from society to society, the gender issue in both these societies is found in different levels.

According to World Bank, (2001)Since the 1990's women have been identified as key agents of sustainable development and women's equality and empowerment are seen as central to a more holistic approach towards establishing new patterns and processes of development that are sustainable. The World Bank has suggested that empowerment of women should be a key aspect of all social development programs.

'Feminism of Poverty': Not an illusion

In recent past there has been a trend to analyse poverty in the context of gender issues. The concept of 'Feminism of Poverty' critically linked poverty with women related issues. Now role of women varies from society to society. But 'Feminism of Poverty' often proves that women empowerment is one of the strong weapons to struggle against poverty. Let's check some statistics. A

1992 UN report disclosed that "the number of rural women living in poverty in the developing countries has increased by almost 50% over the past 20 years to an awesome 565 million — 374 million of them in Asia, and 129 million in Sub-Saharan Africa. ... While poverty among rural men has increased over the last 20 years by 30%, among women it has increased by 48%".

Women in India contribute 89 per cent of the unrecognised sector and it is reality that their participation rate is higher in rural areas. Women responsible for one-third of the labour force of India and 90 per cent of the rural-urban women are unskilled workers. Hence, women empowerment could be a mechanism to pave the way for integrated development. Although, gender discrimination has been banned by the Constitution and women have been guaranteed political equality with men, yet there is a difference between the constitutional rights and the rights enjoyed in reality by women. Even after half a century of independence, barring a few exceptions, women have mostly remained outside the domain of power and political authority.

Bhuyan, (2006) highlighted the issue in different context. He opines that although they constitute about half of the citizen and over the years their participation by way of voting has increased, yet their participation and representation in law making and law implementing bodies are not very satisfactory. No doubt the 73rd and 74th constitutional amendment acts have provided access to women in the decision making process at the grass-root level but their representation in the Parliament and state legislatures is woefully poor. Insecurity does not allow the women leaders to identify leadership at the grass-root level. In politics when a man proposes, they themselves depose. In reality women representatives are ornamental in nature and political consciousness is found lacking among them. They are affected by the caste and class divisions, feudal attitudes, patriarchal nature of the family and village-social, environmental, ethnic, religious separatism and the like. They are members on record only. Allegedly, they are not consulted while taking decision. Thus, women representatives are not free from male dominance in the village administration and no significant change in the power equal is observed in the villages.

SHG: Concept and Context

Indhumati and Palanivelu (2013) stated that Self-Help Groups are increasingly accepted as agent by social changes, development and empowerment among the women. SHG's approach represents a paradigm shift from development toward empowerment. The SHG's are considered as the agents of transformation through the mobilization, organizations, banking in situation and voluntary organizations follow self-help approach to active rural development and empowerment of the women in India. The banking institutions channelize their credit to poor through the SHG's.

Self Help Groups are small homogeneous group of individual members who voluntarily come together and form an association for the purpose of solving their common problems through self-help and mutual help and for achieving a common objective. In most cases SHGs are constituted by persons known to one another and coming from the same village, community or neighbourhood, that is, SHGs are small in size with membership ranging from 10 to 20 are homogeneous (poor, financially weak, same socio-economic background etc.) and have certain pre groups binding factor. These groups start and promote with small saving in a common group-fund and kept in a bank. (Vadivoo and Sekar, 2004).

Research Objectives: With these rationales at the backdrop, the objectives for the study were:

- To analyze women empowerment in the light of pre and post SHG situation
- To find out the role of institutions in women empowerment through SHG
- To trace out major constraints of the SHGs

Methodology

Research design: The study is based on Ex- post facto research design. The methods of data collection were both quantitative and qualitative in nature and were aimed to seek answers to certain research questions. Participatory Rural Appraisal and FGD are being the dominant methods. Keeping a foreside to the main objectives, the study emphasizes on socio-economic, political, entrepreneurial aspects of empowerment. Participatory approach is adopted to analyse the whole situation.

Locale of the study area: The study was conducted in the selected areas of Patamda block of East Singhbhum district and Devipur block of Deoghar, Jharkhand.

Sampling procedure: Multistage sampling technique was adopted for the selection of study area and sample respondents for collection of information required for the study.

District	Blocks	GP	Respondents
East Singhbhum	Patamda	Bangurda Goberghuri	90
	Musabni	Benashole Bhobui	
Deoghar	Devipur	Dhobana Jhundi	90
	Karwon	Badhanadih, Badia	
Total Respondents			180

Findings

Women Empowerment: In the light of Pre and Post SHG situation

Social Empowerment: The present study intends to find out the social empowerment of the women, residing at Deoghar and East Singhbhum district on the basis of social empowerment indicators like employment selection freedom, decision making ability, social equity, employment skill, ability to defend themselves and ability to participate in social issues, autonomy in decision making.

Cobweb Diagram: It is a participation wheel, popularly known as evaluation wheel. The method deals with visual representation of perception of the masses to reveal the relative social status of SHG and Non-SHG members. Here 'Cob Web Diagram' has been utilized as an evaluation tool to measure the social empowerment of women in pre and post SHG situation. The empowerment issues were selected on the basis of pre-determined variables.

Table 1: Empowerment: Social point of view

Variables	Description
1. Skill of Problem solving	The capacity to resolve the problem of the family. Problems are related to social, economic & any other factors.
2. Betterment of social status	The members get social recognition, social understanding and higher social set up and higher place in the community. Anyone of these factors is considered improvement of social status.
3. Employment selection freedom	The women are free to select any employment channel according to their discretion.
4. Decision making ability	Decision making is mental process based on conscious reasoning. In this study, it refers to the participation of women of the family in making decision in all the important matters of home & economic activities.
5. Ability to defend	It is the ability or right to defend herself. The capacity is developed through SHG.

Source: Dr. Shivambhari Mishra viz. 'Empowering Men & Women Through Self Help Group'.

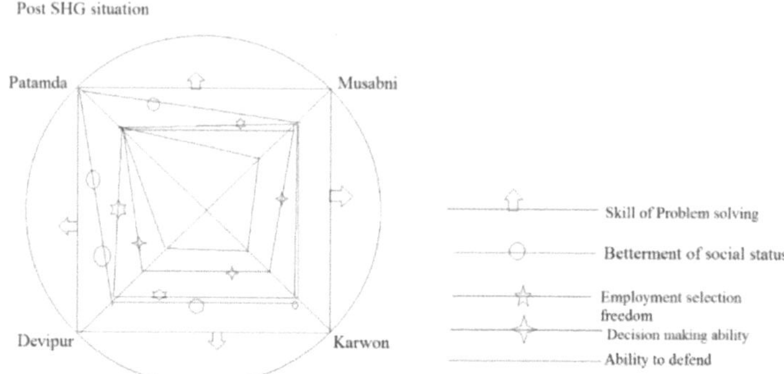

Fig. 1: Cobweb Diagram of Social Empowerment

Firstly the purpose of the PRA exercise was explained. Then participants were requested to provide score against each of the five variables. They agreed to score each of the groups on the five indicators on a five point scale. Scores obtained from separate blocks of Jharkhand, had been transferred into a 'Cobweb Diagram'. Respondents of four blocks have been represented by the four quadrants. These four arms of the diagram are considered as four scales. Performance of each block has been marked using different symbols on the indicators.

The size of the trapezoid thus indicated the empowerment status of women of different blocks on the particular indicator. Thus, bigger the size of the symbolic line (indicator) symbolizes better social empowerment status of the women. Women performed well in skill of problem solving, social status, and selection of employment. Women of Patamda and Musabni (East Singhbhum) are more socially empowered in comparison to other blocks (Deoghar).

Economic Empowerment

Economic empowerment is the ability to access, own and control resources. It can be measured in a variety of ways, using outcome indicators such as income generation, ownership of assets and land, expenditure patterns, degree of participation in paid employment, division of domestic labour and control over financial decision-making.

In post SHG situation, the women respondents were able to meet up financial requirements of family like paying children's fee for school and private tutor, electric bill, medical expenses, household items etc. Some of them were even able to manage regular savings by deciding on priorities heads of expenses. Thus, at times of family need, they were able to contribute from the savings which improved their status and respect in families.

Impact on income

Income is one of the important indicators to investigate the level of living of the members of the society. The data regarding the average house-hold income of beneficiaries was estimated in two point time i.e. 'Pre' and 'Post' SHG situation. Table 2 shows the income of the SHG members before and after joining the SHGs. Table 2: Income of the SHG members before and after joining the SHGs

Rank Test is applied to test the significance of difference between the annual incomes of SHG members before and after joining SHG in table above.

Table 3: Wilcoxon Signed Rank Test

Sl.No.	Income (Rs/Month)	Before SHG		After SHG	
		No. of Respondents	Percentages	No. of Respondents	Percentages
1	Less than 1000	15	8.33	0	0
2	1000-2000	65	36.11	2	1.12
3	2000-3000	55	30.56	65	36.12
4	3000-4000	45	25.00	45	25.00
5	4000-5000	0	0	55	30.57
6	5000-6000	0	0	10	5.58
7	Above 6000	0	0	3	1.71

Test statistics (b)

	Annual Income Before	Annual Income After
Z Asymp. Sig. (2-tailed)	-8.867[a]	.000

a. Based on negative ranks

b. Wilcoxon Signed Rank Test

Since the asymptotic significance value is less than 0.05, we reject the null hypothesis that the there is no difference in income before and after joining SHGs. The number negative ranks for Income after Joining SHG–Income before Joining SHG is given in the following table. It is trivial from the table that for all the respondents Income after Joining SHG is higher than Income before Joining SHG.

Ranks

	N	Mean Rank	Sum Rank
Negative Ranks Annual Income			
Before- Positive Ranks Annual	100[a]	57.50	5750.00
Income After Ties	0[b]	.00	.00
Total	0[c]		
	100		

a. Annual Income Before < Annual Income After

b. Annual Income Before > Annual Income After

c. Annual Income Before = Annual Income After

Chi Square Test for Economic Empowerment

The respondents level of empowerment on the basis of their experi-ence as members of the SHG analysed, among them 10 respondents had a low level of empowerment, 156 respondents had a medium level of empowerment and 14 respondents had a high level of em-powerment. To evaluate the empowerment of women entrepreneur through Self Help Groups, chi-square test has been used. The table value of x^2 at 0.05 level of significance at 6 degrees of freedom was 13.67. As the observed value i.e. 83. 67825 is greater than the table value i.e. 13.67, so it can be said that women were economically empowered through SHG. It revealed that the SHG had been successful in providing economic empowerment to their members.

Economic Empowerment in Post SHG situation: Cobweb Diagram: Here 'Cob Web Diagram' has been utilized as an evaluation tool to measure the social empowerment of women in pre and post SHG situation. The empowerment issues were selected on the basis of pre-determined variables.

Table 2: Empowerment: Economic point of view (Source: Dr. Shivambhari Mishra viz. 'Empowering Men and Women Through Self Help Group'.

Variables	Description
1. Self-financing	The economic activities are carried with the use of own money. The loan given by SHG to its members is included under self-financing of the groups although the members have undertaken individual activities.
2. Development in saving habit	They started to save money for future use
3. Control over property	The control over movable and immovable property makes them empowered. They got ownership of the property, right to use the property or dispose the property for improving economic well-beings.
4. Develop family economic activity	Members of SHG develop their economic activities. They get economic empowerment because of achievement, safety & security.

Firstly the purpose of the PRA exercise was explained. Then participants were requested to provide score against each of the five variables. They agreed to score each of the groups on the five indicators on a five point scale. Scores obtained from separate blocks of Jharkhand, had been transferred into a 'Cobweb Diagram'. Respondents of four blocks have been represented by the four quadrants. These four arms of the diagram are considered as four scales. Performance of each block has been marked using different symbols on the indicators. The size of the trapezoid thus indicated the empowerment status of women of different blocks on the particular indicator. Thus, bigger the size of the symbolic line (indicator) symbolizes better economic empowerment status of the women. Women performed well in Self-financing and saving habit. Women of Patamda and Musabni (East Singhbhum) are more economically empowered in comparison to other blocks at Deoghar.

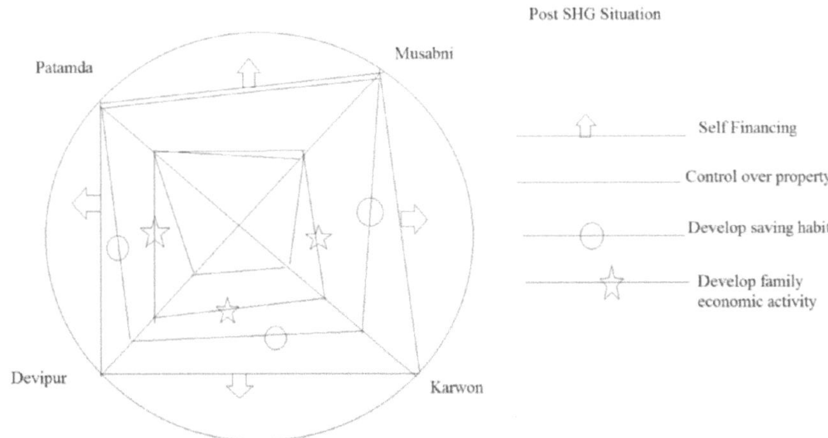

Fig. 2: Cobweb Diagram: Economic Empowerment

Psychological Empowerment

Psychological Empowerment involves a sense of autonomy, competence, efficacy, meaningfulness and trust.

Garret's Ranking clearly depicts that self-employment is the main factor influencing the empowerment of women entre-preneurs through SHGs.

Major findings from Garret's Ranking

The variable 'Self – Employment' obtained highest score .The least score is awarded to 'Serving the Society'.

It is inferred that 'Social Status' and 'Group Domination' are the important factors for the SHG members.

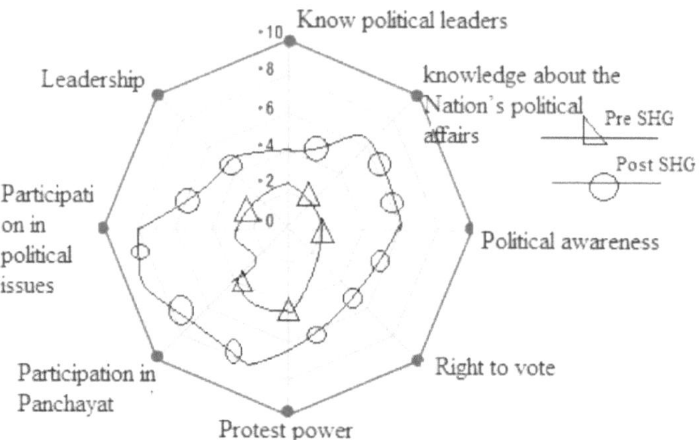

Fig. 3: Spider Diagram: Political Empowerment

Political Empowerment

Traditionally women were considered a passive worker in the society. Hence, political empowerment is needed to melt the ice. Political empowerment has direct relationship with social empowerment. This dimension considers the issues like- Political Awareness, Protest Power, and Participation in election and Panchayat meetings. Although many aspects have been considered, they do not cover all aspects of political empowerment. Moreover, due to variation in mental and socio-economic conditions, the aspects may vary from society to society. Leadership quality is now considered as the major aspect of political dimension.

After joining SHG, the members' level of awareness in political, educational and legal fields had increased. They were well informed about their local leaders and 25% of the respondents have some knowledge about the Nation's political affairs. Almost all the members even knew the legal age of marriage for boys and girls, dates of election as well as the nominated candidates. From the Figure 2, it is clear that the status of political empowerment became better in post SHG situation. 'Protest to Power' has increased significantly.

Role of institutions in women empowerment through SHG

NGOs are sensitizing primary stake holders (women) and they could play a significant role in initiating a SHG and also arranging technical skill development programmes for the SHG members. Women are often able to impart technical knowledge and skill. Some NGOs are also trying to develop marketing chains in the process of establishing on an enterprise by them for more sustainability. If these inputs are not supported strongly then women are least likely to succeed in taking up any income generating activity on a sustainable basis even after having a desire and need for such activity.

Based on Focused Group Discussion (FGD) role of NGOs in women empowerment has been highlighted here.

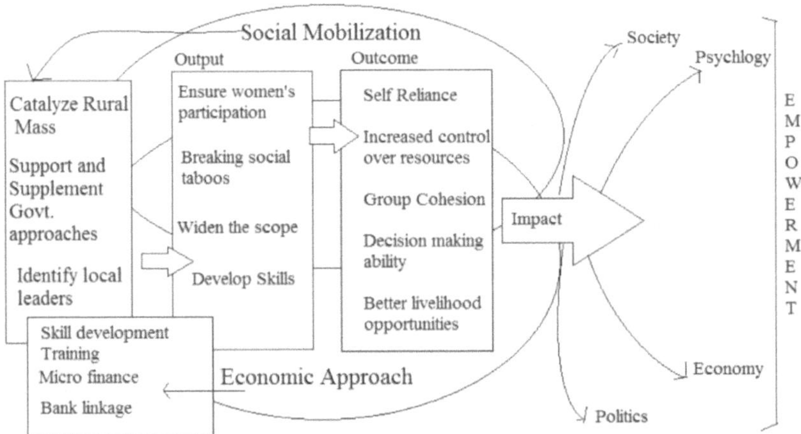

Fig. 4: Role of NGOs in women empowerment

Constraints of SHGs

Keeping in view, the relevance of SHGs in rural development and economic growth, the present study had been tried to trace out the constraints perceived by the members of SHGs.

The study draws attention on increasing importance of women empowerment and also intends to show the role of SHGs on family wellbeing and socioeconomic development as well as on SHG and NGO initiatives which have been proved to be an important tool in such developmental approaches. Analysis of qualitative and quantitative data led to draw some major constraints to comprehend the phenomenon of women empowerment in greater detail. Hence, **Garret Ranking** has been adopted to trace major constraints of SHGs.

The data obtained from the SHG members and NGO workers were tabulated systematically. Garrett's Ranking Method was used to trace and create a priority based ranking of the attributes regarding constraints of the studied SHGs.

Garret's ranking technique provides the change of orders into numerical scores. The prime advantage of this technique over simple frequency distribution is that the reasons and factors are arranged based on their importance from the point of view of the respondents. The per cent position of each rank was converted into scores referring to the Table given by Garret and (Garret and Woodworth, 1969).

Garret's formula for converting ranks into percent:

Percent position = 100* $(R^{ij}-0.5)/N^{j}$

Where, R^{ij} = Rank given for ith factor by j^{th} individual

N^{j} = Number of factors ranked by j^{th} individual

Ranking of Constraints of SHGs: Based on Garret's ranking technique

Sl.No. Constraints	Mean Score	Rank
I. **Cohesion Constraints of SHG**		
1. Non-cooperation from family members	60.8	III
2. Poor leadership quality	64.3	II
3. Lack of trust on Group Leader	64.8	I
4. Women were very busy in their family works	53.2	IV
5. Non-cooperation from group members	45.2	v
II. **Group performance related constraints of SHGs**		
1. Lack of education	58.5	IV
2. Lack of fund	70.6	I
3. No exposure in record keeping	24.6	VI
4. Unskilled members/ Lack of training	59.9	III
5. Inability to loan repayment	43.5	v
6. Problem of marketing channel for the produce	67.4	II
III. Financial Constraints of SHGs		
1. Poor linkage with Bank	54.4	III
2. Inordinate delay in loan from bank	45.3	IV
3. Travel expenses for disbursement of loan	42.5	V
4. Irregular savings	62.6	II
5. Delay in loan repayment	68.3	I

Conclusion

For centuries, women of the nation have stayed in the cave of societal taboos, have maintained their distinctive culture of living, have sustained on their family lives, and now they are struggling to overcome poor socio-economic condition. The women of this nation have stayed outside the realm of the self-motivation and self-reliance. Male dominated society for various needs, ranging from economic to social system has utilized women as the prime unit of child production. These concepts have been gradually integrated and incorporated in the social system of human beings. Domestic activities are often considered as the prime job of women.

With innovative ideas and novel approaches, SHGs could be utilized as a change maker and could be a tool to provide better opportunities to the downtrodden. In this context SHGs have evolved some simple but innovative methodologies which could be a lesson to all. Failure stories could be helpful to revitalize the situation.

By upholding the banner of Women empowerment *vis-à-vis* the success stories, the project reports of several NGOs, speaks of social, economic and political empowerment of women through SHGs. Undoubtedly the civil society movement got huge success, followed by various entrepreneurial success and intends to highlight the success stories. Since the movement started, a bundle of failure stories are still sinking in the darkness. But it's the time when some of them are asking a pertinent question; why should they tolerate this severe darkness?

References

Biju, R. 2005. Women's empowerment, Mohit Publications, New Delhi.

Bhuyan, D. 2003. Empowerment of Indian Women: A Challenge of 21st Century, RBSA, Publishers, Jaipur.

Garret, H.E. and R.S. Woodworth.1969. Statistics in Psychology and Education, Vakils, Feffer and Simons Pvt. Ltd., Bombay: 329.

Indhumati and Palanivelu 2013. Women Empowerment through Self – Help GroupGlobal Research Analysis, Volume (2): ISSN No 2277–8160.

Kabeer, N. 1998. Resources, Agency, Achievements, Reflections on the Measurement of Women's Empowerment, Discussing Women's Empowerment Theory and Practice SIDA Studies No. 3.

Vadivoo Sential, K. and Sekar, V. 2004. Self Help Groups – A Movement for Women Empowerment, Kisan World, July:13–14.

20

Biological Drainage System for Agricultural Sustainability under Salinity and Lowland Conditions

A. Zaman[1] *and Sagar Maitra*[2]

[1]*Emeritus Professor, Centurion University of Technology and Management MSSSoA, Paralakhemundi-761211, Odisha, India*
Email: aftab@cutm.ac.in
[2]*Associate Professor, Department of Agronomy, MSSSoA, CUTM Paralakhemundi-761211, India*
Email: sagar.maitra@cutm.ac.in

Abstract

Biological drainage relies on vegetation, rather than engineering mechanisms to remove excess soil water through evapo-transpiration. It is often considered attractive because it requires only an initial investment in site development (planting of a "bio-drainage crop") and (potentially) returns a benefit when the bio-crop is harvested for fodder, wood or fibre. Presently about one-third of the world's irrigated area faces the threat of waterlogging. Considering long-term viability of bio-drainage, the systems would be subjected to the same scrutiny as other plant-based biological systems with regards to nutrient and salt-balance and soil considerations. Moreover, large-scale adoption of bio-drainage crops could include considerable tracts of land and socio-economic considerations. The driving force behind the bio-drainage concept is the consumptive water use of plants. One of the major factors determining the sustainability of plant productivity (and thus plant water use) processes is *salt balance*. If the salts moving into the root zone are not either (1) taken up by the vegetation and harvested or (2) removed from the root zone by leaching, the vegetation is doomed to succumb to salinity. The deep rooting characteristics of trees make them extremely efficient users of water while shallow-rooted grasses and crops have limited access to underlying water tables, deep-rooted trees can access water tables down to several meters. Added advantages of bio-drainage and bio-disposal systems, depending on the vegetation used are:

production of timber, fuel wood, oil, fruits, fibre, contribution to carbon sequestration, diminishing the effects of wind erosion, provision of shade and shelter, function as windbreaks, yield organic matter for fertilizer, enhancement of biodiversity, as flora and fauna flourishes as well as diminishing air pollution.

Keywords: Bio-drainage; lowland conditions; waterlogged areas; soil salinity; crop productivity

Introduction

The term, 'bio-drainage' is relatively new, although the use of vegetation to dry out soil profiles has been known for a long time. The term 'bio-pumping' is used to describe the use of trees for water table control. Bio-drainage has been designed for extraction of excess water by fast growing tree species and that is removal of excess water by biological drainage system which are eco-friendly and economic for marginal and sub-marginal farmers. In a word, bio-drainage depends on characteristic of the trees to transpire water harnessed to control the rise of ground water table as an alternative drainage measure, which would be most economical and eco-friendly. Bio-drainage concept refers to the copious withdrawal of groundwater by the root system of certain tree species and its subsequent evapotranspiration to the atmosphere in the process, first a reversal of the rising trend of groundwater table, followed by a water table decline make place the crops to be established. The tree once established can withdraw water from deeper layers due to their deep root system. This favours a water balance such that the groundwater table is maintained at a desired level below the active crop root zone. As a result, fields can be cultivated economically in the spaces between the tree rows. Further, the trees that are grown to control water table may be trimmed from time to time to provide green fodder to the cattle and firewood to the resource poor inhabitants of the area. As the water table is sufficiently declined due to high ground water withdrawal by the trees, the land between the rows may even be put under field crops to give additional return to the land holders. Establishment and maintenance of the tree plantation involves comparatively less expenditure. Thus the concept of bio-drainage tunes well with the aspirations of the resource-poor farmers. The biological system of controlling waterlogged hazards and restoring field crop production to a significant level is also affordable, economically viable and accommodates the farmers' perception.

Conventional drainage techniques

An adequate drainage either natural or artificial is necessary for the maintenance of a satisfactory balance of water, air, and salt in the crop root zone for healthy plant growth. Most land has some sorts of natural surface and subsurface drainage. When natural drainage is inadequate, artificial drainage is required to increase the drainage capacity. Artificial drainage is essential to prevent the rising water table and avoid waterlogging and salinization for sustaining irrigated agriculture. The conventional engineering based techniques most commonly used to drain excess water from land are surface drainage, horizontal subsurface drainage and vertical subsurface drainage. When properly designed, installed and maintained, these systems are efficient in lowering water table and controlling salinization in irrigated lands. However, these have two drawbacks, firstly, they are costly and secondly, they generate drainage effluents, which will have to be reused or safely disposed of in safer places.

Non-conventional approaches

The limitations and shortcomings of the conventional drainage techniques call for alternative approaches which are techno-economically feasible, sustainable, cost-effective, affordable to the farming community, socially acceptable and environmentally friendly and do not cause degradation of natural land and water resources. Bio-drainage is one of these alternative options. The absence of generating drainage effluent and small investment in site development makes the system attractive. The new term bio-drainage or bio-pump is although new, however, it is used to describe the use of trees for controlling water table.

Review of literatures

Research works have been conducted extensively in many countries of the world to evaluate the impact of different types of trees, bushes and crops to control water logging. Heuperman *et al.* (1984) found an evidence of water table lowering by trees both in irrigation and dry land areas in Victoria. In a northern Victorian irrigation area an 8 year-old eucalyptus plantation lowered the water table by 2 m or more. Similarly, at several dry land sites, progressively greater and more rapid lowering of the water table was observed as planted trees grew to occupy the site more fully (Morris, 1994). However, the lack of leaching of salts from the root zone in this situation resulted in the development of salt accumulation underneath the tree plantations.

A study on ground water salinity changes beneath four experimental tree planting sites in a 700 mm annual rainfall region in Western Australia, Schofield *et al.* (1989) observed that a salinity decline of 11 % over the first seven years since planting. At a lower rainfall site (500 mm) an agro-forestry plantation lowered the groundwater level by up to 1.9 m but the ground water salinity beneath trees generally increased.

The capacity of productive plantations to extract shallow ground water is a valuable tool for controlling rising water tables and salinity. The results of the experiments reported that plantations established on a suitable scale could be provided useful protection against the twin problems of water logging and salinity and allow rehabilitation of some already salt affected land. Large reductions in water table during wet and dry seasons of the same year showed that trees were able to extract ground water, because faster growing trees provide a large conducting a pathway from soil to air, hence their water use is higher (NIAB, 1997 b).

A number of Australian scientists (George, 1990, Bari and Schofield, 1991) have reported the ability of different trees to influence water tables. They found that if trees tolerant to water logging were introduced in the problematic areas these could easily be assisted in controlling the water logging with minimum expenditure but it needs careful selection of the trees / crop cultivation. It is essential to select and identify suitable situations where this biological drainage is economically, efficiently and usefully adopted.

Smedema (1997) was of the view that bio-drainage is generally not a realistic option when common crops were facing severely waterlogged conditions. However, these biological systems under certain conditions have a merit according to which they might be confined to the growing of tolerant species under not too extreme water logging conditions. The scope of bio-drainage seems to be more favourable in arid zone where drainage surpluses were small in relation to evapotranspiration rates. According to Smedema, the claim of bio-drainage to cope with salinity does not seem to be supported with experimental evidence from Australia and California. In fact bio-drainage without any physical drainage was more likely in the long run lead to a harmful build up of root zone salinity.

In Pakistan attempts have been made to improve the salt-affected land through growing salt-tolerant and water logging tolerant crops, trees, grasses and saltbush. The approach was found to increase the economic return to farmers by improving the productivity of salt affected soil through increased cover on the soil surface which reduces evaporation, salinization rates and enhancing biological activity (Shah *et al.*, 2001). In southern Australia, the replacement of deep-rooted perennial native vegetation with shallow-rooted annual crops has resulted in a rising water table and the development of a major secondary problem. However, data are available to show that stands of halophytes can transpire sufficient water to lower water tables, thereby ameliorating water logging (presumably inundation). One consequence of the use of ground water by salt tolerant perennial plants is the accumulation of salt in the root zone. This has the potential to be highly damaging to plant growth and survival in the longer term (Barrett-Lennard *ct al.*, 2002). This biological system of drainage was not only techno-economically feasible and cost-effective but also socially acceptable and environmentally friendly (Kapoor, 2000), and thus appeared to be more remunerative to the farmers of this low-producing basin area.

In a study on the Indira Gandhi Canal Project in dry-arid region of Rajasthan, Kapoor (1997) observed that the tree plantations had very significant impact on lowering the ground water table. In the irrigated command site tree plantations of 1900 trees /ha lowered the ground water table by 15 meters and the lowering down effect was observed up to a distance of more than 500 meters away from the plantation boundary. The annual tree water use in the plantation area was estimated as 3446 mm, which was about 1.2 times the standard A-pan evaporation. It was estimated that tree plantations over 10 per cent area, in a planned and scattered manner, could help to achieve a water balance such that the ground water table might not be risen to cause water logging and salinization.

In a study on the Indira Gandhi Canal Project in dry-arid region of Rajasthan, Kapoor (1997) observed that the tree plantations had very significant impact on lowering the ground water table. In the irrigated command site tree plantations of 1900 trees /ha lowered the ground water table by 15 meters and the lowering down effect was observed up to a distance of more than 500 meters away from the plantation boundary. The annual tree water use in the plantation area was estimated as 3446 mm, which was about 1.2 times the standard A-pan evaporation. It was estimated that tree plantations over 10 per cent area, in a planned and scattered manner, could help to achieve a water balance such that the ground water table might not be risen to cause water logging and salinization.

In a lysimeteric study, Chhabra and Thakur (1998) evaluated the amount of water bio-drained by eucalyptus (*Eucalyptus tereticornis*) and bamboo (*Bambusa arundinacea)* at the measured water table depths and ground water salinity levels over four years by daily measuring the water needed for maintaining the depth of water table. The trees continued to absorb and transpire water throughout the year, the capacity being more in summer and rainy season than that was in the winter season. The study indicated that due to high transpiration capacity and an ability to extract water from deeper layers containing saline ground water, the trees can control the rise in water table in irrigation command areas and prevent the formation of waterlogged and eventually saved the saline waste lands.

In the command area of Balasamund sub-branch and Balasamund distributaries, Sarma *et al.* (1999) found that a system analysis approach for bio-drainage could control the rise of water table in a land-locked region. The linear programming technique helps to find an optimal cropping pattern and vegetation plan that would increase ET demand for controlling the water table depth.

Chauhan (2000) explained that bio-drainage was found to be a good method of subsurface drainage system. It is not only an economic method of drainage, but also improved the ecology of the area and is environment friendly. It provided costly wood for multifarious purposes and also various range of biomass production. It can transpire water from ground water table in good amounts. Chauhan also opined that change of cropping pattern was also sometimes considered as an alternative to subsurface drainage. A few selected crops only could be grown in a soil with a given low salinity. Most of the crops could not tolerate salinity beyond 4 m mhos/ cm.

The research works conducted by the earlier workers at the national and the International level were mostly centred in the irrigated commands of arid and semi-arid region and the affected lands along the canal banks. The issue of low-cost bio-drainage technology was found to be an alternative option for land amelioration and crop production unless the overall drainage situation in the basin was improved drastically.

Bio-drainage in controlling land degradation

Waterlogging and subsequent soil salinization are the two major causes of degradation of irrigated agricultural lands making it unfit for crop cultivation. Surface drainage where found inefficient to combat the problem adequately. Subsurface Drainage Technology (SSDT) is an effective control measure for both. However, this technology is expensive and has an associated problem of safe disposal of the drainage effluents. Bio-drainage is considered by some as an effective alternative, in which water and salt tolerant plant species are grown to remove the salt solutions from saline and waterlogged lands in an eco-friendly and economical manner. The saplings of *Acacia biflora* to determine the plant's salt uptake capacity from the root zone that was made saline by irrigation with water of different salt concentrations for a period of 6 months likewise *Euclytus* spp can draw substantial amount of groundwater and help land free from inundation. The consequent plant sap salinity was measured and salt balance analysis was carried out to ascertain salt migration from the soil to the plant. The study revealed that the plants irrigated with the water of highest salinity removed 28.7 g of salt in comparison to the plant under control, which removed 8.4 g of salts. Even the highest salt removal by *Acacia biflora* saplings was negligible as compared to that by conventional subsurface drainage system. To portray bio-drainage as a land reclamation option, confirmation of salt uptake from saline lands having established plantation needs to be carried out.

Acacia, Albizzia, Azadirachta and *Eucalyptus* respectively were found useful plantation crops to the direction of lowering down the water table and that also important to control evaporation losses (44-69% of total irrigation) which might be much higher than transpiration. Increased water use uptake by *Eucalyptus* might lead to reduction and scarcity of aquifer resources for irrigated agriculture in arid and semiarid climates. Although water use efficiency of any species was also important, but environmental concerns about the amount of water consumed by these species were also considered most important.

Trees like *Eucalyptus*, poplar, *Albizzia* could be grown well under high water table conditions as compared to other forest species. Evapotranspiration from forest tree species is 2 to 2.5 times higher than agricultural crops. Some Eucalyptus tree species transpire higher than pan evaporation. Pan evaporation in arid climate was found more than 2 m per year. Assuming a specific yield of 0.2 t/ha and evaporation of water from forest plantation area equivalent to pan evaporation, it worked out that plantation in 10 ha area will lower the water table area by 1 m in 1000 ha annually. Proper drainage removes excess water from crop root zone and excess nutrient salts. A suitable soil environment is created for growth of plant roots which results in higher crop production on sustainable basis.

Bio-drainage relied on vegetation, rather than engineering mechanisms to remove excess soil water through evapotranspiration. It was often considered attractive because it required only an initial investment in site development (planting of a "bio-drainage crop") and (potentially) gave returns a benefit when the bio-crops were harvested for fodder, wood or fibre. In addition, under some management scenarios, *viz.* certain cropping systems and slightly saline conditions, it might offer limited scope to achieve nutrient and/or salt-balance through removal of biomass, thus alleviating the problem of the disposal of polluted drainage effluent from the bio-drainage crop area by reducing volumes and improving the quality of the effluent. Presently about one-third of the world's irrigated area faces the threat of waterlogging. But there were doubts and reservations on the long-term sustainability of bio-drainage, because of the apprehension that no salts would be removed.

Kapoor (2001) has described methods of estimating these salt accumulations were whether and to what extent the above requirements could be met depending upon the prevailing field conditions. IPTRID (2002) has presented a knowledge synthesis report describing principles, experiences and applications of bio-drainage. The Indian National Committee on Irrigation and Drainage (INCID) (2003) has brought out a status report on bio-drainage as well.

Consideration prior to bio-drainage installation

Prior to install the bio-drainage as a biological system, the careful considerations are required to follows:

1. The quantity of water removed from the ground water annually should equal the quantity of recharge to maintain proper water balance;
2. The quantity of minerals removed annually should be nearly equal to the quantity of mineral import to keep salt balance;
3. Irrigation method is practised primarily to promote agriculture, horticulture, dairy; Therefore in term of economic returns afforestation or agro-forestry should be comparable with that from other alternative uses of land. If it is not so, afforestation may still be justified, on considerations of the environmental and drainage benefits under the area of plantation;
4. Under ideal situation, trees in afforestation area on full development should be able to draw most of their requirement of water from the ground water table, so that surface irrigation water can be put to other productive uses. If, and so long it is not possible, plantation trees would need some irrigation water. They may also need some water periodically to leach down salts from the root zone, if and when the salinity levels approach threshold limits in maintaining ground water table condition;
5. The quality of ground water, when the water table approaches the root-zone of trees, should be such as can be tolerated by the plant species, otherwise the trees would need to be supplied irrigation water;
6. Trees can lower the ground water table directly underneath the plantation area, to a depth up to which the tree roots can extend. This can be up to 15 m from ground surface or even more. To be effective as a drainage measure, the ground water table must be lowered in the irrigated area to a minimum critical depth (say 2 m below ground level), at the farthest point from the edge of the plantation area.

Positive aspects of bio-drainage system

1. Bio-drainage can effectively contribute to strongly reducing the problems as experienced from water logging in irrigated agriculture and non-irrigated agriculture;
2. The problems associated with a rise in salinity in the root zone can be effectively delayed using bio-drainage systems in semi-arid and arid areas;
3. A more natural environmental setting is achieved in irrigated project areas;

4. A significant contribution is made to integrated rural development and the well-being of rural livelihoods;
5. Added advantages of bio-drainage and bio-disposal systems, depending on the vegetation used, are: production of timber, fuel wood, oil, fruits, fibre, contribution to carbon sequestration, diminishing the effects of wind erosion, provision of shade and shelter, function as windbreaks, yield organic matter for fertilizer, enhancement of biodiversity, as flora and fauna flourishes, diminishing air pollution;
6. As compared to conventional subsurface drain-age systems, bio-drainage systems perform the following four functions together;
7. Dewatering, i.e. inducing water flow through the soil towards a tube-well or pipe drain;
8. Transportation of the drain water to collector drains and thereafter to main drains;
9. Pump lifting effluent to higher levels in the evacuation system (as often required) and/or further conveyance by gravity outflow to disposal sites;
10. Final disposal at selected sites (e.g. by evaporation ponds);
11. Investment might be relatively low and could be borne by the farmer beneficiaries themselves. However, costs could be increased by fencing, wildlife protection, etc., depending on location;
12. Salt balance: No biological system is sustainable without salt balance. It is one of the most important issues to be addressed before biodrainage can be promoted as an appropriate drainage management technology. Salt storage in vegetation components needs to be studied in depth.
13. Involving certain categories of plants that habitually draw water from ground water
14. Control waterlogging and Salinization of irrigated lands
15. Evacuates undesirable and undrainable water with the help of transpiration from plantation trees
16. Controlling of depth of groundwater table
17. Help establish water and salt balance in the groundwater
18. Acts as sub-surface drainage/ shallow tube-well/bio-pump
19. Cost-effective and socially acceptable
20. Environment friendly

Constraints of bio-drainage systems

1. To design the bio-drainage system, it always require a large land are under waterlogged situations;
2. The bio-drainage system is not very equally effective in removing salt;
3. The system does not allow for controlled drainage;
4. Bio-drainage combined with conventional drainage systems has combined positive and negative aspects. These are more positive and less negative than when only making use of conventional drainage systems.

Disadvantages of bio-drainage system

In the areas of over-exploitation of sweet groundwater due to irrigation and other purposes, there is a steep fall in the water table. In such a scenario, monoculture of high bio-drainage potential species like *Eucalyptus* might be adding to the ecological disaster by further contributing to water table decline. Replacement of fast bio-drainers like Eucalyptus with slow bio-drainers is recommended here. Unfortunately quantitative information is not available on the relative bio-drainage potential of different species. The ultimate solution lies in obtaining a steady-state balance between precipitation, irrigation input of water vis-à-vis water requirements of the agro-system and groundwater hydrology. In this system, bio-drainage potential of tree flora can be cardinal component.

The withdrawal of groundwater by trees depend on the following conditions (a) Climatic condition (b)Plant species (c) Density of plantations (d) Area under plantation (e) Water supply (f) Root system and (g) Trees in general lower the groundwater table directly underneath the plantation area, to a depth up to which the tree roots can extend.

Conclusions

Waterlogging is a widespread problem in the inland low-lying areas of both non-irrigated and irrigated lands. Conventional system of drainage produced a lot of drainage effluent which caused environmental degradation. This alarming situation makes the alternative approaches of biological systems with the use of evapotranspirative power of plants to lower groundwater tables and depress the extent of salinity level. However, bio-drainage system may delay the salinization process. In the arid and semi-arid regions, in addition to waterlogging the major problem associated with irrigation was salinization, which caused land degradation and consequently yields reduction. The solution to this problem was the installation of subsurface drainage measures that keeps the water table down and allows for leaching of salts from the crop root zone.

References

Barrett-Lennard, E.G.; Turner Neil, C. and Ward Phillip, R.2002. Restoration of saline land through revegetation. In: Symposium on the role of agroforestry and perennial pasture in mitigating waterlogging and secondary salinity, 53(1- 3): 213 – 226.

Bari, M.A. and Schofield, N.J. 1991. Effects of agroforestry-pasture associations on ground water level and salinity. *Agrofor. Syst.*, 16: 13 – 31.

Chhabra, R. and Thakur, N.P. 1998. Lysimeter study on the use of bio-drainage to control water logging and secondary salinization in (canal) irrigated arid-/ semi-arid environment. *Irrigation* and *Drainage Systems*, 12 : 265-268.

Chauhan, H.S. 2000. Bio-drainage and other alternatives to tile drainage. Proceedings of 8th ICID International Drainage Workshop held from 31st Jan – 4th Feb., 2000 at New Delhi, India. Vol. IV, pp 79 – 95.

George, R.J. 1990. Reclaiming sand plain seeps by intercepting perched ground water with eucalyptus. *Land Degradation Rehabil.*, 2 : 13 – 25.

Heuperman, A.; Stewart, H. and Wilders, R. 1984. The effect of eucalyptus on water tables in an irrigated area of northern Victoria. Rural Water Commission of Victoria. *Water Talk*, 52: 4-8.

Kapoor, A.S. (1997). Bio-drainage of lands under irrigation in dry arid regions with particular reference to the IGNP, Rajasthan.

Kapoor, A.S. 2000. Bio-drainage feasibility and principles of planning and design. Proceedings of the 8th ICID International Drainage Workshop held from 31st Jan to 4th Feb, 2000 at New Delhi, India, Vol. IV, pp 17-32.

Morris, J.D. 1994. The use of trees in salinity control in Victoria. Australia progress and potential In : The Role of Trees in Sustainable Agriculture, pp. 63 – 75. Proc. Nat. Cont. Sept. 30 – Oct. 3. 1991. Albury, Australia.

NIAB. 1997. The Australian Centre for International Agricultural Research (ACIAR). Research Notes Partner in Research and Development RN. 18,8/97.

Patra. S. K., Ray, R. and Das, N.C. 2000. Alternative technology in reclamation of waterlogged areas of Dankuni basin in West Bengal. *Agril. Engineering Today*, 24 (5):1-7.

Sarma, P.B.S., Garg, S.P. and Rao, N.H. 1999. A system analysis approach for biodrainage of waterlogged regions. Proceedings of the 8th ICID International Drainage Workshop held from 31st Jan to 4th Feb, 2000 at New Delhi, India, Vol. IV, pp 47-60.

Schofield, N.J.; Lah, I.C.; Scott, P.R.; Bartle, J.R.; Ritson, P.; Bell, R.W.; Borg,; Anson, B. and Moore, R. 1989. Vegetation strategies to reduce stream salinities of water resource catchments in south-west western Australia. Report of the Steering Committee for Research on Land Use and Water Supply. Water Authority of Western Australia. Rep. No. WS 33, 98pp.

Shah, A.H.; Sufi, A.B.; Bhutta, M.N. and Michael Patta, P. 2001. Prospect in saline agriculture in Pakistan. Proceedings of the 8th ICID International Drainage Workshop held from 31st Jan to 4th Feb, 2000 at New Delhi, India, Vol. IV, pp 61-78.

Smedema, B. 1997. Biological drainage myth or opportunity? GRID Iptrid Network Magazine Issue 9, December 1977, pp 3.

Zaman, A. 2012. Water management technologies to increase crop and income per drop with reference to effective utilization of low land eco-system in eastern region. *J. Water, Energy and Food Security,* 1(1):1-9

21

Natural Calamities, Typical Ecosystems and Environmental Crisis of Sundarbans Areas

Anadi Gayen

Senior Hydrogeologist (Scientist-'D'), National Ground Water Training and Research Institute, Central Ground Water Board, Government of India, Raipur Chhattisgarh-492001, India

Email: anadigayen@rediffmail.com

Abstract

Sundarban is the largest mangrove wetland ecosystems in the world. This World Heritage site along the Bay of Bengal has evolved over the millennia through natural deposition of upstream sediments accompanied by intertidal segregation. Indian part of *Sundarbans* is estimated to be about 4,110 square kilometres of which about 1,700 square kilometres is occupied by water bodies in the form of river, canals and creeks of width varying from a few metres to several kilometres, which together form the complex ecosystem of this Ganges delta.. Rivers in the *Sundarbans* are confluence points of salt water and freshwater. Environmental crisis is caused by severe cyclones sea level rise and storm surge followed by coastal erosion, flood and earthquake phenomenon. River water is saline. Shallow aquifer is brackish/ saline, which is not suitable for drinking. Fresh potable ground water is occurring in deeper aquifer, the exploitation of which is not cost-effective. Sundarbans houses around 500 different species of mammals, fish, bird and reptiles and it is the largest habitat of royal Bengal tigers. Conservation of flora and fauna in the typical ecosystem of *Sundarbans* in India with emphasis on judicious use of ground water resources along with retrospection of historical trend and future sustainable management strategies is a pre-requisite.

Keywords: Mangrove; saline; ecosystem; water; flora and fauna

Introduction

The *Sundarbans* of India lies between 21° 562 593 N, 89° 102 59.993 E (Fig. 1). The history of *Sundarbans* in Indian has been traced back to 200-300 AD. It has been reported that this lower Ganges delta was under water during 6000-7000 years ago. In 1875, a large portion of this mangrove forests was declared as reserved forests under the Forest Act, 1865. The *Sundarban* is considered as the largest mangrove wetland ecosystems in the world. It is named after the *sundari* trees (Fig.1a). The *Sundarbans* along the Bay of Bengal has evolved over the millennia through natural deposition of upstream sediments accompanied by intertidal segregation. Administrative set up of Sundarbans in India is shown in Table 1.

Fig. 1: Location of *Sundarbans*
Source: Wikipedia

Fig. 1a: Sundari trees
Source: Wikipedia

Table 1: Administrative Sub-divisions of *Sundarbans* in India

District	Sub–Division	Blocks/Panchayet Samities
24 Parganas (South)	Kakdwip	Sagar, Namkhana, Kakdwip, Patharpratima
	Diamond Harbour	Mathurapur – I, Mathurapur – II
	Baruipur	Kultali, Joynagar – I, Joynagar – II
	Canning	Canning – I, Canning – II, Basanti, Gosaba
24 Parganas (North)	Basirhat	Hingalganj, Hasnabad, Sandeshkhali – I, Sandeshkhali – II, Haroa, Minakhan

The mangrove forest lies in the vast delta on the Bay of Bengal formed by the super confluence of the Ganges, Padma, Brahmaputra and Meghna rivers across southern Bangladesh. The seasonally flooded *Sundarbans* freshwater swamp forests lie inland from the mangrove forests on the coastal fringe. The forest covers 10,000 square kilometres (3,861 sq. miles) of which about 6,000 square kilometres (2,317 sq. miles) are in Bangladesh. It became inscribed as a UNESCO world heritage site in 1997. The Indian part of *Sundarbans* is estimated to be about 4,110 square kilometres (1,587 sq. miles), of which about 1,700 square kilometres (656 sq. miles) is occupied by water bodies in the form of river, canals and creeks of width varying from a few metres to several kilometres, which together form the complex ecosystem of this Ganges delta.. At present, the total land area of *Sundarbans* is 4,143 square kilometres (1,600 sq. miles) and the remaining water area is 1,874 square kilometres (724 sq. miles). Rivers in the *Sundarbans* are confluence points of salt water and freshwater. Thus, it is a region of transition between the freshwater of the rivers originating from the Ganges and the saline water of the Bay of Bengal.

In this coastal active delta, there are diverse flora and fauna existing within a unique ecosystem. A study in 1991 has revealed that the Indian part of the *Sundarbans* supports diverse biological resources including at least 150 species of commercially important fish, 270 species of birds, 42 species of mammals, 35 reptiles and 8 amphibian species, although new ones are regularly being discovered. This represents a significant proportion of the species present in Bangladesh (i.e. about 30% of the reptiles, 37% the birds and 34% of the mammals) and includes many species which are now extinct elsewhere in the country. Two amphibians, 14 reptiles, 25 aves and five mammals are endangered. The *Sundarbans* is an important wintering area for migrant water birds and is an area suitable for watching and studying avifauna.

Environmental Degradation in *Sundarbans*

Environmental degradation were caused by severe cyclones (Fig. 2), sea level rise and storm surge followed by coastal erosion, flood and earthquake phenomenon in the *Sundarbans* area. The history (since 1582 to 1994) of dreaded and devastating cyclones, storm surge in sea, flood and earthquakes took place in the *Sundarbans* are stated below in table-2.

Table 2: Historical details of natural calamities occurred in *Sundarbans*

Sl. No.	Cyclones	Flood	Earthquake	Famine
1	1582	1823	1737	1770
2	1688	1834	1762	1791
3	1707	1856	1842	1865
4	1737	1864	1895	1921
5	1742	1868	1897	
6	1830	1869		
7	1832	1871		
8	1833	1885		
9	1839	1890		
10	1840	1900		
11	1844	1907		
12	1848	1966		
13	1850	1976		
14	1852	1978		
15	1858	2010		
16	1859			
17	1862			
18	1864			
19	1867			
20	1877			
21	1878			
22	1880			
23	1881			
24	1882			

25	1883
26	1884
27	1887
28	1888
29	1889
30	1893
31	1894
32	1896
33	1898
34	1899
35	1901
36	1904
37	1909
38	1913
39	1916
40	1917
41	1919
42	1922
43	1927
44	1928
45	1929
46	1932
47	1934
48	1935
49	1936
50	1937
51	1940
52	1941
53	1942
54	1943
55	1948
56	1956
57	1960
58	1961
59	1962
60	1965
61	1968
62	1970
63	1973
64	1981
65	1982
66	1985
67	1988
68	1991
69	1994
70	2007
71	2009 (Aila)

Source: Paschim Banger Sundarban: Prakriti o Parichiti, by Kumud Ranjan Naskar. Article in the Paschimbanga Magazine, Information & Broadcasting Dept., Govt. of West Bengal

Islands in the *Sundarbans* area are losing their areas. The areal loss (%) since 1969 to 2001 mentioned in Table 3

Table 3: Areal loss of *Sundarbans*

Sl. No. Name of the Island	Area (Sq.km.) [1969]	Area (Sq.km.) [2001]	Areal Loss (%)	Status of Embankment
1. Dakshin Surendranagar	52.722	44.520	15.55%	125 km, 40 km vulnerable
2. Sagar	253.203	241.042	4.8%	250 km long embankment, 15 km washed off, 90 km vulnerable
3. Namkhana	156.908	149.222	4.879%	150 km long embankment, 50 km vulnerable.
4. Moushuni	35.244	30.067	14.6%	30 km long embankment, 15 km vulnerable.
5. Lothian	36.041	34.986	2.92%	Mangrove Forest, no embankments.
6. Jharkhali	162.00	161.328	Stable	150 km long embankment
7. Ghoramara	9.083	5.349	41.109%	5 km washed off
8. Dulibhasani	190.515	185.338	2.72%	Mangrove Forest, no embankments.
9. Dhanchi	39.041	34.174	12.465%	- do -
10. Dalhousie	78.233	66.713	14.725%	- do -
11. Bulchery	32.700	26.227	19.80%	- do -
12. Ajmalmari West	25.189	25.096	0.37%	- do -
13.Ajmalmari East	75.664	70.923	6.266%	- do -
14. Bhangaduni	45.112	30.855	31.6%	- do -
15. Jambudwip	17.543	5.519	68.541%	- do -

Source: Studies on embankments and vulnerability of Sundarbans, West Bengal, by Dr.Sugata Hazra, Prof. & Director, School of Oceanographic Studies, Jadavpur University.

At least 8 lakh people in eight blocks in the Sunderbans area are vulnerable to cyclones and floods (report submitted by the South 24 Parganas district administration to the State government). In Namkhana, Patharprotima, Sagar, Gosaba, Mathurapur I, Mathurapur II, Basanti and Kultali blocks, the infrastructure to face floods and cyclones is lacking in alarming extent.

Fig. 2: Cyclone devastated *Sundarbans*
Source: Wikipedia

Occurrence of Groundwater in *Sundarbans*

Apart from the natural disasters, human intervention has been affecting the Sundarbans green cover. Exponent increase of population in the area resulting deforestation, which may affect the core area like dense jungles of Sundarbans like Neti Dhopeni, Haldi, Baghmara and Mayadeep in future. The fresh ground water occurs in very deep aquifers under confined conditions. In Sagar block area fresh potable aquifers exist within 300 – 306 & 313 – 319 mbgl depth range; in Mathurapur area within 311 – 320 mbgl; in Kultali block 185 – 209, 271 – 280, 314 – 332 and 393 – 400 mbgl; in Basanti block 332 – 356 mbgl; in Patharpratima block area within 201 – 204, 235 – 255 & 260 – 275 mbgl; in Kakdwip block area within 204 – 215 & 241 – 250 mbgl; in Namkhana block area fresh water occurs within 156 – 162, 167 – 173, 206 – 212, 215 – 227 & 247 – 252 mbgl depth range (CGWB). Ground water withdrawal in this area is not abundant. In some places people are using pond water for drinking as well as domestic purposes. Soils in *Sundarbans* are mainly saline to saline alkali in nature. Maximum cultivable land is used for mono-crops like paddy during monsoon. Few tube wells for irrigation are constructed in the area so that more than one crop may grow in the same land. Mainly these irrigation tube wells are pumping ground water during lean spell period. In South 24 Parganas of *Sundarbans,* out of 243320 ha, net cultivable area is 76203 ha, area in which more than one crop grows, whereas in North 24 Parganas *Sundarbans* area out of 61514 ha, 25270 ha has more than one crop in a year. In multiple (double /triple crops) instead of ground water, rainwater storage structures like ponds, canals and tanks play a key role. During monsoon period, the rainwater in heavy showers drain into sea as surface run – off, small percentage of which are being stored in those structures and being utilized for growing non-monsoon crops. In order to maintain the fertility of the soil, control soil erosion and simultaneously to obstruct surface run–off of rain water, the rain water harvesting is an inevitable solution. Tanks and other surface water sources support irrigation of 34259 ha and 8637 ha area in *Sundarbans*. Shallow tube wells (STW) irrigate 16108 ha area. Environmental degradation were caused by severe cyclones (Fig. 2), sea level rise and storm surge followed by coastal erosion, flood and earthquake phenomenon Canal area covers 8366 ha. Deep tube wells are irrigating 275 ha area.

Suitable rainwater harvesting structures present in the Sundarbans area are drainage channels, protective bandhs and land shaping cum fresh water reservoir. Nearly 30 years ago, the storage of rainwater through the above conservation structures were started.

Sagar, the biggest island of Sundarbans area is witness of serious environmental degradation. Within 70 years (1922 to 1992). The Sagar Island has decreased in its' area upto 58 Sq.km. area and loss of area is 4.80% in last 32 years. Tidal impact causes river banks erosion. Bishalakshipur, Lohachara and khashimara villages of the Sagar Island has already washed out into the sea. A lot of human settlement has got damaged. Frequent flood and cyclones made people homeless. Fresh water reservoirs get saline during floods causing severe drinking water scarcity during those periods. Existing water supply arrangements through pipeline supply and hand pump tube wells for 212,037 (2011 Census) population is not sufficient. Sagar Island has 250 km long embankment, out of which 15 km is already washed off already and 90 km is vulnerable at present.

Endangered and Extinct Species of *Sundarbans*

The **endangered species** (Fig. 2a) that live within the Sundarbans and extinct species that used to be include the Royal Bengal tigers, estuarine crocodile, northern river terrapins (*Batagur baska*), olive ridley sea turtles, Gangetic dolphin, ground turtles, hawksbill sea turtles and king crabs (horse shoe). Some species such as hog deer (*Axis porcinus*), water buffalos (*Bubalus bubalis*), barasingha or swamp deer (*Cervus duvauceli*), Javan rhinoceros (*Rhinoceros sondaicus*), single horned rhinoceros (*Rhinoceros unicornis*) and the mugger crocodiles or marsh crocodiles (*Crocodylus palustris*) began **to become extinct** in the Sundarbans towards the middle of the 20th century, because of extensive poaching and man hunting by the erstwhile monarchs. There are other threatened mammal species, such as the capped langurs (*Semnopithecus pileatus*), smooth-coated otters (*Lutrogale perspicillata*), Oriental small-clawed otters (*Aonyx cinerea*), and great Bengal civets (*Viverra zibetha*), which may soon become extinct. Conservation of flora and fauna in the typical ecosystem of *Sundarbans* in India with emphasis on water issues along with retrospection of historical trend and future sustainable management strategies is a pre-requisite.

Fig. 2a: Endangered species of *Sundarbens*
Source: Wikipedia

The river water salinity rate slowly increases with time and more rivers were affected by the salinity in the Sundarbans. The reduction in the flow rate of Ganges River has resulted in the inflow of high saline water in upstream areas. The high salinity zone is situated in the South-Western corner.

Climate change

Along with the temperature, precipitation is also showing an increasing trend in the *Sundarban* region. Variability abnormal in the climatic parameters has increased in the past few years. The warm climate would cause an increase in the cyclone speed and amount of rainfall.

Oil Spill

Around 350 tonne of oil was spilled in the river of *Sundarban* on December 9, 2014. The spill occurred at a protected area of the river where the rare dolphins were present. The oil spill disturbed the aquatic life and resulted in the death of certain aquatic animals especially the dolphins. The oil spill is generating a major threat to the forest habitant's food cycle. The primary level of the food cycle is dying to the effect of the excessive oil spill. Due to the oil spill, dolphins, saltwater crocodile, Bengal tigers, leopards, chitals are facing a severe life threat.

Cyclones

Despite Sundarbans being a cyclone prone area, there is a lack of disaster shelters. There has been an increase in the cyclonic storms hitting the *Sundarbans* in the past many years. During the Cyclone Aila in 2009, the waves crossed the heights of the embankment and entered the plains. Many people were affected due to the shortage of water supply and destruction of the agriculture land.

Environmental impact of Aila

As reported, the endangered Bengal Tigers (265 nos.) in the *Sundarbans* area was inundated with 6.1 m (20 ft) of water rise during the Aila and also suspects that dozens of the tigers were drowned in *Aila's* storm surge along with deer and crocodiles. Moreover, the forest remains under an estimated 2.4 m (7.9 ft) height of water. A team of conservationists searched for the tigers throughout the forest after *Aila* havoc in 2009. The search teams supplied fresh drinking water to the tigers as their natural water source was inundated with salt water from Aila's devastating storm surge.

Pollution

Due to a heavy disposal of solid waste from the nearby cities, the Ganges river does not carry fresh water to the Sundarbans. A huge amount of domestic and industrial effluents are carried by the tributary rivers. Loads of mud disposals from the harbours also causes pollution

Man-made disasters

Oil spills are one of the biggest man-made disasters that affect the people as well as the biodiversity environment of the fragile *Sundarbans*. The oil spill of 2014 which was caused by the Southern Star VII, a Bangladeshi ship created a lot of destruction especially to the aquatic life. Modernization of agriculture activity and farmers adopting the techniques of using pesticide and chemicals are affecting the flora and fauna of the region. The dumping of various industrial and domestic wastes into the river systems is also possessing a great threat to the environment. Added to these, there are concerns regarding the construction of 1320 megawatt coal-fired power plant near the *Sundarbans.*

Fishing

The issue of over fishing has put the aqualife under great pressure. The population of the fish is decreasing due to the combination of overexploitation and climatic changes. The density of the fish in shallow waters reduced tremendously.

Wood cutting

There is an increase in illegal cutting down of trees and woods in the Sundarbans. People living nearby the forests in the South 24 Paraganas are directly involved in the trading of the wood. Deforestation is taking a rise in most of the areas of *Sundarbans,* which will ultimately lead into the severe loss of the biodiversity.

Water issues

1. Sundarban mangrove wetland spreads over the Ganges delta with an average elevation of 0.9 to 2.1 meter above the mean sea level (m amsl).
2. The lower Ganges delta was under the water before 6000 – 7000 years.
3. The highest water salinity values are found at the Malancha – Mandarbaria river in the South.
3. The global sea surface temperature is 0.06 degree Celsius.
4. The major oil spill in Sundarbans occurred on December 9, 2014 at the Shela river.

6. The proposed coal-fired plant by NTPC would be situated at the Bagerhat district of south-west Bangladesh.

Suggested Future Sustainable Management Strategies

1. Eco-conservation.
2. Eco-development.
3. Training.
4. Awareness.
5. Education.
6. Research.
7. Plantation of mangrove plants.
8. Construction of fresh water ponds for drinking water to the wild animals.
9. Periodic monitoring of polluted river water samples to understand the gravity and extent of the problems caused by Water component.
10. Ant poaching campaign.
11. The report on 'District Disaster Management Plan of South 24 Parganas, 2015' indicates that the flood and cyclone shelters in many blocks of *Sundarbans* are in very sorry state. Therefore, it needs attention on top priority to face flood situations.

Conclusion

Sundarbans hoses around 500 different species of mammals, fish, bird and reptiles and above it is the largest habitat of royal Bengal tigers. Therefore, sustainable environmental management plan is a pre-requisite to conserve the flora and fauna. Mangrove forest of *Sundarbans* acts as natural barrier and can save the India and Bangladesh from natural calamities like flood, cyclone and tsunamis. As the mangroves have very high potential of carbon capturing, the protection of mangroves would help to combat the danger of increased global warming. During peak monsoon, every year earthen embankments used to get erode into rivers. As a result, areal loss takes place throughout *Sundarbans*. To address this burning issue, more number of trees like *Sundari* and *Goran* are to be planted along the rivers/estuaries banks in the area. *Sundarbans* hold the unique coastal ecology and complicated ecosystem, wherein shallow aquifer is saline to brackish in nature. Hence, the economics involved in ground water extraction from deeper aquifers is a big issue for poor villagers. Fresh ground water should be used for drinking only, not for irrigation.

The available household ponds may be deepened for increasing water holding capacity and conserve rain water for domestic uses. The conservation structures like drainage channels, protective bunds and reservoir ponds may be adopted to enhance the judicious use of rain water in monsoon. Moreover, due to the increased salinity, less salt tolerant mangrove species of the *Sundarbans* may be eliminated by the more salt tolerant species to arrest the loss of biodiversity.

References

Anon, Central Ground Water Board Report

www.wwfindia.org

https://en.wikipedia.org/wiki/Sundarbans_National_Park

https://www.wordpress.com

22

Watershed Management for Agricultural Sustainability

A. Zaman[1], Md. Hedayetullah[2] and Parveen Zaman[3]

[1]Emeritus Professor, Centurion University of Technology and Management M.S. Swaminathan School of Agriculturen, Paralakhemundi-761 211 Odisha, India
Email: profazaman@gmail.com
[2]Assistant Professor, Department of Agronomy, Directorate of Research BCKV, Kalyani-741235, West Bengal, India
Email:hedaye.bckv@gmail.com
[3]Assistant Director of Agriculture, Pulse and Oilseed Research Station Beldanga-742133, India
Email: parveenzaman1989@gmail.com

Abstract

The effect of watershed management programme was studied with introduction of new crop-cultivars including six rainy season crops namely direct seeded rice (drilled rice), transplanted rice, arhar, groundnut, sesamum and maize and two winter season crops namely mustard and wheat. The crop demonstration programme were undertaken for consecutive four years during 1988 to 1991 to utilize different land situations and for effective use of created irrigation potential at a drought prone tract of West Bengal (India). The results of the crop demonstration revealed that with selection of crops and its suitable varieties, adopting improved agronomic practices including scientific management of fertilizers and irrigation water, plant protection measures, the productivity of both land and crops markedly increased. The crop production strategy on the basis of land capability classification changed the land use, cropping pattern and crop intensity in the watershed area. The scenario of rainfed agriculture got a dramatical change with creating irrigation potential, harvesting and reclying the excess rain water. Watershed approach appeared to be most effective and integrated management of all natural resources like soil-plant-water-animal and human to uplift to socio-economic status of the farmers disseminating the location specific proven agro-technologies in massive scale for sustainable agriculture in rainfed areas. The scope left for development of agro-forestry, silvipasture, agri-horticulture, animal husbandry, fisheries and building up several agro-based industries with abundant supply of raw materials. The programme was found cost effective, generated employment opportunities of the rural people throughout the year and increased individual family income of each beneficiaries of the adopted villages. The success gave encouragement and impetus to the fellow farmers to keep the process on towards further improvement the different activities through formation of various Watershed Level Committees.

Keywords: Crop productivity; integrated resource management; watershed development

Introduction

Radhamanpur watershed in drought prone red and lateritic tract of West Bengal having a total geographical area of 364 hectare with rolling and undulating topography, had no chance of growing second crop(s) after harvest of rainy season tall indica rice cultivar under below optimum production level less than 1 ton/hectare before commencement of the project. The watershed was one, out of the twenty watersheds identified initially during early eighties to undertake experimentation for development of rainfed agriculture under National Watershed Development Projects in India. The watershed was located in the district of Bankura at the western end of West Bengal, continuation of Chottonagpur plateau lies between 22°38' to 23°30' N latitude and 86°36' to 87°46' E longitude having an elevation of 90 m to 120 m above the mean sea level. The soil contained 0.35% to 0.38% organic carbon, 60 kg to 65 kg available P_2O_5/hectrare and 100 kg to 105 kg exchangeable K_2O/hectare having pH of 5.5 to 6.0 (1:2 soil:water suspension).

In India, National Watershed Development Project for Rainfed Areas (NWDPRA) was launced to generate all available and possible natural resources and its effective utilization for sustainble agriculture and the effective management of soil-water resources to uplift the socio-economic status of the rural people wherein emphasis was applied on rain water conservation for development the ares under rainfed conditions. With a view to develop suitable agro-technology for soil water conservation, cropping systems, agro-forestry and livestock production and for its effective transfer under massive scale, the Operational Research Projects for Resources Development and Management on Watershed Basis was formulated to meet the location-specific need along witth need based research programme.

At present, out of 141 million hectare net area sown, about 100 million hectare of land was remained rainfed (Tripathi *et. al.*,1990) which contributed about 42% of the total food grain production in the country. Most of the land occupied under rainfed conditions was predominantly based on monocropping where after harvest of rainy season crop(s), nothing could be grown and remained fallow during winter months due to lack of irrigation water or adequate rainfall in these areas (Zaman,1993). Growing second crop(s) with necessary soil and water conservational practices was possible with varying degree of success in such areas (Zaman and Mallick, 1987). Watershed was appeared as integrated management of all possible natural resources like soil-water-plant-animal and human to maximize agricultural production on sustainable basis (Zaman,1993).The performances of various watershed located at different agro-climatic zones in India revealed that the production of food, fodder, fuel, fibre and fruits along with the animal components increased considerably (Bhardwaj and Singh,1990). The achievement of all these watersheds operated in the country was quantified on the basis of results obtained under the Operational Research Projects for Resource Development and Management. With a view to increase the agricultural production on sustainable basis and to bring more area under crop cultivation specially introducing second crops during winter months under appropriate soil and water conversational measures along with increasing or creating irrigation potential collecting run-off, harvesting and recycling of excess rain water, the programme was undertaken at location-specific watershed under red and laterite tract of West Bengal.

Methodology

Out of twenty such operational projects run experimentally during eighties in the country, Radharamanpur watershed was one, comprising the villages like Radharamanpur, Bara-kalaijharia and Bankati in the district of Bankura in West Bengal of Eastern India. Bankura district was located at the western end of West Bengal, continuation of Chottonagpur plateau lies between 22°38' to 23°30' N latitude and 86°36' to 87°46' E longitude having an elevation of 90 m to 120 m above the mean sea level. The crop demonstration programme was carried during 1988 to 1991 for a period of four years. The soil contained 0.35% to 0.38% organic carbon, 60 kg to 65 kg available P_2O_5/hectare and 100 kg to 105 kg exchangeable K_2O/hectare having pH of 5.5 to 6.0 (1:2 soil:water suspension). Some of the important physical-chemical properties of the soils were given in Table 1. The climate of the region was sub-tropical humid having an annual rainfall of about 1,200 mm confined mainly during rainy season that occurred June to August. The uneven distribution of annual rainfall throughout the year caused a drough spell usually starting from the month of November. The maximum temperature during summer months went up to 45° C and minimum temperature came down in winter upto 3° C.

Table 1: Physical-Chemical properities of the soil

Particulars	Range (0 cm—30 cm soil depth)
Bulk Density (g/cc)	1.47—1.67
Particle Density (g/cc)	2.50—2.65
Field Capacity (0.1 bar, %)	25.0—30.0
Permanent wilting point (%)	6.4—11.2
Infiltration rate (cm/hr)	1.12—1.52
Hydraulic Conductivity of undisturbed soil (cm/hr)	0.25—0.45
CEC (meq/100g)	12.35
EC (milimhos/cm)	
Average Slope (%)	0.04
Soil Class	3.0—5.0
Soil Type	Laterite
Soil texture	Light in nature
Soil Order	Sandy-loam
	Oxisol

The total geo-graphical area of the watershed was 364 hectare. Out of which 139 hectare was under crop cultivation mainly with tall *indica* rice following traditional practices with below optimum production level of less than 1 ton/hectare before commencement of operational project. The rainy season crops such direct seeded rice (drilled rice), arhar, groundnut and maize were sown under medium to upland situations whereas sesame and transplanted high yielding rice cultivars were given under medium to low land situations with recommended agronomic practices during July-August and the crops like mustard and wheat were sown during winter months in medium to low land situations with minimal irrigation facilities created under the programme. The soil and water conservational measures to improve land situations and to create irrigation potential renovating farm ponds were taken as continuous process. Introduction of high yielding varieties, new crops and adaptability of different cultivars were carried on. The programme was undertaken on the basis of land capability classifications. All these attempts were taken to motivate the farmers at massive scale to adopt location specific improved agro-technologies as a part of effective transfer of proven technology to the farmers, which wre found encouraging building up and increase the confidence level of the farmers be associated themselves with the modern hi-techs.

Results and discussion

Land productivity

Thc Land usc pattern before and after commencement of operational projects were shown in Table 2. The results of the programme revealed that crop production was increased due to increase in gross cropped area and cropping intensity. The results also revealed that there was an appreciable increase in crop yield under demonstration programme adopting improved agro-technologies and soil-water conservational measures (Table 3).

Table 2: Change in land use, crop pattern and cropping intensity as influenced by the watershed programme

Particulars	Before Project		After Project	
	Area (ha)	(%)	Area (ha)	(%)
Land use pattern				
Area under crop cultivation	139.0	100	194.0	140.0
Area under irrigation	0.0		33.0	20.0
Area under double crop	0.0		30.0	18.3
Gross cropped area	139.0		224.0	161.0
Crop Pattern				
Cereals (rainy season)	139.0	100	118.0	72.0
Oiseeds (rainy season)	0.0		20.0	12.2
Pulses (rainy season)	0.0		12.0	7.3
Others	0.0		14.0	8.5
Cererals (winter)	0.0		24.0	18.6
Oilseeds (winter)	0.0		20.0	12.2
Other	0.0		25.0	26.6
Cropping intensity		100		150

Table 3: Crop yield under demonstration, non-demonstration and District average with per cent increase

Crop(s)	Yield (q/ha) in ORP area				
	Demo	Non-demo	% increase	District	% increase Average
Rainy Season					
Drilled rice	29.0	22.0	32.0	17.0	71.0
Transplanted rice	32.0	25.0	28.0	19.5	64.0
Arhar	10.0	8.3	20.5	5.8	72.0
Groundnut	18.0	12.0	50.0	10.0	80.0
Sesamum	12.0	9.4	27.7	6.0	100.0
Maize	28.0	23.0	21.7	14.5	93.0
Winter Season					
Mustard	13.0	11.5	13.0	10.0	30.0
Wheat	20.0	16.7	19.7	15.0	33.0

Crop productivity

The impact of watershed programme particularly land levelling, field channelling, bunding, terracing and improved agronomic practices including proper time of sowing, seed rate and application of fertilizers, introduction of irrigation, weed control along with other appropriate plant protection measures increased the crop yields under demonstration (Table 3). Mittal and Dhruvanarayana(1989) also reported the similar results from integrated watershed management in Himachal Pradesh and Hazra (1989) reported the successful implementation and dramatically increase in crop yield in Tejpura watershed in Jhansi district of Uttar Pradesh. All rainy season crops gave the higher yield at least 20% to 50% under demonstration over non-demonstration which was obviously higher than that of the district average.

Crop pattern

The alternative crops like arhar, maize, groundnut, sesame were introduced to utilize medium to upland situations were found effective though these measures decreased the rice area during rainy season. The replacement of tall rice varieties with high yielding rice cultivars and introducing these entire new crop which increase the cropping intensity from 100% to 150%. The crop demonstration programme gave impetus to the farmers to grow pulses, oilsedds and fodder. The area under these crops increased from 0 to 19%, 0 to 20% and 0 to 15%, respectively during winter months.

Land use

The rolling and undulating topography that was enhanced the soil erosion and caused land degradation prior to commencement of the project. Land levelling, bunding, terracing, gully puging and other conservational land treatment works carried out in the projected area resulted in increased area about 40% under crop cultivation from 139 to 194 hectare wherein under created irrigation potential on harvesting the excess rain and its reclying helped to grow second crop(s) in the watershed area.

Conclusion

Thus the programme on watershed management was found successful and accepted as pivotal for improvement of crop yields in rainfed areas that left the tremendous expansion possibilities including development of agro-forestry, agri-horticulture, silvipasture and animal husbandry along with fisheries. Selection of suitable crop(s) and its appropriate varieties, introduction of new crops with proper sequences at different land situations the areas through proper utilization of all possible natural resources increased the crop as well as land productivity.

The watershed management programme was proved both labour and capital intensive and benefit cost ratio was found as high as 7:1. So the programme offered high avenues for employment generation in the rural areas which augmented the average income of all the beneficiaries of the adopted villages.

References

Bhardwaj, S.P. and Singh, G.1990. Achievements and performances of twenty operational research projects on integrated watershed management. Proc. National Seminar on Watershed Development and Management, Kanpur Feb 2-4,1990. pp 263-269.

Hazra, C.R., 1989. Crop yields increased dramatically in Tejpura watershed. *Indian Farming*, 39: 37-39.

Mittal, S.P. and Dhruvanarayana V.V.,1989. Watershed management increases crop yields in Himachal Pradesh. *Indian Farming*, 39:14-15.

Tripathi, S.K., Dhotre, R.S. and Yoganarsimham G.N.,1990. A model schedule for agricultural watershed management. Proc. National Seminar on Watershed Development and Management, Kanpur Feb 2-4,1990. pp 314-319.

Zaman, A., 1993. Effect of watershed management program on land and crop productivity in drought prone region of West Bengal. *Environment & Ecology*, 11(2): 442-445.

Zaman, A. and Mallick, S. 1987. Feasibility of growing second crop in winter after rainy season crop at rainfed condition in semi-arid region of West Bengal. *Environment & Ecology*, 5(2): 239-243.

23

Lowland Ecosystem Management in Agricultural Sustainability

A. Zaman[1] *and Parveen Zaman*[2]

[1]*Centurion University of Technology and Management*
MS Swaminathan School of Agriculture, Paralakhemundi-76211 Odisha, India
Email: profazaman@gmail.com
[2]*Assistant Director of Agriculture, Pulse and Oilseed Research Sub-Station, Beldanga-742133, Murshidabad, West Bengal, India*
Email: parveenzaman1989@gmail.com

Abstract

Rainfed agriculture plays and will continue to play a dominant role in providing food and livestock for an increasing world population. Rainfed agriculture practiced on 80% of the world' agriculture land area and generates about 70% of the world's staple food including most of the food in poor communities in developing and least favoured areas. The most recent estimates have put global rainfed croplands at 1.75 billion ha (Bha) at the end of the last millennium, or about 5.5 times the irrigated area in the world . India ranks 1st among the countries that practice rainfed agriculture both in terms of extent 86 (mha) & value production. Agriculture water management deals with the administration of a key input to agricultural production & offers to hope that land and water use will be intensified .Scarcity of water resources and growing ofm competition for water many sectors reduces its availability for irrigation. Effective management of water for crop production in water scare area requires efficient approaches for increasing crop productivity and drought tolerance by genetic improvement and physiological regulation may be the means to achieve efficient & effective use of water. A single approaches would not be able to tackle the forthcoming challenge of producing more food and measures with engineering solutions (water saving ,irrigation method, deficit irrigation, proper deficit sequencing, modernization of irrigation system etc) and agronomic & soil manipulation (seed priming\ seedling age manipulation, direct or wet seeded rice. proper crop choice, integrating agriculture & aquaculture increasing soil fertility & addition of organic matter, tillage & soil mulching etc.) The term enhancing water productivity implies how we can most efficiently improve the outcome or yield of a crop with water currently in use, the answer lies in three main pathways (Passioura, 2006): Transpire most of the supplied water (minimization of unwanted loss),exchanged water for carbon dioxide more effectively in\ producing biomass, and convert most of the biomass into grin or other farm of harvestable product. Many technologies to improve water productivity and the management of scare water resource are available .among the most promising and efficient proven technique are-limited supplemental irrigation for optimize the use of limited water and water harvesting for improved farm income in drier environment. Improving (water productivity, however requires exploiting not only water management but also other input such as improved cultivars, fertility management & cultural practice which influence yield.

***Keywords*:** Lowland utilization; rainfed agriculture; crop productivity; cropping sequence

Introduction

India is the seventh largest having 329.73 million ha of geographical area and second most populous country having more than a billion population in the world. There is no other substitute other than to use every square inch of land and every drop of water effectively and efficiently. Though there was markedly increase in annual food grain production from 51 (1950-51) million tones t0 206 million tones (2000-01), still the death on hunger is visualized.

The demand for food is increasing with ever-increasing population growth rate. So there is no option other than to increase food production. There is very little or no scope of horizontal expansion due to land limitation, the question of vertical expansion is the only way to fit with the situation. Genetically improved crop variety, hybrids, mechanized agriculture, higher rate of fertilizers and chemicals application may give one or two more green revolution, but the question lies how far it can sustain greenness?

Adoption of improved crop cultivars, cropping system, increasing cropping intensities, precision farming can hold good to get out of this situation where whole system is to be considered.

The cropping intensities in the states like Assam, Bihar, MP, Orissa, UP and West Bengal are 137.7, 137.8, 125.6, 154.6, 148.6 and 159.6 percent, respectively (FAI,1994-95). This indicated that a vast majority of lands remain fallow during dry season due to lack of proper technologies, which are under rainfed conditions, neglected. Except West Bengal, more than half of the arable land remains fallow during rainless period. Increasing cropping intensities in rainfed uplands is not easily achievable but there is great potential of intensification of cropping in rainfed lowland by utilizing natural resources and conserved residual moisture. Pulses like black gram, green gram, lentil, lathyrus, chick pea and oilseeds like linseed, groundnut, rapeseed could not only increase the cropping intensities but also come forward to increase national productivity.

Table 1: Increasing demand of food at the end of 9th and 10th 5-year plans along with production trends

Crops	Production (in million tones)		
	1998-99	Ending 2001-02	Ending 2006-07
Rice	84.74	94.00	103.50
Wheat	71.01	75.07	84.30
Coarse Cereals	30.90	32.60	34.40
Total Cereals	186.65	202.30	222.20
Pulses	15.89	18.20	21.00
Food grains	203.54	220.50	243.20
Oilseeds	25.60	28.00	33.80
Sugarcane	295.70	287.00	352.80
Vegetables	87.53	93.60	110.70
Fruits	44.04	53.70	70.50

Source: Kumar, 1996 and The Hindu Survey of Indian Agriculture 2000

Rainfed lowland ecology

Topographically, the cultivated lands are divided in to three broad categories viz: up, medium and lowlands. These classifications are based on the relative elevations o the fields and determine the depth and duration of submergence of these lands during a year. Rainfed lowlands offer very limited scope for growing crop where rice is compulsory options to the farmers for this situation and environment.

Table 2: Total lowlands rice area in different states of the country

States	Total rice area (in million ha)	Rainfed lowland rice area (in million ha)
Assam	2.32	0.99
Bihar	5.17	2.90
Madhya Pradesh	4.83	3.12
Orissa	4.37	2.30
Uttar Pradesh	5.54	2.34
West Bengal	5.20	2.96
East (Total)	**24.43**	**14.61**
Haryana	0.56	
Himachal Pradesh	0.09	
Jammu & Kashmir	0.28	
Punjab	1.65	
North (Total)	**2.58**	
Andra Pradesh	3.53	0.01
Karnatak	1.16	0.23
Kerala	0.74	0.44
Tamil Nadu	2.52	0.05
South (Total)	7.95	0.73
Gujrat	0.57	0.13
Maharasthra	1.52	0.62
Rajasthan	0.16	
West (Total)	**2.25**	**0.75**
Other small areas	0.95	0.36
All India (Total)	**41.16**	**16.45**

About 16.45 million ha of rainfed lowland, rice is grown under several constraints. In Eastern India, comprising of Assam, West Bengal, Bihar, Orissa, eastern MP and eastern UP, rainfed lowland rice occupies more than 50% of the total area of 27.43 million ha with an average yield of 1.2 to 1.5 tones/ha as against national average of 1.9 tones/ha. The flood plains of Brahmaputra and its tributaries in Assam, the Ganges and its tributaries in eastern UP, Bihar and WB and the Mahanadi river system in Orissa constitute about 50% of the total rainfed lowlands in this zone.

Generally, three physiographic-hydrologic typesf rice lands are recognized by pluvial, phraetic and fluxial (Mooremann and Breeman, 1978). Since pluvial lands have free drainage, the lowlands are either phraetic or fluxial. In phraetic lands, the water requirement of rice crop is met both by rainwater and groundwater. The water table remains shallow at least during most part of the growing/cropping season. Water availability in the root zone during the growing cycle is influenced by the nature and distribution of natural rainfall, inherent soil characteristics and its management. In phreatic lands, rice could be grown with less drought risk. Fluxial rice lands receive water fully or partly from surface flow runoff or floodwater from rivers, streams. These lands remain inundated during the cropping season. The duration of flooding preludes the growing of other upland crops in the rainy season. In fluxial rice lands, floodwater availability depends on the rainfall pattern and the runoff from the catchment. Floodwater is not dependable source of water for growing rice in drought years. Both shallow and deep-water rice also suffers. In wet years, excessive rains in the catchment area cause widespread flooding, damaging the standing crop, Rice may be drowned or uprooted and land damaged by scouring.

Agro-Climatic Zones and Soil Characteristics

Out of 15 agro-climatic zones of the country delineated by the Planning Commission, five zones (II) eastern Himalayas comprising Assam, West Bengal, Manipur, Meghalaya, Nagaland, Arunachal Pradesh, Tripura and Sikkim, (III) lower Gangetic plains region comprising West Bengal, (IV) middle Gangetic plains with eastern UP and Bihar, (VIII) eastern plateaus and hills comprising MP, Orissa and part of West Bengal and (IX) the northern part of east coast plains and hills comprising coastal Orissa constitute eastern India. The climates in these zones varies from humid to sub-humid (rainfall: 1840-3528 mm), dry sub-humid to moist sub-humid (1302-1607 mm), dry sub-humid to moist sub-humid (1271-1436 mm) and semi-dry to dry sub-humid (780-1287 mm), respectively (Annoymous, 1998; Vaswani, 1990). The average minimum and maximum temperature in these zones are in the ranges of 6-10 and 33-41 degree C. Alluvial soils (Entisols and inceptisols), coastal and deltaic alluvial soils (Inceptisols and Alfisols), red soils (Alfisols) and laterite sois (Ultisols) occupy a considerable area of this part of the country (Murthy, 1978). In addition to these, salt affected soils, mixed red & yellow and mixed red & black soil also occurs in this tract.

Soils of lowlands areas of eastern India are usually finer in texture, massive in structure, slightly acidic to neutral in reaction and relatively more fertile than that of other land situations. Rice is harvested during mid November to end of December depending upon the land physiography and rainfall pattern. Immediately after rice harvest, there remains adequate residual soil moisture and a shallow groundwater table in these lowlands, which can sustain a short duration pulse or oilseed crop (Misra and Panda, 1994).

In a typical lowland at CRRI, Cuttack clay content in soils ranged from 35 to 41% at 0-4 cm depth, which slightly declined with increase in depth. The total available water holding capacity of soil in 1 m profile was 17-22 cm. In lowlands at Balasore, clay content was in the range of 22-23% in 0-20 cm soil layer, which suddenly drop down to 6% at 20-30 cm and then increased to 49% at 1 m depth. The total available water in 1 m profile was 17.3 cm.

Rice farming under different subsystems of lowland ecology

(1) Rainfed lowland with limited water at seeding and excess water at later growth stages

In rainfed lowland areas, the rice is direct seeded in dry soil in April-May before onset of monsoon (June) and seeds germinate with the showers received after sowing. Drought at the seedling stage kills seedlings and results in poor stand establishment. Furthermore, heavy rains in catchment areas of river basins result in water logging and flooding at late growth stages. The water hinders tillering and results in poor panicle number less than 200/m^2 as against 350 /m^2 of high yielding rice cultivars.

(2) Rainfed lowlands with limited/excess water at transplanting and excess water at later stages

When on-set of the monsoon is delayed, transplanting delayed because the farmer has to wait for adequate rains for land preparation. Monsoon rain submerges the transplanted crop, resulting in poor stand. In some areas, the land gets inundated before transplanting and farmers generally transplant aged and tall seedlings to keep the plant parts above the water. Plant survival in deeper water depends on rise in water level and on the elongation ability and submergence tolerance of the plant. Varieties that elongate fast develop weak culms and may lodge when the water level recedes. In this situation tillering is poor and panicle per unit area are less.

(3) Assured irrigated areas with excess water during crop growth due to poorly drained condition

Rice grown in the river deltaic areas of Krishna, Godavari and Indo-Gangetic rivers suffer from water stagnation even though some areas are having irrigation facilities. Heavy rains in the catchment cause flooding and stagnation in rice fields because of congested drainage or lack of drainage caused by topographically saucer-shaped land. These conditions develop during peak monsoon (July-September). The late planted crops, being at an early stage of growth, suffersfro water stagnation. In these areas, transplanting early in the season is necessary for plants to grow enough to escape submergence.

(4) Rainfed areas with excess water throughout the crop growth

Some lowland areas as get inundated with the first few monsoon breaks before seedlings and transplanting operation take place. Such areas fall in to two categories: (i) those in which water depthdoes not exceeds 1 m and (ii) those in which depth is exceeds 1 m. In former case, seedlings of 75 cm are normally used for transplanting. In later case, deep-water rice varieties are grown and the crop is seeded well in advance of the rains.

(5) Rainfed lowlands with flash flood

The lowland areas where water depth varies from 0-15 cm during most of the crop growth period, transplanted rice may be submerged for a few days to 2 weeks due to flash floods caused by over-flood from river or due to excess rainfall in shorter period of time. Some cultivars, which could tolerate floods to an extent and recover, might be suitable if floodwater recedes within 2 weeks.

(6) Rainfed coastal saline areas with excess water at later growth stage

Large part of Orissa, West Bengal, Karnataka, Tamil Nadu, Goa and Maharasthra suffer from tidal inundation. The coastal saline areas having varied characteristics, the most common being low-lying topography, inadequate drainage, fairly high water table of poor quality (5-8 m mhos/cm) and moderate to heavy soil texture. About 2 million ha of rice lands are saline and waterlogged (30-70 cm) and in these areas rice yields vary from 0.5 to 1.0 tones/ha. Cyclonic storms bring seawater to coastal areas through rivers and estuaries during May and June. The lands, even not inundated, it does not support crops because of high salt content. The salt rises to the surface through capillaries and accumulates.

(7) Rainfed lowlands with acid sulfate soils and tides

In Kerala and West Bengal, most of the acid sulfate soils are found. Crops failure due to salinity and acidification during droughts, intrusion of salt water and deep flooding are common. Observed problems include low pH, toxicity from Al, Fe and H_2S partial pressure of CO_2, salt injury, toxicity of organic acids, deficiency of P and micro-nutrients and low nutrient status (Bloomfield and Coulter,1973); Ponnamperuma *et al.*, 1973). The potential acid sulfate soils in tidal swamps will not acidify as long as the tidal effects are strong enough to prevent prolonged aeration. The soil remains productive only if wet conditions are maintained (Sinh *et. al.*, 1982).

Available water resources

The annual global flow of water ranges from 4000 to 4700 million ham, out of which 1400 m ham is utilizable for irrigation and other purpose. The present level of utilization of this available potential is just around 25%., out of which irrigation constitute about three-forth (Harswarup Singh, 1990). As per estimate of National Commission of Agriculture (1976), the average annual rainfall over the geographical area of the country is 370 m ham (112 cm). An amount of 20 m ham could be added, which is brought in form of catchment areas lying outside the country. It is estimated that about 85 m ham of water soaks into the soils (contributing to profile storage/groundwater). And 130 m ham evaporates into atmosphere. The remaining 180 m ham constitutes the average annual flow in different rivers, of which only 70 m ham is utilizable owing to the topographical, hydrological, climatic, soil and other limitations (Harswarup Singh, 1990). The available data on groundwater potential in AP shows the groundwater balance for development to be round 2.7 million ha. Distribution of groundwater reveals that the canal command districts of coastal areas as well as Rayalseema and Telengana, 33 in Rayalseema and 15 on the coastal districts largely in the non-ayacut areas (Siddique,1995), if policy decision is taken on groundwater development in command areas, an additional areas of 0.5 to 1.0 million ha in Andra Pradesh, Tamil Nadu, West Bengal, Orissa, Bihar and Assam could be brought under crop cultivation during *rabi* season, and summer rice could be grown. Even other crops like groundnut, maize, potato and linseed could successfully be grown in this rainfed lowland ecology after rainy season rice harvested on availability of profile moisture.

Cropping system under rainfed lowland ecology

The cropping system in these areas is highly influenced by rainfall pattern and amount of precipitation, soil type, topography within lowlands, time of rice harvesting and resourcefulness of the farmers. Despite their adequate skill for managing crops, farmers could be provided tools at the levels of crop germ plasm, crop management and cropping system to minimize losses due to vagaries of nature. With the objective to produce more quality food grain with less water in the forthcoming decade, rainfed agriculture has its pivotal role in meeting the daunting challenge.

Profile moisture in various soils

The lowland rice growing imposes high demand on time and energy for conversion of lowland soil to freely drained upland soil. This turn around time required for conversion to upland environment discourages the use of rice fallows. The vast majority of rainfed lowland rice of eastern India is kept fallows due to this reason. The raising of non-rice crops on residual moisture in lowland rice fallows depends on;

1. The rapidity with which the soil dries up and become amenable for tillage in a crumb soil structure of a garden soil;
2. Availability of irrigation after rice crop harvested;
3. Severity of aridity and mildness of temperature in post-rice season;

4. Sensitivity of the crops to a saturated soil condition; and
5. Length of time the soil remains moist.

Table 3: Profile moisture storage of different soils:

Soil type	Storage (cm)
Clay	76.20
Silt loam	60.96
Sandy loam	45.72
Fine sand	30.48
Coarse sand	15.24

Harvesting time of *kharif* rice

Utilization of residual moisture under rainfed lowland through various crops depends on the time of *kharif* rice harvest. Chandra and Manna,1988 reported that rice crop occupied in lowland field even up to December and thus crop planning could be done to sow other crop in December-January.

Table 4: Time of *kharif* rice harvest in different states

State	Time of harvest
Orissa	Mid-December
Bihar	Last week of November
Eastern UP	First week of December
Eastern MP	Second week of December
West Bengal	Second week of November
Assam	Last week of October

Tillage operations

Under rainfed lowland situation, after *kharif* rice harvest, it is most important to establish next crop at earliest opportunity. There would be problem for land preparation in light and medium textured soils than that with heavy clay soils, which generally occurs in lowlands. Several pulses like green gram, black gram and lentil and oilseeds like linseed and sunflower could be raised under heavy soil condition through zero tillage and dibbling of seeds in rows. The time consuming and energy demanding tillage operations is totally dispensed with and the seed broadcast either in standing crop ready for harvest pr in the stubble soon after harvest is followed. The crop so sown used to thrive on receding moisture regime. The non-rice crop is a bonus in such farming system. Surprisingly, this opportunity could not be availed of in many situations, particularly in eastern India.

Crop options for the no-till farming on residual moisture are: sun hemp for hay where soils are heavy and the non-rice growing season is short, black and green gram when whether is mild, and rape seed when ambient temperature is low for grain legume. Research agenda for no-till residual moisture farming in lowland would entail, selection of benchmark sites, development of seeding techniques, screening of crops and suitable varieties and evaluation of soil moisture conserving techniques. Besides, there is need for screening the crops towards sensitivity to tilth regarding their germinibilty and growth. It was observed that sesame required fine tilth, sunflower cloddy seddbed and black gram and *khesari* required muddy fields. The crops may also be evaluated for root penetrability. Experiments conducted at CRRI shown that the reduced tillage or shallow tillage at 0-5 cm surface conserved adequate soil moisture to grow successful crop of groundnut. Study on soil moisture profile revealed that under non-tillage treatment, the profile progressively dried, whereas under reduced tillage such drying was effectively checked due to soil mulch effect. Sharma *et al.,* (1985) observed the shallow tillage of surface soil (3-5 cm depth) during dry season and use of 1.5 – 2 cm thick layer of wheat straw mulch on wet black soil reduced evaporation by 30% and signicantly increased water use efficiency and grain yield of non-irrigated wheat , wheat, safflower, chick pea, lentil, linseed, *taramira*, mustard and toria could be established well on the residual moisture in black soil with applied minimum tillage or animal drawn improved blade harrow, bullock-drawn sweeps with tool frame.

Sowing methods and prospect of post rainy season crops

In rainfed lowland areas, deep sowing (5-7 cm) provides good opportunity for better germination and stand establishment of pulse crops. Saraf *et al*., (1990) reported that for sustaining the productivity of chick pea sowing should be done taking opportunity of first available shower when soil would have sufficient moisture at seeding depth to ensure good germination. Pulses like chickpea, lentil, lathyrus, could be advocated in rice fallows under such situations. Seed treatment, such as soaking of seeds in potassium dihydrogen orthophosphate (KH_2PO_4) before sowing or even seed soaking in water facilitates the germination and early establishment. Chatterjee and Maiti (1983) reported that during period of November-February, the temperature is low and provided good scope for growing a number of winter crops under rainfed lowland conditions like lentil, gram, safflower, sunflower, linseed, niger and peas depending upon topographical and agro-climatic conditions. They also reported that where winter in mild and short with possibilities of receiving some rainfall during February-March, the sequence of rice-groundnut and rice-sesame appeared to be very profitable.

Soil water retention and movement

At relatively low suction (less than 1 bar), soil aggregation and at higher suction (10-15 bars) the clay minerals regulate soil water retentivity in soil (Chibber, 1964; Tamhane and Dutta Biswas, 1965; Pharande *et al*., 1969.) The water content at any particular suction followed the decreasing order: black soil (smectite), alluvial soil (illite) and laterite soil (kaolinite) (Ali and Biswas, 1971). Water retention and release for crop use in soil can be improved by application of FYM, straw, lime or SSP (Biswas *et al*., 1967; Biswas *et al*., 1971; Bhagat and Acharya, 1989). The values of organic carbon, silt + clay and CEC correlated significantly with water retention (Gupta *et al*., 1984).

Soil water movement processes such as infiltration and percolation of rain or irrigation water into soils and n upward flux from ground water containing plant nutrients and salts are govern by hydraulic conductivity K of soil, being its water content on volume basis. Darcy's law, which governs the movement of water in soil, takes the form

Q = -k (-). DH/dZ

Where 'q' is the flux of water, 'H' the hydraulic head (pressure head + gravity head) and 'Z' is the position coordinate. Saturated hydraulic conductivity K (s) as relevant to waterlogged rice soil, is invariant at a given temperature but depends on soil properties. Its values may be in the range of 1-10 cm h^{-1} in coarse textured soils and 1-0.1 cm ha^{-1} or even less in heavy soils. But in *rabi* season, water moves under unsaturated conditions and such movement is guided by K (-) depending on water content. Using soil water pressure head values (H) measured by tension meter and soil water content (-) by gravimetric or neutron probe method, K(-) can be directly determined in the field (Nielsen *et al*., 1973; Patra and Misra, 1985). K(-) for coarse and fine textured soils of Orissa has been formulated as :

K = 0.362 expo 48.17 (0-0.335) and

K = 0.205 expo 67.15 (0-0.468)

Respectively, where K is expressed in cm h^{-1}.

Dongale (1987) reported that under zero tillage, infiltration rate after 180 minutes increased from 4.4 cm h^{-1} in rice mono-cropped soil to 11.0 cm h^{-1} in rice - legume double-cropped soil. Tillage increased the intake of water further.

Mulching

Mulching is an important technique for conserving residual soil moisture especially in the no-tillage plots under rainfed lowland conditions. Besides conserving moisture, it also suppresses weed growth. Chatterjee and Sen (1977) reported that mulching the crop fields with 5 cm thick straw could increase the yields of winter crops. They observed yield improvement with mulching in case of linseed, safflower and lentil because of increased water use efficiency.

MandaI *et at.* (1990) observed that a good crop of safflower can be harvested under late sown condition after the harvest of lowland paddy even without giving any irrigation with the help of residual soil moisture and with a lower dose of each of N, P_20_5 and K_20 @ 30 kg h^{-1} provided that the soil moisture is conserved with the help of straw mulch @ 7 th^{-1} at sowing.

Paira Cropping

Paira cropping is another useful practice for raising winter crops in rainfed lowlands. Roysharma et al. (1982) reported that lentil give grain yield of 15.9, 16.3 and lOA q h^{-1} at Dholi (1976-80), Kanpur (1974-76) and Barielly (1974-75), respectively with inoculated seeds and top-dressing of 20 kg N and 30 kg P20S h^{-1} (twice at 40 and 80 DAS) with paira cropping. Gangasaran (1990) reported that paira cropping *of* linseed is certain parts of U.P. and West Bengal. Linseed is broadcast in the standing crop *of* rice 10-20 days before harvest.

In an experiment on *paira* cropping, rice seedlings were transplanted on 15th July and chickpea, oilseed flax and lentil were broadcast into the standing rice crop on 25th November at normal and one and half times the normal seed rate without and with 20 kg N + 18 kg P h^{-1} for chickpea and lentil and 30 kg N + 9 kg P h^{-1} for flax. Oilseed flax intercrop with rice gave the highest rice yield equivalent of 5.1 th^{-1} when fertilizers were applied. Sowing rates had no significant effects on yield (Thakur,1990).

Physical improvement in soil structure

Physical soil health is an important consideration in seed germination and conservation. Utilization ofresidual soil moisture in *rabi* pulses and oilseeds after rice in rainfed lowlands is depended on physical health of the soil. Biswas (1982) has conclusively suggested three practices for improvement of soil physical condition leading to increased soil productivity *viz.* (1) Judicious use of organic manure, (2) use of phosphates and (3) legume cultivation in crop rotation. Biswas and Khosla (1971) have earlier reported that long term manuring with FYM or sun-hemp green manuring in red and laterite soils at Bhubaneswar and Ranchi improved water stable aggregates percentage, water storage capacity and available water content. The results obtained from All India Coordinated Project on long-term Fertilizer Experiments revealed that available water holding capacity ofsoil increased from 18.8% with N alone to 19.5% with combined use ofNPK

+ FYM for 11 years in 8.8% was reported in laterite soils of Bhubaneswar (Nambiar and Ghosh, 1984). Long-term fertilizer experiments conducted at CRRI for17 years showed that continuous combined application ofinorganic N (44.8 kg/ha) and compost (9.2 t/ha/year) significantly improved soil aggregation (Bandyopadhya *et al.,* 1969). For sustainability in production, it is neither chemical fertilizers nor the organic manures alone but their integrated use is highly beneficial (Patnaik *et al.,* 1989).

Soil fertility improvement

All India Coordinated Project on Long-term fertilizer experiments clearly showed declining trend of productivity even with application of N, P and K fertilizer due to emergence of deficiency of Sand Zn, Available S (Morgan's extractable S) decreased to around 3-5 mg/kg in plots where DAP had been applied but increased in plots where SSP had been continuously applied. This effect was quite spectacular at Barrackpore, Jabalpur and Bhubaneswar (Nambiar and Abro, 1989). In the context of growing pulses and leguminous oilseeds after rice in rainfed lowlands, supply of P, S and Ca is very important to improve crop productivity.

Effective utilization of water

In lowland areas of eastern India, because of high rainfall there is a rise reserve of ground water at shallow depth. But, except in Uttar Pradesh and some parts of West Bengal, in rest of the eastern region, the ground water utilization is very poor (Verma and Agarwal, 1990; Rao, 1990; Patnaik, 1994). Although in semi-arid and arid regions of the country, indiscriminate use of ground water lowers the water table and is expected to eventually create the problems of drought, there is less chance of and deep tube wells should be developed depending on the position of the aquifers and quality of water. This may help in providing life saving irrigation to the crops grown after rice. Lowland areas of eastern India also provide great scope for storage of rainwater in ponds, ditches or trenches at one site or at comer of the field for adoption of the water-harvested structures (Sinhababu and Venkateswarlu, 1995).

Coastal saline soil management

Coastal soil salinity generally develops due to seawater inundation, capillary rise of saline ground water and its evaporation during the dry season and deposits of windblown salts on land surface. This can be minimized by providing sea dykes with one way sluice gate to allow drainage from land to sea but to check the reverse flow of sea water into land (Yadav *et at.,* 1979). These dykes need to be stabilized by growing *Casuarina*, cashew nut and salt tolerant grasses on them. Shelterbelt and windbreak plantations along seacoast are also recommended to check deposit of wind-blown salts on arable land. Salt tolerant rice variety could be grown during wet season. Lunishree recently developed at CRRI, Cuttack and CSR 6 is more suited to coastal salt affected lowlands. Salt tolerant crops such as sugar beet, linseed, barley and even wheat have been found to be promising during dry season. Most of the land in coastal saline areas remains fallow during dry season because of high salinity or non--availability of good quality irrigation water. Bandyopadhya and Sen (1975) have found that at few locations of Sunderban area, quality of ground water is good only at a depth of 180 m or below. A study on assessment of quality of surface water available in Orissa coast canal and at few definite sites along the Suvarnarekha and Salandi rivers closer to seas coast indicate that there is scope of adopting multiple cropping in some pockets of Balasore by using water from these sources during the first half of the dry season (Panda *et at.,* 1980).

Important cropping pattern followed in rainfed lowlands of eastern India

An exhaustive compilation on agro-climatic zone specific research under National agricultural Research Project indicates the feasibility and prospects of the following rice based cropping systems in rainfed lowlands of different agro-climatic zones of eastern India (Ghosh, 1991)

Assam	Jute-rice-rapeseed, Jute-rice-lentil, Rice-black gram
Bihar	Rice-linseed, Rice- Khesari (*Piara*crop)
	Rice-chickpea, Eastern MP, Rice-khesari, Rice-linseed, Rice-Black gram, Rice-chickpea andRice-rapeseed –mustard/kulthi
Orissa	Rice-green gram, Rice-black gram, Rice-kulthi, Rice-sesamum, Rice-groundnut, Rice-khesari and Jute-rice-green gram/black gram and Rice-field pea (*Piara* crop)
Eastern UP	Rice-Chickpea, Rice-peas, Rice-linseed Rice-lentil
West Bengal	Rice-lentil Rice-khesari (*Paira* crop) Rice-rape seed/mustard Rice-linseed and Rice-safflower/sunflower

Crop response under residual soil moisture ecology

The various crops have been identified for cultivation under residual soil moisture condition under different agro-climatic regions.

Table 5: Rice based crop yield under residual soil moisture in irrigated ecology

Crop	Variety	Yield t/ha		Place of work	Source
		Residual moisture	Irrigated Ecology		
Mustard	Pus a Bold	0.37	0.37	Cuttack	Dinesh Chandra
	Pusa Bahar	0.37	1.03	Cuttack	1993
Raimash	PDR 14	0.53	1.07	Cuttack	-do-
Peanut	ICGS 44	0.71	1.36	Cuttack	Chandra *et al.* 1988
Mungbean	PDM 11	0.11	0.10	Cuttack	-do-
Urdbean	PDU 54	0.15	0.18	Cuttack	-do-
Wheat	Swati	3.72	4.16	Bilaspur	Parihar *et al.*, 1955
Chickpea	JG74	2.02	1.88	Bilaspur	-do-
Indian mustard	Baruna	1.05	1.22	Bilaspur	-do-
Sunflower	Modem	1.91	2.11	Bilaspur	-do-

Irrigating these crops sometimes reduced the yields e.g. mung bean safflower and horse gram, Mung bean variety Pusa Baisakhi, safflower varieties JSF-l and APRR-3 and horse gram (local varieties) were well suited under the residual moisture in rainfed lowland ecology at Cuttack (Table 5).

Future Strategies of research and development

The extent of occurrence of rainfed lowland, particularly in eastern India is large. But cropping intensity in this ecosystem is low, most probably due to the lack of technology of efficient utilization of residual soil moisture during dry season. In fact, intensification of cropping with the utilization of this rice-natural reserve in rainfed lowlands of eastern India is possible and can provide food security to the growing population of this region. To achieve this objective, the following research and development strategies may be pursued.

1. Improvement in soil moisture storage and regulation of upward flux of water through soil structural manipulations;
2. Multi-locational evaluation of cropping systems, introduction of new crop species and varieties of different pulses and oilseeds in dry season under rainfed lowland situations along with their screening for sensitivity to soil tilth and soil/air temperature and for root penetrability;
3. Genetic improvement in pulses, oilseeds and other crops with particular reference to residual soil moisture utilization;
4. Identification of specific limiting factors of crop production (bio-physical and socio-economic factors) and their mitigation options.
5. Integrated approaches of nutrient management, water management and plant protection measures.

References

Ali, M.H. and Biswas, T.D. 1971. *J Indian Soc Soil Sci,* 19: 363.

Bandyopadhya, A.K. and Sen, H.S. 197 6. Technology, *Sindri*, 12: 151.

Bandyopadhya, A.K., Sahoor, R, Bhadrachalam, A and Patnaik, S.1969. *J Indian Soc Soil Sci,* 17 : 369.

Bhagat, R.M. and Acharya C.L. 1989. *J Indian Soc Soil Sci,* 37: 229.

Biswas, T.D. 1982. *J Indian Soc Soil Sci,* 30 : 427.

Biswas, T.D. and Khosla B.K. 1971. *Proc Int Symp Soil Fert Evaluation*, 1 : 831.

Biswas, T.D. Jain B.C. and Mandal S.C. 1971. *J Indian Soc Soil Sci.,*19 : 31.

Biswas, T.D., Pharande K.S. and Naskar G.C. 1967. *J Indian Soc Soil Sci,*15 : 289.

Bloomfield, C. and Coulter, T.K. 1973. Genesis and management of acid sulfate soils. Adv. Agron 25. 265-326.

Chandra, D. and Manna, G.B. 1988. Rice going deep. *Intensive Agriculture,* 26(4 & 5):12-18.

Chatterjee, B.N. and Sen, H.S. 1977. *J Soil and Water Conservation in India* 27 (1 & 4):101.

Chhiber, R.K. 1964. *Bull Natn Inst Sci India* 26 : 148

Dangale, J.H. 1987. *J Indian Soc Sloil Sci.,* 35: 274.

Ganga Saran 1990. In Agronomic Research towards Sustainable Agriculture (ed) Singh K.N. and Singh R.P., p., 153-62.

Gangwar, M.S., Sachan, R. S. and Rathore, T.R. 1982. In Review of Soil Research in India. Indian Soc Soil Sci, New Delhi, p. 537.

Ghosh, S.P. 1991. Agro-climatic zone specific research. ICAR, New Delhi.

Gupta, R.K., Kumar, S. and Singh, T. 1984. Consolidated Report. AICRP Soil Physical Conditions to increase Agricultural Production of Problematic Areas, IARI .

Harswarup, Singh 1990. Management of soil and water resources for sustainable agriculture. In Proceedings of the International Symposium on Natural Resource management for a Sustainable Agriculture, February 6-10, 1990 held at New Delhi, pp. 40-68.

Mandal, B.K., Mandal, B.K. and Das S.K. 1990. *Indian J Agron* 35(3): 270-274.

Misra, C. and Panda, D. 1994. In Management of Land and Water Resources for Sustainable Agriculture and Environment. Indian soil Sci, New Delhi, pp 121-144.

Moormann, F.R., van Breman N. 1978. Rice: soil, water, Land. International Rice Research Institute, Manila, Philippines.

Murthy, R.S., Hirekerum L.R., Deshpande, S.B., VenkataRao B.V. and Shankarnarayana H.S. 1982. Benchmark soils of India. National Bureau of Soil Survey and Land Use Planning (ICAR), Nagpur.

Nambiar, K.K.M. and Abrol, I.P. 1989. *Fertil News* 34 (4) : 11.

Nambiar, K.K.M. and Ghosh, A.B. 1984. in L.T.F.E Res. Bull.No.1, IARI , New Delhi.

Nielsen, D.R., Biggar J.W. and Erth, K.T. 1973. Hilgardia42: No.7

Patanaik, J. 1994. Proc. Workshop Integrated Development of Irrigated Agriculture (east zone). Central Board of Irrigation and Power, New Delhi, April 20-22, 1994, pp. 10-15.

Patanaik, S., Panda, D. and Dash, R.N. 1989. *Fertil News* 34 (4): 47.

Patra, T.G. and Misra, C. 1985. *J Indian Soc Soil Sci* 33: 469.

Pharade, K.S., Jain, B.L. and Biswas, T.D. 1969. *J Indian Soc Soil Sci.* 17: 240.

Ponnamperuma, F.N., Attanaudana and Beye, G.1973. Amelioration of three acid sulfate soils for lowland rice. Pages 391-393 in Proceedings of the International Symposium on Acid sulfate soils, 13-20 August 1972, Wageningen, Netherlands.

Rao, Y.P. 1990. Proc Workshop on Research needs on land and water management for enhancing agricultural production in eastern region, October 4-5, 1990. Water Technology Centre for Eastern Regions, Bhubaneswar, pp. 1-15.

Roysharma, C. 1985. In Agronomic Research Towards Sustainable Agriculture (ed) Singh K.N. and Singh, R.P., pp 163-179.

Saraf, C.S., Shinde, V.S. and Hegde, R. 1990. In Agronomic Research Towards Sustainable Agriculture (ed) Singh K.N. and Singh R.P., pp 153-166.

Sharma, R.A., Upadhyay, M.S. and Tanon, R.S. 1985a. *J Indian Soc Soil Sci* 33:383 Hindu Publication, Madras, pp. 35-43.

Singh, B.N., Sahu, S.P. and Banerjee, S. 1982. Varietal requirement of rain fed rice in north Bihar. Paper presented at the Task Force meeting on Rainfed lowland rice, 12-13 November, 1983. Agricultural Research Institute, Mithapur, Patna, India.

Sinhababu, D.P. and Venkateswariu, B. 1995. Rice-fish in rainfed lowlands. Central Rice Research Institute, Cuttack, pp.I-14.

Tamhane, R.V. and Dutta Biswas, N.R. 1965. *J Indian Soc Soil Sci* 13:205.

Thakur, R.B. 1990. *Int Rice Res Newsl.* 15 (3):34.

Verma, V.P. and Agrawal, D.K. 1990. Proc Workshop on Research needs on land and water management for enhancing agricultural production in eastern region October 4-5, 1990. Water Technology Centre for Eastern Region, Bhubaneswar, pp. 85-89.

Yadav, J.S.P., Bandyopadhya, A.K., Rao, K.V.G.K., Sinha, T. S.and Biswas, C.R. 1979. Coastal Saline Soils of India. Central Soil Salinity Research Institute, *Kamal Bull* 5:1-34.

Section 5

Crop Management and Productivity

24

A Review on Approaches of Phyto-Extracts on Plant Disease Management

***Diganggana Talukdar*[1], *Utpal Dey*[2], *Dipika Sarmah*[3] *and Sudip Das*[4]**

[1]Department of Plant Pathology and Microbiology
College of Horticulture, CAU, Ranipool, East Sikkim-737135, India
[2]Division of Plant Production, ICAR Research Complex for NEH Region
Umiam - 793103, Meghalaya, India
[3]Department of Floriculture, College of Horticulture, CAU, Ranipool
East Sikkim-737135, India
[4]Department of Plant Biochemistry, College of Horticulture, CAU, Ranipool
East Sikkim-737135, India
Email: talukdardigangana@gmail.com

Abstract

The increase incidence of plant diseases have created challenging troubles in commercial agriculture and posed economical threats to both conventional and organic farming systems. To combat these problems there has been indiscriminate use of the synthetic chemicals that created different types of environmental and toxicological problems. Therefore attention has been directed towards exploitation of plant products or phytoextracts as novel chemotherapeutants in plant protection. Numerous botanical pesticides have been used globally and more has been discovered in due course of time. Pyrethroids, neem, etc are well established commercially as botanical pesticides and recently some essential oils of higher plants have also been used as antimicrobials as they are safe and widely accepted by the consumers. Use of phytoextracts may thus offer a new source of antibacterial, antifungal and antiviral agents against microorganisms. Botanical pesticides are best fitting for organic food production in industrialized countries and can plays greater role in the production and post-harvest protection of food products in developing countries.

Keywords: Antibacterial; phyto extracts; antifungal; antiviral; antimicrobials; Botanical pesticides; plant protection;

Introduction

The global concern regarding safety and security of food, environmental feature and pesticide residue free food have raised the general public's interest in the use of alternative disease and pest control measures (Singh and Saratchandra, 2005) resembling use of phytoextracts (botanicals) and organic amendments (biopesticides and biocontrol agents) which manages the diseases and pest in an eco friendly way. Holistic plant disease management approach is a sustainable and wide-ranging approach of pathogen control utilizing combined means to reduce the population of pathogens while minimizing health, environmental and economic risks. Plant pathogens are responsible for reducing losses both in quantitative and qualitative way including increased production cost, threat to animal health and environment, limiting the type of cultivars grown, loss of natural resources and less remunerative alternatives adopted. More than 800 million people in the developing countries do not have adequate food supplies and 10% food loss due to plant diseases (Strange and Scott, 2005). In order to combat the crop losses caused by plant diseases, holistic, cost effective and eco-friendly management practices which are cheap, locally available, non-toxic, easily biodegradable are to be adopted as alternative methods. Numerous antifungal compounds such as *azadirachtin*, organosulfur, phenolics, phenolic acids, coumarins, pyrones, flavonoids, isoflavonoids, steroids, steroidal alkaloids and other miscellaneous compounds have been found in the plants naturally which exhibits antibacterial, antifungal and antiviral activity. Utilization of organic amendments for controlling plant pathogens has been considered as best technique as mentioned by Lazarovits, (2010). Organic amendments like poultry manure, meat and bone meal, and soy meal contains high nitrogen and significantly reduces wide range of soil-borne plant pathogens. Disease control with organic amendment takes place when soil and biological factors are in conducive state that reduces the survival of the pathogens. Thus, they are potential ecofriendly approaches for disease management strategy. The prospect of using phytoextracts which are attributed to release plant essential oils, adehydes, antibiotics and phenolics are easily available, environmentally safe, less risky for developing resistance in pathogens, less hazardous to non target organisms and has no adverse effect on plant growth and economic.

Over the past decade there has been a prominent attention in excavating the antimicrobial agents of plant origin (or phytoextracts), isolating and identifying active components having potential use in integrated crop protection and pest management programs (Eksteen *et al.*, 2001). Control of plant pathogens requires employment of alternative techniques because traditional handling with synthetic chemicals has caused various problems such as toxicity to users (Whalen *et al.*, 2003) and impairment of beneficial organisms (Anderson *et al.*,

2003). Another important aspect is that pathogenic organisms have generated resistance to the active ingredient of some synthetic fungicides in response to selection pressure due to high dose and continuous applications, causing great economic losses (Cooke *et al.*, 2003; Leroux, 2003). An economical and efficient alternative for disease control is the use of natural products derived from plants (secondary metabolites) (Wilson *etal.,* 1999), since it does not affect environment and their residues are easy to degrade. On the other hand, vegetal biodiversity in Mexico is available to be exploited, especially by their natural non-toxic, biodegradable compounds (Hernandez *et al.*, 2007). The potential use of plant extracts to control plant pathogens has been reported in different laboratory (Hernández *et al.*, 2010; Castillo *et al.,* 2010; Osorio *et al.*, 2010), greenhouse (Bergeron *et al.*, 1995) and field studies (Hernandez *et al.,* 2006, 2008). Based on the popular uses of plants in the Coahuila Southern region, botanical resources can be identifying by their antimicrobial potential against plant pathogens. However, still there is a lack of research in this field. As previously mentioned, there is a need for new plant disease management options with lower environmental and economic impact, which expresses a similar or greater effect in controlling pathogens.

Chemical compounds in bio-fungicides

The botanical bio pesticides represent an alternative for pest control with low environmental impact and high food safety. Several products derived from plants have shown an antimicrobial effect. Among the main compounds present in these extracts are: flavonoids, phenols, terpenes, essential oils, alkaloids, lectins and polypeptides. Some plant extracts containing these metabolites has been extracted in water or other solvents, depending on its polarity, and in powder form (Bautista *et al.*, 2003). The enormous diversity of secondary metabolites and biological properties present in plants, are still subject of study. The limited knowledge that currently prevail about plant extracts, is an interesting point to begin studies with plants of almost any kind. However, there are some chemical derived from the knowledge base of the plants (Montes, 1996). Some families of plants may be more feasible for study, such as: *Solanaceae*for its high alkaloid content, or *Mimosaceae*t hat's present species rich in tannins, or *Lamiaceae* and *Meliaceae* because their wide diversity of terpenoids.

Effectiveness of phytochemicals

The phytochemicals derivatived from the plants act as an alternative methods for controlling plant diseases caused by different phytopathogens. These phytochemicals have anti fungal and anti bacterial properties as most of the extracts inhibited mycelium or bacterial growth. This effect varies according to

Table 1: Source, phyto-pathogen, solvent and references

Source	Phytopathogen	Solvent	Reference
Eucalyptus lameldulensis	*Xanthomonasoryzae,*	Ethanol	Shehu *et al., 2013*
Moringao leifera	*Corynoebacteriumxerosis*		
Azadirachtaindica,	*Fusariumoxysporum*f.sp.		
Allium sativum	*lycopersici,Rhizoctoniasolani*		Hadian, 2012
Azadirachtaindica,	*Phomopsisvexans, Fusariumoxysporum,*		
Putranjivaroxburghii	*Aspergillusflavus, Aspergillusnigar,*		
	Curvularialunata, Penicillium spp.		Kuri*et al.*, 2011
Flourensiacernua	*Rhizoctoniasolani*	Methanol	Castillo *et al.*, 2011
F. microphylla	*Phytophthorainfestans*	Chloroform	Castillo *et al.*, 2011
Eugenia aromatica	*Agrobacterium tumefaciens*, *Erwiniacarotovora*pv. *carotovora*, *Pseudomonassyringae*pv. *syringae*, *Ralstoniasolanacearum*, and *Xanthomonascampestris*pv. *pelargonii*, *Rhodococcusfascians*and *Streptomyces* spp.		Huang and Lakshman, 2010
F. retinophylla	*Alternariasp*	Hexane	Castillo *et al.*, 2011
Origanummajorana	*Fusariumoxysporum*	Diethyl	Castillo *et al.*, 2011
Bouvardiaternifolia	*Colletotrichumcoccodes*	Ethanol	Castillo *et al.*, 2011
Aloe vera	*Colletotrichumgloeosporoides*	Water	Castillo *et al.*, 2011
Larreatridentata	*Pythium* sp.	Lanolin	Castillo *et al.*, 2011
Agave lechuguilla	*Botrityscinerea*	Cocoa butter	Castillo *et al.*, 2011
Larreatridentata	*Botrytis cinerea, Fusariumoxysporum, Colletotrichum Coccodes*		Lira *et al.*, 2006
Origanummejorana	*Phytophthorainfestans*	Methanol	Gamboa*et al.*, 2003a
Agave lechuguilla	*Clavibactermichiganensis*subsp. *michiganensis*		Castillo *et al.*, 2011
Camellia sinensis, Acacia arabicae, Azadirachtaindica	Erwiniacarotovora subsp.Carotovora		Bhardwaj and Laura, 2008; Newton *et al.*, 2002
Phyllanthusamarus	*Curvularialunata*		Akinbode, 2010
Ocimumbasilicum	*Drechsleraheveae*		Ogbebor and Adekunle, 2008
Vitexnegundo	*Colletotrichumfalcatum*	Chloroform/ Ethanol/ Aqueous	Prince and Prabakaran, 2011

plant species and solvent used for phytochemicals extraction (Gamboa *et al.*, 2003b; Hernández *et al.*, 2006, 2008, 2010; Castillo *et al.*, 2010).

Factors involved in phytochemicals

The diversity of compounds in plant tissues is widespread from the chemical point of view which resulted in the identification of thousands of these compounds having unknown biological functions, which represents a challenge in the search for compounds that are highly efficient for crop plant diseases control. The main groups of secondary metabolites present in plant tissues are polyphenols, terpenes and nitrogen compounds. Tannins (polyphenols) are phytochemicals widely distributed in the plant kingdom, found in roots, leaves, seeds and fruits. These compounds are secondary metabolites with a molecular weight range of 300-20000 D (Gonzalez *et al.,* 2009). The tannins are non nitrogen compounds with astringent flavor, amorphous, mostly soluble in water and alcohol (Medina *et al.,* 2007). Its hydrophilic nature allows them associate with carbohydrates, alkaloids and proteins by hydrogen bonds, covalent bonds and hydrophobic interactions.

Plant extracts effect on microorganism's growth inhibition

Fungal growth inhibition

Plant based metabolites appear to be one of the good alternatives for fungal disease management and in this context, Tapwal *et al.*, (2011) screened the crude extracts of five plants having potential antifungal properties against *A. solani, A. zinnia, Fusarium oxysporium, Rhizoctonia solani* and *Curvularia lunata.* Most polyphenolic compounds and resins derived from plants inhibits mycelia growth plant pathogens and the inhibition ranges from 0 to 100% with a enhance effect with increase in concentration, *i.e.* as concentration of resin or polyphenols increases, the mycelia growth of plant pathogen is significantly reduced (Hernández *et al.*, 2010; Castillo *et al.*, 2010). Inhibition of fungal mycelium growth by plant extract may be as effective as that of synthetic molcules. having no negative impact on the environment (Hernández *et al.*, 2006; Hernández *et al.*, 2008). Botanical extracts have a very good potentiality against seedborne pathogen of brinjal (Usha*etal.* 1993; Devi *et al.* 1999; Koirala *et al.* 2005; Alberts *et al.*, 2006).

Bacterial growth inhibition

Plant extracts has an excellent effect on growth inhibition of plant pathogenic bacteria like *Clavibacter michiganensis* sub sp. *Michiganensis* (Castillo *et al.*, 2011). Application of clove oil (*Eugenia aromatica*) has been a potential alternative control measure to combat bacterial wilt of tomato and geranium. Clove oil is a mixture of different compounds including mainly eugenol, as well as eugenyl acetate, caryophyllene and other minor constituents (The Merck Index, 2001) which contribute to its antibacterial properties. Ji *et al.* (2005) have evaluated the application of essential plant oil thymol significantly reduced bacterial wilt incidence and increased plant yield of a susceptible tomato cultivar. Many scientist like Bowers and Locke, 2000, 2004; Momol *et al.*, 2000; Oka *et al.*, 2000, depicted that the clove oil as well as other essential plant oils possess broad-spectrum activity against a wide range of plant pathogens including bacteria, fungi and nematodes in soil. Pretorius *et al.*, (2003) collected plant extracts from 26 plant species in the Free State Province of South Africa, and conducted in vitro test for their potentially to inhibit the growth and found four plant species namely *Eucleacrispa*, *Acacia erioloba*, *Buddlejasaligna* and *Senna italic* showed highest level of antibacterial activity upon five plant pathogenic bacteria (*Agrobacterium tumefaciens*, *Clavibacter michiganense*, *Erwinia carotovora* pv. *carotovora*, *Pseudomonas solanacearum* and *Xanthomonas campestris*. Leaf extracts of *Camellia sinensis*, known for its antimicrobial activity of different components like caffeine, tannins and other polyphenolic compounds particularly gallocatechin (Kubo *et al.*, 1992). These plant essential oils, therefore, have the potential to be developed into an environmentally friendly and potential alternative to the use of methyl bromide in integrated management of soil-borne plant diseases.

Viral growth inhibition

Effective viricides are lacking for the control of plant virus diseases. Certain higher plants posses endogenous virus inhibitors of which proteinaceous antiviral substances are of particular interest (Balasubramanyam *et al.*, 2000) However, in recent years, safer botanical extracts are gaining importance in crop protection against diseases. *Bouganvallie spectabilis* extract was found to be effective against Tobamovirus infecting bell pepper and Tomato plants (Madhusudhan *et al.,* 2011). Plant seed oils were mainly used to control plant viruses (Jayashree *et al.*, 1999; Mandal and Singh, 2001; Kulkarni and Byadagi, 2004; Madhusudhan *et al.*, 2005). Spray treatment is more effective than seed treatment in reducing the virus concentration. Along with lower virus concentration the higher peroxidase activity and lower catalase activity can be observed in spray treated plants.

Mode of action

Plants contain wide array of secondary metabolites or natural products that facilitate interactions with the environment. Plant Kingdom could be considered a rich source of the most diverse structures (Radulovi, *et al.*, 2013). There are more than 12,000 known alkaloids, more than 8,000 phenolic compounds and over 25,000 different terpenoids, many of which were proven to possess strong antimicrobial properties (e.g. thymol, eurabienol, etc.). Phenolic substances containing one carbonyl group and have been reported to possess antiviral activity against a wide range of viruses. Some phytochemicals have been reported to exert antimicrobial activity against bacteria, fungi yeast having different mechanisms. Toxicity of phenols for microorganisms is attributed to enzyme inhibition by oxidation of compounds. The mode of action of terpenes and essential oils has not been fully elucidated, but is postulated to cause cell membrane disruption by the lipophilic compounds. The postulated effect of alkaloids is that these compounds may alternate with DNA, while lectins and polypeptides are known cause competitive inhibition of adhesion of microbial proteins to host polysaccharide receptors (Cowan, 1999).

Conclusion

Plant secondary metabolites are the excellent starters for the development of new antimicrobial agents. It depends on the medicinal chemists, pharmacologists, phytochemists and researchers from diverse fields how to extract the enormous potential hidden facts from the plants and utilize them in the best possible way. Exploitation of naturally available chemicals from plant protection would play for a major role in development of future commercial pesticides for crop protection strategies, with special reference to plant disease management (Varma and Dubey, 1999; Lale and Abdulrahman, 1999; Islam *et al.*, 2003; Yazdani *et al.*, 2011). Natural plant products are important sources of new agrochemicals for the control of plant diseases (Kagale *et al.*, 2004; Maya and Thippanna, 2013). Their uses in controlling diseases is considered as an interesting alternative to synthetic fungicides due to their less negative impacts on the environment, as they do not leave any toxic residues and therefore can effectively replace synthetic fungicides (Cao *et al.*, 2001; Munoz-Mingaro *et al.*, 2003). Plant preparations have been used for centuries in medicine and pest control. Farmers in India use neem leaves to protect their stored grain from insects. Herbs and spices, such as basil and clove, have been used to protect food from spoilage, as both have antimicrobial properties (Arora and Kaur, 1999; Manohar *et al.*, 2001). In recent times, focus on plant research has increased all over the world and a large body of evidence has been collected to show immense potential of plants used in various traditional systems.

References

Akinbode, O.A. 2010. Evaluation of antifungal efficacy of some plant extractson *Curvularialunata*, the causal organism of maize leafspot. *African Journal of Environmental Science and Technology,* 4(11): 797-800.

Alberts, J. F., Engelbrecht, Y., Steyn, P. S., Holzapfel, W. H., and Vanzyl, W. H. 2006. Biological degradation of aflatoxin B1 by Rhodococcusery thropolis cultures. *International Journal of Food Microbiology*, 109:121–126.

Anderson, B.S., Hunt, J.W., Phillips, B.M., Nicely, P.A., Vlaming, V. de, Connor, V., Richard, N., &Tjeerdema, R. S. 2003. Integrated assessment of the impacts of agricultural drain water in the Salinas River. *Environmental Pollution*, 124(3):523-532.

Arora, D. S. and Kaur, J. 1999. Antimicrobial activity of spices.*International Journal of Antimicrobial Agents,* 12(3):257-262.

Bautista, B.S., García, E., Barrera, L., Reyes, N. and Wilson, C. 2003. Seasonal Evaluation of the postharvest fungicidal activity of powders and extracts of huamúchil (*Pithecellobiumdulce*): action against *Botrytis cinerea*, *Penicilliumdigitatum*and *Rhizopusstolonifer*of strawberry fruit. *Postharvest Biology & Technology*, 29(1): 81-92

Bergeron, C., Marston, A., Hakizamungu, E., & Hostettmann, K. 1995. Antifungal constituens of *Chenopodiumprocerum. International Journal of Pharmacognosy*, 33(2):115-119.

Bhardwaj, S. K. and Laura, J. S. 2008.antibacterial activity of some plant extracts against pathogenic bacteria *Erwiniacarotovora* sub sp. *carotovora*. *Potato J.*, 35 (1-2): 72-77.

Boehm M.J., Wu T., Stone A.G., Kraakman B., Iannotti D.A., Wilson G.E., Madden L.V., Hoitink H.A.J., 1997. Crosspolarized magic-angle spinning (Sup13) nuclear magnetic resonance spectroscopic characterization of soil organic matter relative to culturable bacterial species composition and sustained biological control of Pythium root rot. *Applied and Environmental Microbiology,* 63: 162-168.

Borrero C., Trillas M.I., Ordovás J., Tello J.C., Avilés M., 2004. Predictive factors for the suppression of Fusarium wilt of tomato in plant growth media. *Phytopathology*, 94:1094-1101.

Bowers J.H., Locke J.C., 2000. Effect of botanical extracts on the population density of *Fusariumoxysporum* in soil and control of Fusarium wilt in the greenhouse. *Plant Disease* 84: 300-305.

Bowers J.H., Locke J.C., 2004. Effects of formulated plant extracts and oils on population density of *Phytophthoranicotianae* in soil and control of Phytophthora blight in the greenhouse. *Plant Disease* 88: 11-16.

Cao, K. Q. and Forrer, H. R. 2001b.Current status and prosperity on biological control of potato late blight. *Journal of Agricultural University of Hebei*, 24, 51-58.

Castillo, F., Hernández, D., Gallegos, G., Méndez, M., Rodríguez, R., Reyes A. &. Aguilar, C.N.2010. *In vitro* antifungal activity of plant extracts obtained with alternative organic solvents against *Rhizoctoniasolani* Kühn. *Industrial Crops and Products*, 32(3):324–328.

Castillo, F.D.H., Reyes, F.C., Morales, G.G., Herrera, R.R. and Aguilar, C.N. 2011. Plant extracts from mexican native species: an alternative for controlof plant pathogens, Research in organic farming, Dr. Raumjit Nokkoul (Ed.), ISBN: 978-953-307-381-1, InTech, available from:http://www.intechopen.com/books/research-in-organic-farming/plant-extracts-frommexican-native-species-an-alternative-for-control-of-plant-pathogens

Cooke, D.E.L., Young, V., Birch, P.R.J., Toth, R., Gourlay, F., Day, J.P., Carnegie, S.F., & Duncan, J.M. 2003.Phenotypic and genotypic diversity of *Phytophthorainfestans* populations in Scotland (1995-97). *Plant Pathology*, 52(2):181-192.

Cowan, M. M. 1999. Plant products as antimicrobial agents.*Clinical Microbiology Review*, 12(4):564-582.

Devi, K. T., Mayo, M. A., Reddy, K. L. N., Delfosse, P., Reddy, G., Reddy, S. V.1999.Production and characterization of monoclonal antibodies for aflatoxin B1.*Letters in Applied Microbiology*, 29: 284–288.

Eksteen D, Pretorius JC, Nieuwoudt TD, Zietsman PC 2001.Mycelial growth inhibition of plant pathogenic fungi by extracts of South African plant species. Annals of Applied Biology 139: 243–249.

Gamboa, A.R., Hernández, C.F.D., Guerrero, R.E. & Sánchez, A.A.2003a. Inhibition of mycelial growth of *Rhizoctoniasolani*Kühn and *Phytophthorainfestans* Mont (de Bary) with methanolic plant extracts of Hojasén (*Flourensiacernua* DC), Marjoram (*Origanum majorana* L.) and trumpet [*Bouvardiaternifolia* (Ca.) Schlecht.]. *Mexican Journal of Phytopathology,* 21(1):3-18

Gamboa, A.R., Hernández, F.D., Guerrero, E., Sánchez, A., Villarreal, L.A., López, R.G., Jiménez, F., & Lira, S.R.H.2003b. Antifungal effect of *Larreatridentata* extracts on *Rhizoctoniasolani* Kühn and *Phytophthorainfestans* Mont. (De Bary). PHYTON-*International Journal of Experimental Botany*, 72:119-126

González, G.E., Rodríguez, H.R.& Aguilar, G.C.N.2009. Biodegradación de Taninos, CIENCIACIERTA No.17, http://www.postgradoeinvestigacion.uadec.mx/CienciaCierta / CC17/cc17taninos.html

Hadian, S.2012. Antifungal activity of some plant extracts against some pathogenic fungi in Iran, *Asian J. Exp. Biol. Sci.* 3(4):714-718.

Hernández, C. F.D., Lira, S.R.H., Cruz, Ch.L., Gallegos, M.G., Galindo, C.M.E., Padrón, C.E. & Hernández, S.M. 2008. Antifungal potential of *Bacillus* spp. strains and *Larreatridentata* extract against *Rhizoctoniasolani* on potato (*Solanumtuberosum* L.) crop. *International Journal of Experimental Botany*, 77:241-252

Hernández, C.F.D., Aguirre, A.A., Lira, S.R.H., Guerrero, R.E. & Gallegos, M.G.2006.Biological efficiency of organic biological and chemical products against *Alternariadauci* Kühn and its effects on carrot crop. *International Journal of Experimental Botany,* 75:91-101

Hernández, L., A. Bautista, B., S. &Velázquez-del Valle, M. 2007. Prospective of plant extracts for controlling postharvest diseases of horticultural products. *MexicanJournal of Phytopathology*, 30(2):119-123

Hernández, C.FD., Castillo, R.F., Gallegos, M.G., Rodríguez, H.R. & Aguilar, G.C.N. 2010. *Lippiagraveolens* and *Caryaillinoensis* organic extracts and there *in vitro* effect against *Rhizoctoniasolani* Kuhn. *American Journal of Agricultural and Biological Sciences*, 5(3): 380-384.

Huang, Q. and Lakshman, D.K. 2010. Effect of clove oil on plant pathogenic bacteriaand bacterial wilt of tomato and geranium. *Journal of Plant Pathology*, 92(3):701-707.

Islam, M. R., Alam, S., Rahman, M. Z., Chowdhury, S. P., Begum, M. F., Akhter, N., Alam, M. S., Han, K. D. and Lee, M. W. 2003. Effects of plant extracts on conidial germination, mycelial growth and sporulation of fungi isolated from poultry feed. *Mycrobiology*, 31(4): 221-225.

Jayashree, K., Pun, K.B. and Doraiswamy, S. 1999. Effect of plant extracts and derivates, buttermilk and virus inhibitory chemicals on Pumpkin vein mosaic virus transmission. *Indian Phytopathol.*, 52:357-361.

Ji P., Momol M.T., Olson S.M., Pradhanang P.M., Jones J.B., 2005. Evaluation of thymol as biofumigant for control of bacterial wilt of tomato under field conditions. *Plant Disease,*89: 497-500.

Kagale; T Marimuthu; B Thayumanavan; R Nandakumar; R Samiyappan.2004, *Physiol. Mol. Plant,* 65:91-100.

Koirala 2005 Prevalence of fluconazole-resistant strains of *Candida albicans* in otherwise healthy outpatients. *J. Oral Pathol Med.*, 31:99–105.

Kubo, I., H. Murei and M. Himejina. 1992. Antimicrobial activity of green tea flavor components and their combination effect. *J. Agric. Chem.*, 40: 245-48.

Kulkarni, V.R. and Byadagi, A.S. 2004. Evaluation of plant extracts against safflower mosaic virus disease. *Karnataka J. Agric. Sci.*, 17:838-840.

Kuri, S. K., Islam, R.M. and Mondal, U. 2011. Antifungal potentiality of some botanical extracts against important seed bornd fungal pathogen associated with brinjal seeds, *Solanummelongena* L. *Journal of Agricultural Technology,* 7(4):1139-1153.

Lale, N.E.S. and Abdulrahman, H.T. 1999. Evaluation of neem (*Azadirchtaindica* A. Juss) seed oil obtained by different methods and neem powder for the management of *Callosobruchusmaculatus* (F.) (Coleoptra Burchidae) in stored cowpea. *J. Stored Prod. Res*, 35:135-143.

Lazarovits, G. 2010. Management of soil-borne plant pathogens with organic soil amendments: A disease control strategy salvaged from the past. *Can. J. Plant Pathol.* 23: 1–7.

Leroux, P. 2003. Mode of action of agrochemical towards plant pathogens. *Comptes Rendus Biologies*, 326(1):9-21.

Lira, S.R.H., Hernández, S.M. and Hernández, C.F.D. 2006.Activity of *Larreatridentata* (D.C.) Coville L. extracts and chitosan against fungi that affect horticultural crops. *Rev. Chapingo Serie Horticulture,* 12(2):211-216.

Madhusudhan, K.N. Vinayarani, G., Deepak, S.A., Niranjana, S.R., Prakash, H.S., Singh, G.P., Sinha, A.K. and Prasad BC 2011. Antiviral activity of plant extracts and other inducers against tobamovirus infection in bell pepper and tomato plants. *International journal of Plant pathology*. 2:35-42.

Madhusudhan, K.N., Nalini, M.S., Prakash, H.S. and Shetty, H.S. 2005. Effect of inducers against to bamo virus infection in tomato and bell pepper. *Int. J. Bot.*, 1:59-61.

McKellar M.E., Nelson E.B., 2003. Compost-induced suppression of Pythium damping-off is mediated by fatty-acid metabolizing seed-colonizing microbial communities. *Applied and Environmental Microbiology,* 69: 452-460.

Manohar, V., Ingram, C., Gray, J., Talpur, N. A., Echard, B. W., Bagchi, D. and Preuss, 2001. Antifungal activities of origanum oil against Candida albicans. *J. Mol. Cell Biochem.*, 228: 111-7.

Maya, C. and Thippanna, M. 2013. *In vitro* evaluation of ethno-botanically important plant extracts against early blight disease (*Alternariasolani*) of tomato. GJBB 2: 248–252

Medina, M.M.A., Belmares, C.R.E., Rodriguez, H.R & Aguilar, C.N. 2007.Valorizacióncomercial de los subproductos de la nuezpara la producción de antioxidantes. Cienciacierta No. 9 *http://www2.uadec.mx/pub/pdf/cienciacierta9.pdf*

Momol M.T., Mitchell D.J., Rayside P.A., Olson S.M., Momol E.A., 2000. Plant essential oils as potential biofumigants for the management of soilborne pathogens of tomato. *Phytopathology* 90: S127.

Montes, B. R. 1996. Natural plant products to combat pathogens. *Mexican Journal of Phytopathology*, 14(1): 9-14.

Newton, S.M., C. Lau, S.S. Gurcha, G.S. Besra and C.W. Wright. 2002. The evaluation of forty-three plant species for in-vitro anti mycobacterial activities: Isolation of active constituents from *Psoraleacorylifolia* and *Sanguinariacanadensis*. *J. Ethnopharmacol.*, 79: 57-67.

Munoz-Mingaro, D., Acero, N., Llinares, F., Fozuelo, J.M., Galan de Mera, A., Vicenten, J.A., 2003. Biological activity of extracts from *Catalpa bignonioides* walt (Bignoniaceae). *J. Ethnopharmacol.*, 87: 163-167.

Ogbebor, O. N. and Adekunle, A. T. 2008. Inhibition of *Drechsleraheveae* (Petch) M. B. Ellis, causal organism of Bird's eye spot disease of rubber (*Heveabrasiliensis* Muell Arg.) using plant extracts. *African Journal of General Agriculture*, 4(1):19-26.

Oka Y., Nacar S., Putievsky E., Ravid U., Yaniv Z., Spiegel Y. 2000. Nematicidal activity of essential oils and their components against the root-knot nematode. *Phytopathology,* 90: 710-715.

Osorio, E., Flores, M., Hernández, D., Ventura, J., Rodríguez, R. and Aguilar, C. N. 2010. Biological efficiency of polyphenolic extracts from pecan nuts shell (*Carya illinoensis*), pomegranate husk (*Punica granatum*) & creosote bush leaves (*Larrea tridentata* Cov.) against plant pathogenic fungi. *Industrial Crops and Products,* 31(1):153-157.

Pretorius, J.C., Magama,S. and Zietsman, P.C. 2003. Growth inhibition of plant pathogenic bacteria and fungi by extracts from selected South African plant species. *South African Journal of Botany,* 69(2): 186–192.

Prince, L. and Prabakaran, P. 2011. Antifungal activity of medicinal plants against plant pathogenic fungus *Colletotrichumfalcatum. Asian Journal of Plant Science and Research*, 1 (1):84-87.

Radulovi, N.S., Blagojevi, P.D., Stojanovi, Z.Z., Radi and Stojanovi, N.M. 2013. Antimicrobial Plant Metabolites: Structural Diversity and Mechanism of Action. *Current Medicinal Chemistry*, 20: 932-952.

Shehu, K., Abdullahi, A., Abiala, M.A. and Odebode, A.C. 2013. *in vitro* bioprospecting of botanicals towards inhibition ofmicrobial pathogens of rice (*Oryza sativa* L.). *World Applied Sciences Journal*, 22 (2): 227-232.

Singh, R.N. and Sharatchandra, B. 2005. The development of botanical pesticides with special reference to seri-ecosystem. *Capsicum Journal of Environmental Science*, 3(1):1-8.

Strange, R.N. and Scott, R.P. 2005. Plant diseases: a threat to global food security. *Annu. Rev. Phytopathol.*, 43: 83-116.

Tapwal A., Gargi S., Gautam N and Kumari R. 2011. In vitro antifungal potency of plant extracts against five pathogens. *Brazilian Archieves of Biology and Technology*, 54 (6):1093-1098.

The Merck Index, 2001. 13th Ed. Merck & CO, Whitehouse Station, NJ, USA.

Usha, C. M., Patkar, K. L., Shetty, H. S., Kennedy, R., and Lacey, J. (1993). Fungalcolonization and mycotoxin contamination of developing rice grain. *Mycological Research*, 97(7): 795–798.

Verma, J. and Dubey, N. K. 1999.Prospectives of Botanicals and Microbial products as Pesticides of tomorrow.*Current Science,* 76: 172-179.

Whalen, M.M., Wilson, S., Gleghorn, C. & Loganathan, B.G. 2003. Brief exposure to triphenyltin produces irreversible inhibition of the cytotoxic function of human natural killer cells. *Environmental Research*, 92(3): 213-220.

Wilson, C.L., Ghaouth, A.E., & Wisniewski, M.E. 1999. Prospecting in nature′s storehouse for biopesticides. *Mexican Journal of Phytopathology*, 17(1):49-53.

Yazdani D, Tan Y.H., Zainal Abidin M.A. and Jaganath IB 2011. A review on bioactive compounds isolated from plants against plant pathogenic fungi. *Journal of Medical Plant Research*, 5 (30):6584-6589.

25

Strategies for Cultivation of Chickpea in the Rice Fallows Under Residual Soil Moisture

Md. Hedayetullah[1], Parveen Zaman[2] and Raghunath Sadhukhan[1]

[1]AICRP on Chickpea, Directorate of Research
Bidhan Chandra KrishiViswavidyalaya, Kalyani, West Bengal -741235, India
E-mail: hedaye.bckv@gmail.com
[2]Assistant Director of Agriculture, Pulse and Oilseed Research Sub-Station
Beldanga, West Bengal, India

Abstract

Rice fallows are those land which remain uncultivated during *rabi* after harvesting of *kharif* paddy. Several reason for not taking second crops such as early withdrawal of monsoon rains leading to soil moisture stress at planting time of *rabi* crops, water logging and excessive moisture in the months of November and December, lack of appropriate varieties of winter crops for late planting, and socio-economic problems like stray cattle, blue bulls etc. India accounts for 11.65 million ha (79%) of the total rice fallows of South Asia. The rice fallows offer good scope for area expansion of chickpea. The residual soil moisture left in the soil after rice harvest is often sufficient to grow chickpea in low to medium land soil. It is also beneficial for chickpea where short duration paddy grown which vacate fields in September-October. In long duration paddy, chickpea could be grown under *Utera* cultivation (relay cropping) in the states of Chhattisgarh, Jharkhand, Bihar, West Bengal, Orissa and Assam. Optimum plant population can only be ensuring in well levelled land, sufficient soil moisture and good seed under *Paira* (*Utera*) cropping. Overnight imbibed seed is broadcast at the rate of 20-25 % higher seed rate of recommended seed rate. Manual hand weeding is followed at 25 to 30 days after sowing to control broad spectrum of weed. Fertilization is not possible under *Utera* cropping system. Under nutrient stress condition nutrient solution like urea and DAP at the rate 2% is foliar sprayed before flowering. Seed inoculated with suitable strain of rhizobium at the rate of 20 g per kg of seed after seven days of fungicides treatment. Insect-pests and diseases should be promptly controlled. Seed treatment with fungicides like carbendazim at the rate of 2 g per kg seed should be done. Efforts should be made to choose disease-resistant varieties to control root rot and blight.

***Keywords*:** Rice fallow; Residual moisture; *Utera;* Chickpea; Management

Introduction

The strategies for cultivation of chickpea under rice fallow have multidimensional techniques for higher production in *rainfed* areas. Before sowing of chickpea crop in standing crop is very much crucial for germination of the seed. Soil moisture has to check for proper germination. The rice crop field should be well levelled for maintaining soil moisture in uniform mode in which uniform soil moisture can obtained, it facilitates uniform germination of seed. The planting geometry of rice crop also plays an important role for higher production of succeeding *paira* chickpea crop. Machine transplanting or line transplanting rice crops gives higher yield itself rather it also gives higher yield of second crop. During sowing of chickpea it facilitates to germination, proper establishment and minimal disturbance of chickpea crop during harvesting of rice crop (Mahmood *et al*. 2003). In *paira/relay/utera* cropping system land are not prepared only under wet soil seed are sown. In low to medium land soil characterised by clay loam soil with high water holding capacity is suitable for chickpea crop period. The residual soil moisture of this soil is sufficient for chickpea plant growth and development (Pande *et al*., 2012).

Chickpea (*Cicerarietinum* L.) is widely grown winter pulse crop in India. Chickpea is popularly known as Bengal gram or Brown gram or White gram or Kabuli gram. Chickpea is used as different purposes i.e. consumed as leafy vegetable, chana dal, flour (Besan) as well as fodder. Variety of snacks, sweets and dishes can be made out of chickpea flour. Seed contains around 25% proteins and 60% carbohydrates. The desi type of chickpea contributes to around 80% and the Kabuli type around 20% of the total production. India is the largest producer of chickpea contributing to around 70% of the world's total production. In India, Chickpea production was 7170 thousand tonnes in 2015-16 whereas in 2013-14 it was 9530 thousand tonnes (DES, 2015). Chickpea production also fluctuates year to year because of adverse weather during winter season and recorded a sharp decline over years (Pandey *et al*., 2012). Chickpea (Bengal gram) is a suitable winter pulse crop for farmers to grow on residual soil moisture, i.e. without irrigation, following the harvest of transplanted main season *kharif* paddy in medium to low land situation. A good combination of agronomic package and practices have been developed but the principal constraint remained the difficulty in establishing a reasonable crop stand once the surface layers of the soil had dried out. In rice fallow situation chickpea intervention shows as a low-cost compare to other interventions like wheat and potato which is highly profitable crop that can be cultivated without irrigation under residual soil moisture. Well soaked seed is beneficial for ensuring germination percentage of chickpea crop. The seed-priming technology gives better results than non-primed fields (Hedayetullah *et al.*, 2017).

What is rice fallow?

- Rice fallows are those land which remain uncultivated during *rabi* after harvesting of *kharif* paddy
- *Utera*cultivation (relay cropping)/paira cropping, Zero till drill are generally followed in rice fallow areas
- It is also called relay crop /zero tilled crop /residual crop

Rice fallow in India

Total rice fallow area in India is 11.65 mha. Eastern region area covered about 4.22 million ha in India comprises districts are eastern UP, Bihar, Jharkhand and West Bengal. North east hill region namely Assam area covered 0.54 million ha. Central region of India 5.01 million ha area comprises districts are Madhya Pradesh, Chhattisgarh and Maharashtra. Rice fallow area in coastal region is about 1.52 million ha; districts are Andhra Pradesh, Odisha and Tamil Nadu.

Reason of Rice fallow

- Lack of knowledge of *rabi* cropping
- Early withdrawal of monsoon rains
- Unavailability of inputs particularly good variety of seed
- Water logging and excessive moisture in the months of November and December
- Lack of appropriate varieties of winter crops for late planting
- Socio-economic problems like stray cattle, blue bulls etc. India
- Poor marketing facilities for sell their produce with minimum support price.
- Unavailability of modern seed sowing and harvesting machinery.

What is *paira/relay/utera*cropping?

***Paira/relay/utera*cropping** is sowing of winter crop before harvesting of the main crops (*Kharif* rice). *Paira*cropping is followed where late transplanting long duration rice harvested in late with high soil saturation and water is scarcity to save the second crop at a later stage. Due to high soil saturation during rice harvesting, fields take another 15 to 20 days for proper soil moisture for sowing of second crop in low to medium land ecosystem of West Bengal. The strategies have to adopt for utilization of vast fallow land in low to medium land. In conventional agriculture, farmers keep fallow the land instead of second crop. In some areas *paira* cropping of grass pea and lentil is followed under rice fallow situation after harvesting of rice in residual soil moisture but chickpea also possible in low to medium land ecosystem.

Production constraints of rice fallows

1 Biotic and abiotic stresses

The Alternaria blight disease tolerant lines found were 24034-4-1, 24003-2-1, C-91, C-1143, C-1159, C-1160. Susceptible lines found were C-83, C-58, C-83, C-71. Varieties identified for late sown condition High temperature tolerance. Abiotic stress like drought (water deficit), excessive watering (water-logging/ flooding during winter rains), extreme temperatures (cold, frost and heat), salinity (sodicity) and mineral (metal and metalloid) toxicity negatively impact growth, development, yield and seed quality of chickpea crop.

2 Poor crop management practices

Lack of knowledge of scientific cultivation package and practices i.e. quality seed of high yielding varieties. No use of integrated pest management. Poor linkage to market and government support price policies.

Technological interventions for improving productivity

1. Land preparation

1 Land preparation for relay/ paira/ uteracropping

In this cropping system land preparation is not required. Only imbibed seed are sown in standing rice crop with proper soil surface moisture before 10 to 15 days of harvesting. After harvesting of rice, chickpea seedlings already started browsing soil moisture from the deeper layer of soil for its survival.

2 Land preparation for zero till drill

Just after harvest of rice, chickpea seeds are placed with the help of zero till drill machine. The residual soil moisture promotes its growth at early stage and its later stage too. In this process land are not ploughed thoroughly; so primary tillage is avoided and secondary tillage is restricted here.

3 Land preparation for short duration medium to upland rice

Short duration rice variety generally harvested in the months of October and November. In this situation, sufficient time is available for land preparation with primary tillage implements i.e. country plough, disc plough followed by secondary tillage implements i.e. disc harrow, cultivator, rotavator and planker which takes about 5 to 7 days. In this practice, crop is faced with terminal drought stress and life saving irrigation is essential to overcome heat stress.

2. Suitable varieties

- Using high-yielding varieties with appropriate maturity duration
- Pusa 372', 'PG 186, JG 315, JG 16, JG 14 and 'Udai' are suitable
- Cultivation of early maturing and heat tolerant varieties for eastern India

3. Crop management

After rice harvest with soil residual moisture that could sustain a short-duration crop such as chickpea. Unfortunately, the surface layers of the soil dry out rapidly so crop establishment is the key objective. Few thumb rules are essential: (1) *Relay/paira/utera* cropping: sowing of chickpea seeds in standing crop 15 days before of harvesting of rice; it ensure germination of seeds where soil moisture is the key factor for its germination (2) Minimum/ zero tillage to cover the seeds (and minimal loss of moisture); (3) Imbibed seed for 4-6 hours in water before surface-drying them to facilitate good germination.

4. Seed treatment and inoculation

Before sowing, seed should be treated with fungicides followed by rhizobium and trichoderma inoculation for disease free plant and better nodulation. The seed should be treated with fungicides bavistin @ 2 g/kg seed at least seven days before rhizibium and trichoderma inoculation @ 20 g/kg seed with suitable strain. The disease-resistant varieties of chickpea are more suitable to control root rot and blight in *paira* cropping situation.

- Seed treatment with fungicides- Bavistin @2 g/kg
- Seed inoculated with rhizobium @ 20 g/kg seed for better nodulation
- Seed inoculated with rhizobium PSB @ 20 g/kg seed for soil phosphorus solubilization.

- Seed inoculated with trichoderma @ 20 g/kg seed tp control soil born pathogen

5. Seed Sowing

The seed is broadcast at the rate of 100 to 120 kg/ha, 20-25 % higher seed rate of recommended seed rate. To Tillage machines- Zero till drill, Seed cum ferti drill & minimum tillage etc. Sowing methods- Broadcasting and line sowing. Seed priming- for better germination

- Higher seed rate- 20% more than normal seed rate
- Timely sowing- September-October

6. Nutrient management

Basal application of fertilizer is not possible under *paira* cropping system. Under nutrient deficient condition nutrient solution like urea and DAP at the rate 2% is foliar sprayed at vegetative stage or before flowering. Foliar spray of urea/ DAP to supplement N and P @ 2 % in case of relay/ paira/ utera cropping is recommended. Machine (Zero till drill, Seed cum ferti drill) sowing crop, fertilizers are automatically placed below seed. In conventional sowing, organic manure like farm yard manure @ 5 t/ha are used and inorganic fertilizers (N: P_2O5: K_2O) used @ 20:40:20 kg/ha.

7. Micronutrient deficiencies

Mo, B, Zn as seed pallets

8. Weed Management

Manual one or two hand weeding is followed at 25 to 30 days or 45 to 50 days after sowing depending upon weed seed bank to control broad spectrum of weed. Pre-emergence like pendimethaline @ 4-5 litre with 400-500 litre of water per hectare. Post-emergence herbicides like Quizalofop ethyl.

9. Diseases and pests

To control soil borne pathogen liquid formulation of *trichoderma harzianum* may spray in wet soil. Effective control of all fungal disease can be achieved through seed treatment with *trichoderma harzianum or Pseudomonas fluorescens* @ 20 g/kg seed.

Pod borer (*Helicoverpa armigera*) are the serous insect of chickpea. Insect-pests are control through application of pesticides according to types of pest for effectiveness. Mostly pod borer (*Helicoverpa armigera*) at branching and pod development stage is infested more.

10. Terminal moisture/heat stress

Chickpea heat tolerant line ICCV 92944, while least germination was observed in a heat sensitive line ICC 10685. ICC 5912 and ICC 10685, the two sensitive genotypes, showed the lowest pollen viability. ICC 1205 and ICC 5912 have highest and lowest germination percentages respectively. Residue mulching is effective for other chickpea varieties to avoid terminal heat stress. It facilitates sufficient soil moisture during crop period. Live saving irrigation is beneficial where water is available.

11. Mechanization

- Use erect type varieties for mechanical harvesting
- Tillage machines,
- Zero-till planter and
- Harvester

12. Transfer of technology

- Innovative farmer's participatory approach

References

Bitew, Y., Asargew, F. 2014.Rice (*Oryza sativa*) and chickpea (*Cicer arietinum* L) relay intercropping systems in an additive series experiment in rain fed lowland ecosystem of Fogeravertisols. *Science Research* 2(6): 179-184.

DES. 2015.http://eands.dacnet.nic.in/Agricultural_Statistics_At_Glance-2015%20E-book/book.swf. Accessed on 9th March, 2017, 10: 12 PM.

Mahmood, A., Hussain, Anjum, F.H., Ali, A. 2003. Rice planting geometry facilitating relay cropping at zero tillage. *International Journal of Agriculture & Biology* 15(4): 435–437.

Pande, S., Sharma, M., Ghosh, R., Rao, S.K., Sharma, R.N. and Jha, A.K. 2012. Opportunities for chickpea production in rainfed rice-fallows of India – Baseline survey report. International Crops Research Institute for the Semi-Arid Tropics Patancheru 502 324, Andhra Pradesh, India p. 56.

Pande, S., Sharma, M., Ghosh, R., Rao, S.K, Sharma, R.N., Jha, A.K.2012. Opportunities for chickpea production in rainfed rice-fallows of India – Baseline survey report. International Crops Research Institute for the Semi-Arid Tropics Patancheru 502 324, Andhra Pradesh, India.p. 56.

26

Evaluation of Jute Varieties (*Corchorus capsularis* L) under Different Fertilizer Management Schedules in Indo-Gangetic Plains

Parveen Zaman

Assistant Director of Agriculture, Pulse and Oilseed Research Sub-Station, Beldanga 742 133, Murshidabad, West Bengal, India
Email: parveenzaman1989@gmail.com

Abstract

Jute is known as golden fibre of Bengal and the most important bast fibre of the world. Jute is a natural fibre crop and is second in the world after cotton in terms of global production, consumption and availability. It is a completely biodegradable, recyclable and eco-friendly lingo-cellulose fibre. The genus *Corchorus* belongs to the family *Tiliaceae*, which is composed of approximately 100 species. Two species (*Corchorus capsularis* L. and *Corchorus olitorius* L.) are widely cultivated for natural fibre. The field experiment was conducted in Pre-Kharif season of 2014 in Mondouri Research Farm, of Bidhan Chandra Krishi Viswavidyalaya, Mohanpur, Nadia, West Bengal. This farm is located very close to the tropic of cancer having approximately 89° E longitudes, 23 ° N latitude and about 9.75 in altitude above the sea level. The experiment was in a Factorial Randomized Block Design replicated thrice with first factor variety i.e. V_1-JRC 321, V_2- JRC 698, V_3- JRCJ 2 and V_4-NDJC 2011 and Second factor fertilizers with three levels i.e. F_1-60:13:25, F_2-80:17.5: 33.3 and F_3- 100:21.8:41.7. The plant height was not significant at the harvest stage. The significantly maximum LAI (9.58) was recorded in T_{12} treatment (NDJC 2011× F_3) followed by 9.39 in T_{11} (NDJC 2011× F_2) at harvest stage (120 DAS). The significantly maximum Basal Diameter (1.72 cm) was recorded in T_{12} treatment (NDJC 2011× F_3) followed by 1.64 cm in T_3 (JRC 321× F_3) at harvest stage (120 DAS).The significantly maximum dry matter (162.22 g/ m^2) accumulation was recorded in treatment T_3 (JRC 321× F_3)followed by158.75 g/m^2 in T_{12} treatment (NDJC 2011× F_3) at harvest stage. The crop growth rate trend followed suit. The maximum green jute fibres yield (656.07 q/ha) was recorded in NDJC 2011× F_3 treatment followed by 630.74 q/ha was recorded in JRC 321× F_3 treatment. The lowest green jute fibres yield (404.54 q/ha) was obtained in JRCJ× F_1 treatment. The significantly maximum jute fibres yield (39.56 q/ha) was recorded in NDJC 2011× F_3 treatment followed by 37.89 q/ha was recorded in NDJC 2011× F_2 treatment. The lowest jute fibres yield (23.82 q/ha) was obtained in JRCJ× F_1 treatment. Important findings followed from reduction of retting period of the different varieties with NDJC 2011 and JRCJ 2 having better response with the microbial consortia. The grading of white jute, analysed in NIRJAFT, showed that NDJC 2011 and JRC 321 have better grades in W_4 and W_4+ 25%. The response curve equations showed that JRC 321 has better fertilizer responsiveness than NDJC 2011. The economics was best with NDJC 2011 for both F_3 and F_2 levels of fertilizers.

Keywords: Jute; varieties; fertilizers, quality; fibre yield

Introduction

Jute (*Corchorus* spp.) is a natural fiber crop and is second in the world after cotton in terms of global production, consumption and availability. It is a completely biodegradable, recyclable and eco-friendly lingo-cellulose fiber (Mir *et al.*, 2008). The jute fibers are derived from the bark of the plant. The genus *Corchorus* belongs to the family *Tiliaceae*, which is composed of approximately 100 species (Saunders, 2001). Of these, two species (*Corchorus capsularis* L. and *Corchorus olitorius* L.) are widely cultivated for natural fiber in areas distributed throughout the tropical and sub-tropical regions of the world, particularly in Asia, Africa and Latin America (Hossain *et al.*, 2002). The most common use of jute fiber is in packaging materials such as hessian, sacking and ropes. A variety of products, such as floor coverings, home textiles, agro-textiles, blankets, handicrafts and fashion accessories, are also made from jute. In recent years, jute has been used for making pulp and papers in the paper industry (Mohiuddin *et al.*, 2005).

The fibre of commerce is extracted from the stem of two cultivated species of jute–*tossa* jute (*Corchorus olitorius* L.) and white jute (*C. capsularis* L.). Though there has been a significant increase in productivity, there is still a wide disparity in the yield level of the crop between the states as well as between the agro-climatic zones of the same state. Jute and jute-based production systems under the present system of cultivation is labour intensive and costly. Around 65-70% of the total cost of production in jute is due to weed management and retting. Lack of large volume of clean and slow moving water required to produce good quality.

The fact that the fertilizer was the key input in augmenting food grain production after the availability of the seeds of high yielding crop varieties is evident from the increase in fertilizer consumption from 1.98 Mt in 1969-70 to 18.07 Mt by 1999-00. A simple regression analysis between the food grain production and fertilizer consumption during 1960-61 to 1999-00 showed that the partial factor productivity of fertilizers has been continuously declining. This is supported by the fact that the farmers in the rice-wheat cropping system belt are forced to apply more and more fertilizer to obtain the same crop yields as in the preceding years. The efficiency of fertilizer nitrogen is only 30-40% in rice and 50-60% in other cereals, while the efficiency of fertilizer phosphorus is 15-20% in most crops. The efficiency of K is 60-80%, while that for S is 8-12%. Keeping this in mind the present experiment on "Evaluation of jute varieties (Corchorus capsularis) under different fertilizer management schedules in Indo-Gangetic Plains of West Bengal" has been taken up to screen out capsularis jute varieties to understand the nutrient need in the present day context.

Methodology

The field experiment was conducted in Pre-Kharif season of 2014 in Mondouri Research Farm, of Bidhan Chandra Krishi Viswavidyalaya, Mohanpur, Nadia, West Bengal. This farm is located very close to the tropic of cancer having approximately 89° E longitudes, 23 ° N latitude and about 9.75 in altitude above the sea level. The experiment was in a Factorial Randomized Block Design replicated thrice with first factor variety i.e. V_1-JRC 321, V_2- JRC 698, V_3-JRCJ 2 and V_4-NDJC 2011 and Second factor fertilizers with three levels i.e. F_1-60:13:25, F_2-80:17.5: 33.3 and F_3- 100:21.8:41.7. Plant population of individual plots from one square metre area was recorded at the time of harvest. Ten plants were selected at random and tagged in each plot according to treatments. The plant heights (cm) at the basal girths (cm) were recorded at 30 DAS, 60 DAS, 90 DAS and 120 DAS, i.e. at harvest. The dry fibre yield was recorded from the net plot harvest from each plot and this was later converted to per hectare basis. Grading of dry fibre was done with the help of NIRJAFT ICAR Laboratories. Plot wise fibre samples were sent to the NIRJAFT for standard grading for grading of white jute. Composite samples were made across nutrient levels for the different varieties/entries. Soil chemical analysis: A composite soil sample collected from the entire field before sowing, was chemically analysed for pH, available nitrogen, available P_2O_5 and available K_2O. The data of yield and yield attributing characters were subjected to statistical analysis by the analysis of variance method. The significance of different sources of variation was tested by Error Mean Square by Fisher Snedecor's F test at probability levels of 0.05. For comparison of "F" table and for computation of critical difference (C.D.) at 5% level of significance, Fisher and Yate's tables were consulted.

Results and discussion

Growth attributes

Plant population (m^{-2}) of Jute at 30 days after sowing was maintained after thinning operation which ranges between 49.671 to 50.18 plants per m^2 to avoid overcrowding and nutritional deficiencies. The effect of Varieties and Fertilizer Schedules on Plant Height (cm) of Jute at harvest is depicted in Table 3. Amongst the varieties the highest mean plant height (273.67 cm) was recorded in NDJC 2011 variety followed by JRC 321 and JRC 698 at harvest whereas the lowest mean plant height (245.44 cm) was found in JRCJ-2 variety. The highest mean plant height (291.34 cm) was recorded with F_3 fertilizer schedule at harvest followed by F_2 fertilizer schedule whereas the lowest mean plant height (213.83 cm) was found in F_1 fertilizer schedule. The interaction effect of Varieties and Fertilizer Schedules on leaf area index of Jute at harvest is showed in. Amongst the varieties the maximum leaf area index (8.58) was observed in NDJC 2011 variety followed by JRC 321 and JRC 698 at harvest whereas the minimum leaf area index (6.15) was found in JRCJ-2 variety. The maximum leaf area index (7.82) was recorded with F_3 fertilizer schedule at harvest followed by F_2 fertilizer schedule whereas the minimum leaf area index (5.91) was found in F_1 fertilizer schedule.

Amongst the varieties the maximum dry matter accumulation (1421.97 g m^2) was observed in JRC 321 variety followed by NDJC and JRC 698 at harvest stage whereas the minimum dry matter accumulation (1273.10 g m^2) was found in JRCJ -2 variety. The maximum dry matter accumulation (1550.13 g m^2) was recorded with F_3 fertilizer schedule at harvest followed by F_2 fertilizer schedule whereas the minimum dry matter accumulation (1092.45 g m^2) was found in F_1 fertilizer schedule. The effect of varieties and fertilizer Schedules on dry matter accumulation of Jute at harvest showed significant response. The maximum dry matter accumulation (193.20 g m^2) was recorded in NDJC 2011 with F_3 fertilizer schedule. The *capsularis* varieties when subjected to acute moisture stress at early stage accumulated 67% more biomass when compared with *olitorius* ones which indicated that *C. capsularis* could withstand natural drought better than *C. olitorius* in its early growth stage. The maximum crop growth rate (19.07 g/m^2/day) was observed in JRC 321 variety with F_3 fertilizer schedule during 90-120 days after sowing (DAS) followed by NDJC 2011 variety and the minimum crop growth rate (7.02 g/m^2/day) was observed in JRC 698 variety with F_1 fertilizer schedule. The above findings confirm the observations reported by Mitra *et al.*, (2004).

Yield attributes

The significantly maximum green yield of jute fibre (656.07 q ha^{-1}) was recorded in NDJC 2011 variety with F_3 fertilizer schedule followed by JRC 321 (630.74 q ha^{-1}) and JRC 698 (609.89 q ha^{-1}) at harvest stage (120 days after sowing) whereas the minimum green yield of jute fibre(404.54 q ha^{-1}) was found in JRCJ -2 variety. Similarly, maximum green weight without leaves (60.02 tha^{-1}), green weight with leaves (68.56 t ha^{-1}) were recorded in T_{10} ($N_{90}P_5K_{30}S_{30}$) treatment (Table 1).

The yield of jute fibre was varied with varieties and fertilizers schedule is shown in table 12. The maximum yield of jute fibre (39.56 q ha^{-1}) was recorded in NDJC 2011 variety followed by JRC 321 (36.92 q ha^{-1}) and JRC 698 (34.39 q ha^{-1}) whereas the minimum yield of jute fibre (33.69 q ha^{-1}) was found in JRCJ -2 variety. The maximum yield of jute fibre (36.14 q ha^{-1}) was recorded with F_3 fertilizer schedule followed by F_2 fertilizer schedule whereas the minimum yield of jute fibre (26.94 q ha^{-1}) was found in F_1 fertilizer schedule (Fig 1). Optimum yield for each of the varieties, JRC 698 & JRC J2 will be obtained with further increments in nutrient levels. Agronomic optimum is reached within the applied nutrient level for JRC 321 & NDJC 2011 where as JRC 321 is more nutrient responsive with a value of 136.76 kg nutrients & NDJC 2011 is less responsive requiring 183.77 kg nutrient to reach agronomic optimum. In NDJC 2011 variety 62.38 % more fibre yield can be achieved with F_3 fertilizer schedule compare to F_1 fertilizer schedule but other varieties showed less fibre yield compare to NDJC 2011 variety with the same fertilizer schedule. Similar finding was reported by Ghosh, (2008); Singh *et al.* (2014); Das and Kumar (2012) and Alamgir *et al.* (2008).

Table 1: Effect of Varieties and Fertilizer Schedules on Green Yield of Jute Fibres in q ha^{-1}

F×V	JRC 321	JRC 698	JRCJ-2	NDJC 2011	Mean
F1	465.89	424.37	404.54	495.70	447.63
F2	592.70	562.63	540.79	608.85	576.24
F3	630.74	609.89	583.01	656.07	619.93
Mean	563.11	532.30	509.45	586.87	547.93
	(V)	(F)	(F X V)		
SEm±	19.79	17.14	34.28		
$CD_{(P=0.05)}$	NS	50.27	100.55		

F_1=60:13:25 (N: P: K, Kg ha^{-1}); F_2=80:17.5:33.3 (N: P: K, Kg ha^{-1}), F_3= 100: 21.8: 41.7, (N: P:K, Kg ha^{-1}); V= Variety; F= Fertilizers; DAS= Days after sowing, NS= Not significant

Nutrients response on variety

The NDJC 2011 variety showed 63.40% more nitrogen uptake compare to JRCJ-2 variety. The maximum nitrogen uptake (96.82 kg ha^{-1}) was found in NDJC 2011 variety with F_3 fertilizer schedule and minimum nitrogen uptake (56.04 kg ha^{-1}) was found in JRCJ-2 variety with F_1 fertilizer schedule. The maximum phosphorus uptake (48.71 kg ha^{-1}) was found in NDJC 2011 variety with F_3 fertilizer schedule and minimum phosphorus uptake (24.02 kg ha^{-1}) was found in JRC-698 variety with F_1 fertilizer schedule. In NDJC 2011 variety 66.97 % more phosphate uptake observed compare to JRC-698 variety. Similarly, the maximum potassium uptake (220.48 kg ha^{-1}) was found in NDJC 2011 variety and minimum potassium uptake (131.67 kg ha^{-1}) was found in JRC-698 variety. In NDJC 2011 variety 62.61 % more potassium uptake showed compare to JRC-698 variety (Table 2). The above findings confirm the observations reported Olaniyi *et al.* (2008) and Singh *et al.* (2007). Similar finding was reported by Saha *et al.* (2008) and Paikary *et al.* (2006).

Table 2. Nutrient Uptake in the Jute Crops at Harvest

Treatments	Nitrogen	Phosphorus	Potassium
JRC 321 (F1)	60.66	24.92	134.45
JRC 321 (F2)	88.88	25.84	170.82
JRC 321 (F3)	95.00	44.34	212.64
JRC 698 (F1)	58.88	24.02	131.67
JRC 698 (F2)	82.88	34.20	171.32
JRC 698 (F3)	80.95	42.68	192.43
JRCJ - 2 (F1)	56.04	27.22	134.40
JRCJ - 2 (F2)	79.77	37.56	172.89
JRCJ - 2 (F3)	84.90	40.80	185.65
NDJC 2011 (F1)	65.56	27.44	146.78
NDJC 2011 (F2)	87.23	44.24	198.46
NDJC 2011 (F3)	96.82	48.71	220.48
Sem ±	C O M	P O S	I T E
CD $_{(P=0.05)}$	S A	M P L	E S

F_1=60:13:25 (N: P: K, Kg ha^{-1}); F_2=80:17.5:33.3 (N: P: K, Kg ha^{-1}), F_3= 100: 21.8: 41.7, (N: P: K, Kg ha^{-1}); V= Variety; F= Fertilizers

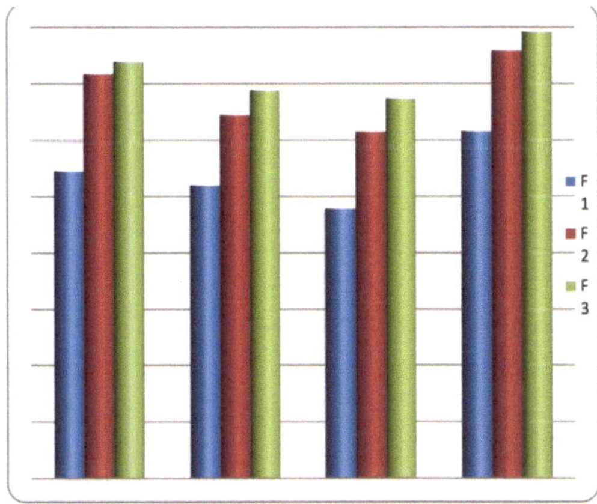

Fig. 1: Effect of Varieties and Fertilizer Schedules on Yield of Jute Fibres in q/ha

Microbial formulation on retting on the different varieties

Delayed retting process results defects in fibre which is related to market price. Reducing retting process is one of the noble techniques for good quality fibre and good net profit. Among the varieties JRC-698 is taking more time (13 days) with consortia which is almost 5 days reduction compare to the other capsularis varieties. The NDJC 2011 variety is taking less time (12 days) with consortia which is 7 days reduction compare to other varieties. Similarly, JRCJ-2 also found the same reduction (7 days) with consortia NDJC 2011 variety was harvested in 120 days after sowing and fibre quality secured with grade W_4 without losing any fiber yield followed by JRC 321 variety. Among the Capsularis varieties the NDJC 2011 variety showed best quality fibre compare to others varieties. The grading of white jute, analyzed in NIRJAFT, showed that NDJC 2011 and JRC 321 have better grades in W_4 and W_4+ 25% respectively.

Amongst the capsularis varieties NDJC 2011 showed maximum NPR (1.41) with F_3 fertilizer schedule followed by same variety with F_2 fertilizer schedule and JRC 321 (1.25) with F_3 fertilizer schedule and minimum NPR (0.51) by JRCJ-2 with F_1 fertilizer schedule. The costs of production of NDJC 2011 variety not too much high compare to the others capsularis varieties with different fertilizers schedule but net return was much more. Alamgir *et al.* (2008) reported that fibre grading and quality were tested of capsularis varieties in the laboratory of Fibre Quality Improvement Division and in Textile Physics Division of BJRI, Dhaka.

Recommendations

Amongst the test varieties, NDJC 2011 proved to be most high yielding in fibre yield and was the frontrunner in capsularis jute fibre production. The variety out yielded the check JRC 321 by highest yielder in jute fibre by 7.68%. The check variety of JRC 698 and the test variety in JRCJ-2 were poor performers. Amongst the fertilizer schedules, 100:21.8:41.7 of N: P: K, respectively was instrumental in obtaining the highest response registering yield increment of 25.45 % over 60:13:25 but it was at par with 80:17.5: 33.3 of NPK application. Amongst the good performing varieties, JRC 321 is more responsive to nutrient than NDJC 2011. The test variety NDJC2011 was responsive to consortia application by reduction of retting period in and days in the pronounced manner. The same variety could attain a commendable W4 grade. JRC 321 further improved in W4 + 25 %. The variety NDJC2011 was most economic with a net profitable ratio of 1.44 and 1.35 for 80:17.5: 33.3 and 100:21.8:41.7 of NPK, respectively.

References

Alamgir, M.A; Islam, M.M; Hussain, M.; Moniruzzaman, S.M; Ahmed, I. 2008.Effect of population density and harvesting time on fibre yield, quality and economics of white jute varieties. *International-Journal-of-Sustainable-Agricultural-Technology*, 4(4): 71-75.

Das, A. And Kumar, D. 2012. Genetic evaluation and characterization of jute (*Corchorus* spp. L) genotypes using DUS parameters. *SAARC-Journal-of-Agriculture*,10(2): 147-153.

Ghosh, A. 2008.Integrated nutrient management by using target yield equations for rice-jute system in eastern India.*Journal of Sustainable Agriculture,* 32: 149–60.

Mitra, Sabyasachi, Sinha, M.K. and Nath, R. 2004. Evaluation of growth and yield performances of jute genotypes under salinity and moisture stress condition. In: *Annual Report,* CRIJAF, Kolkata, pp. 68–73.

Olaniyi, J.O. and Ajibola, A.T. 2008.Growth and Yield Performance of *Corchorusolitorius* Varieties as Affected by Nitrogen and Phosphorus Fertilizers Application.*American-Eurasian Journal of Sustainable Agriculture*, 2(3): 234-241.

Paikary R.K., Mahanta D. and Swanil S.K. 2006.Effect of nutrient management in white jute- rice cropping system under rained condition.*Indian Journal of Agronomy.*51(4): 256-258.

Saha A.R., Mitra D.N., Mazumdar B., Saha S. and Mitra S. 2008.Effect on integrated nutrient management on rosselle (*Hibiscus sabdariffa*) productivity its mineral nutrition and soil properties. *Indian Journal of Agricultural Sciences,*78(5):418-421.

Singh, M.V., Neeraj Kumar, Singh, R.K. and Vinay Kumar.2014. effect of nitrogen management on fibre yield of capsularis jute in eastern uttarpradesh. *Annals of plant and soil research,* 16(1): 72-74.

Singh, M.V; Kumar, N.; Misra, B.N. 2007.Response of integrated nutrient management in capsularis jute (*Corchoruscapsularis*). *Haryana-Journal-of-Agronomy*, 23(1/2): 73-75.

27

Influence of Integrated Nutrient Management (INM) on the Availability of Nutrients in Soil in relation to Yield and Nutrition of Rape (*Brassica campestris* L.)

Prasenjit Adhikary, Soumya Sundar Tewary, S.K. Sadikur Rahaman Priyanka Naskar and D.K. Das

Department of Agricultural Chemistry and Soil Science
Institute of Agricultural Science, University of Calcutta
51/2, Hazra Road, Kolkata-700019, India
Email: prasenjitad1993@gmail.com

Abstract

The field experiment was conducted during *rabi* season at the Baruipur Research Farm of the Calcutta University using treatments as T_1-NPK 100%, T_2- Compost 100%, T_{3-} Compost 50%+NPK 50%, T_4-Vermicompost. 100%, T_5- Vermicompost 50%+NPK 50%, T_6-FYM 100%, T_7-FYM 50%+NPK 50%, in a Randomised Block Design (RBD) replicated thrice. The results show that the amount of available N content in soil has been found to be increased with the progress of crop growth upto 45 days of crop growth and thereafter, the amount of the same decreased with the crop growth irrespective of treatments. The amount of P content in soil has been found to be increased with different treatments, being maintained highest amount (18.30 kg/ha) in the treatment Vermicompost 50%+ NPK50% followed by the treatment where 100% NPK and 100 % vermicompost were applied. The results show that the amount of K content in soil was increased initially and thereafter, the amount of the same decreased with the progress of crop growth. The magnitude of such increase, however, varied with treatments, being maintained highest amount (310.18 kg/ha) in the treatment Vermicompost 50%+NPK 50% followed by the treatment (276.62 kg/ha) where FYM 50%+NPK50% was applied.

The highest amount of S content (32.14 g/kg) in the treatment T_5 where vermicompost 50% + NPK 50% was applied where 100% vermicompost was applied followed by the treatment T_4 (18.95 g/kg) where 100% vermicompost was applied at 75 days of crop growth. The overall results suggested that the amount of available N, P, K and S content in soil and contents in plant as well as yield of rape was recorded an increase in most treatment where combined application of organic manures and chemical fertilizers were applied, being maintained higher amount with the application of vermicompost along with chemical fertilizers excepting S content in seed.

Keywords: Availability; of nutrients; INM; nutrient contents in rape; organic manures; yields

Introduction

Agriculture is a soil-based industry that extracts nutrients from the soil. Therefore, effective and efficient approaches to minimise nutrient removal and returning of nutrients to the soil is required for maintaining soil fertility vis-a-vis increasing crop production. Organic manures play an Important role on the maintenance of physico-chemical properties as well as maintain carbon pool in soils. Of late, due to steady increase in the cost of chemical fertilizers, extensive nutrient mining from soils, environmental and health hazards and expected premium prices for organically produced crops, the use of organic manures in the present farming systems are gaining importance (Ramesh *et al.*, 2005). However, integrated approach of applying organic manures and chemical fertilizers not only improves the soil health and fertility but also helps to continue agricultural practices for sustainable for long time (Das, 2016). Keeping these in view, the present investigation is being undertaken to study the combined effect of different sources of organic manures and chemical fertilizers on the changes in soil nutrient status vis-a-vis yield and nutrition of oil seed crop rape (*Brassica campestris* L.) with the following objectives:

i) To study periodic changes in N, P, K and S contents in soil growing rape

ii) To study the yield and nutrition of N, P, K and S contents in rape seed.

Materials and Methods

1. Site of the experimental field

The experiment was conducted at the agricultural experimental farm, University of Calcutta, situated at Baruipur, 24 Parganas (South), West Bengal during the winter season of 2016-17. The farm is located at about $22^0 22^1$ N latitude and

$88^0 28^1$E longitudes with an average altitude of 9.75 meters above the mean sea level in the South genetic delta region of West Bengal. On station experiment on rape (*Brassica campestris* L.) CV. B – 9 was conducted during rabi season during the year 2016-2017 using following different treatments in a randomised block design (RBD) replicated thrice. T1=NPK100%(80:40:40)kg/ha, T2=Compost100% (2.5t/ha), T3=Compost50%(1.25 t/ha)+NPK50% (40:20:20)kg/ha, T4=Vermicompost100%(1t/ha)T5=Vermicompost50%(0.5t/ha)+NPK50%(40:20:20)kg/ha, T6=FYM100%(5 t/ha), T7=FYM50%(2.5 t/ha)+NPK50%(40:20:20)kg/ha. Determination of available nutrient contents and relevant soil properties were made by following the method described (Jacksan, 1973).

Results and Discussion

Available nitrogen in soil

The results (Table 1) show that the amount of available N content has been found to be increased with the progress of crop growth upto 45 days of crop growth and thereafter, the amount of the same decreased with the crop growth irrespective of treatments. The magnitude of such changes, however, varied with treatments. The treatment T5 (Vermicompost 50%+NPK50%) maintained a substantially highest amount of N (384.57 kg/ha) followed by the treatment T7 (FYM 50%+NPK 50%) where it maintained 353.98 kg/ha at 75 days of crop growth. Pathak and Yadav (2014) also reported similarly.

Table 1: Combined application of different sources of organic manures and N, P,K on the change in available N content (kg/ha) in the Soil..

Treatment	15DAS	30DAS	45DAS	60DAS	75DAS	Means
T_1-NPK 100%	407.68	501.93	470.97	344.10	282.02	401.34
T_2- Compost100%	232.20	313.44	250.43	219.49	188.78	240.86
T_{3-} Compost50%+NPK50%	313.60	360.31	390.70	282.46	250.30	319.47
T_4-Vermi.100%	282.24	407.85	375.40	330.95	344.79	348.24
T_5-Vermi.50%+NPK50%	376.32	471.28	501.98	407.27	384.57	428.28
T_6-FYM 100%	250.88	371.46	360.50	313.71	309.86	321.28
T_7-FYM 50%+NPK50%	319.47	392.36	439.76	376.52	353.98	376.41
C.D at 5%	14.339	15.809	8.208	3.974	11.010	
SE(m)$^+_-$	4.602	5.074	2.634	1.276	3.534	
C.V	2.556	2.183	1.145	0.680	2.032	
R	0.441	0.515	0.767	0.914	0.967	

T_1:NPK(80:40:40),T_2:Compost (2.5t), T_3: Compost (1.25) +NPK (40:20:20), T_4: Vermicompost (1t) T_5:Vermicompost (0.5t)+NPK (40:20:20), T_6:FYM(5t), T_7: FYM (2.5t) +NPK (40:20:20). r-correlation with yield DAS-days after sowing

Such increase in N content might be explained by the availability of greater amount of N in the soil solution resulting from the higher rate of mineralization in presence of organic manures, being more with vermicompost.

Available phosphorus in soil

The results (Table-2) reveal that the amount of available P has been found to be increased with the progress of the crop growth upto 30 days of crop growth and thereafter, the amount of the same decreased upto 75 days of crop growth irrespective of treatments.

Table 2: Combined application of different sources of organic manures and N.P.K on the change in *available P content (Kg/ha)* in the Soil

Treatment	15DAS	30DAS	45DAS	60DAS	75DAS	Means
T_1-NPK 100%	16.53	27.28	25.29	19.80	17.34	21.24
T_2- Compost100%	10.32	25.63	23.03	18.35	14.47	18.36
T_{3-} Compost50%+NPK50%	14.62	26.33	26.46	18.23	14.95	20.11
T_4-Vermi.100%	10.24	29.31	24.86	19.17	16.33	19.98
T_5-Vermi.50%+NPK50%	13.02	32.87	27.30	21.15	18.30	22.52
T_6-FYM 100%	9.94	19.92	17.68	13.69	12.40	14.72
T_7-FYM 50%+NPK50%	12.96	27.41	21.24	15.78	13.20	18.11
C.D at 5%	1.606	1.857	2.690	1.396	2.066	
SE(m)$^+_-$	0.515	0.596	0.863	0.448	0.663	
C.V	7.130	3.828	6.310	4.304	7.486	
R	0.086	0.503	0.155	0.094	0.255	

T_1: NPK (80:40:40), T_2: Compost (2.5t), T_3:Compost (1.25) + NPK (40:20:20), T_4: Vermicompost (1t)T_5:Vermicompost(0.5t)+NPK (40:20:20),T_6:FYM(5t), T_7:FYM (2.5t)+ NPK (40:20:20). r-correlation with yield DAS-days after sowing

The magnitude of such increase, however, varied with treatments, being maintained highest amount (18.30 kg/ha) in the treatment Vermicompost 50% +NPK50% followed by the treatment where 100% NPK and 100% vermicompost were applied.

Exchangeable potassium in soil

The magnitude of such increase, however, varied with treatments , being maintained highest amount (310.18 kg/ha) in the treatment Vermicompost 50%+NPK50% followed by the treatment (276.62 kg/ha) where FYM 50%+NPK50% was applied. The mean value of exchangeable K in soil is present table –(3)

Table 3: Combined application of different sources of organic manures and N. P.K on the changes in exchangeable K content (kg/ha) in the Soil.

Treatment	15DAS	30DAS	45DAS	60DAS	75DAS	Means
T_1-NPK 100%	407.95	420.98	374.47	293.50	245.50	348.48
T_2- Compost100%	270.16	360.41	295.72	240.70	191.54	271.70
T_{3-} Compost50%+NPK50%	310.84	394.12	344.74	282.88	255.18	317.55
T_4-Vermi.100%	295.75	378.58	338.53	272.56	264.83	310.05
T_5-Vermi.50%+NPK50%	389.46	401.62	388.52	332.27	310.18	364.43
T_6-FYM 100%	285.01	372.97	322.40	255.42	230.18	293.19
T_7-FYM 50%+NPK50%	331.52	376.58	349.55	305.35	276.62	327.92
C.D at 5%	10.731	6.603	6.448	5.636	9.595	
SE(m)$\pm$	3.445	2.119	2.070	1.809	3.080	
C.V	1.823	0.950	1.039	1.106	2.104	
R	0.423	0.280	0.696	0.797	0.945	

T_1:NPK(80:40:40), T_2:Compost (2.5t), T_3:Compost (1.25)+NPK (40:20:20), T_4: Vermicompost (1t)T_5: Vermicompost (0.5t)+NPK (40:20:20), T_6:FYM(5t), T_7:FYM(2.5t)+NPK(40:20:20). r-correlation with yield , DAS-days after sowing

Available sulphur in soil

The results (Table 4) reveal that the amount of available sulphate sulphur content has been found to be increased with the progress of the crop growth upto 30 days of crop growth and thereafter, the amount of the same decreased upto 75 days of crop growth irrespective of treatments. The magnitude of such increase, however, varied with treatments, being maintained highest amount (32.14 mg/kg) in the treatment vermicompost 50%+ NPK 50% followed by the treatment (18.95kg/ha) where 100% vermicompost was applied. Chand *et al.* (2006) also reported similarly.

Table 4: Combined application of different sources of organic manures and N.P.K on the change in available S content (mg/kg) in the Soil.

Treatment	15DAS	30DAS	45DAS	60DAS	75DAS	Means
T_1-NPK 100%	9.54	23.39	21.39	16.14	7.23	15.53
T_2- Compost100%	16.48	29.28	22.85	19.29	10.91	19.76
T_{3-} Compost50%+NPK50%	11.75	26.03	25.17	20.88	13.95	19.55
T_4-Vermi.100%	17.65	27.65	24.10	22.40	18.95	22.15
T_5-Vermi.50%+NPK50%	26.56	36.53	28.56	25.37	32.14	29.83
T_6-FYM 100%	14.71	28.58	26.87	23.36	15.73	21.85
T_7-FYM 50%+NPK50%	16.49	30.94	25.50	18.35	17.32	21.72
C.D at 5%	2.989	4.779	N/S	N/S	4.266	
SE(m)$\pm$	0.959	1.534	1.618	2.695	1.369	
C.V	10.275	9.189	11.247	22.411	14.283	
R	0.569	0.558	0.695	0.502	0.786	

T_1:NPK(80:40:40), T_2:Compost (2.5t),T_3: Compost (1.25) + NPK (40:20:20), T_4: Vermicompost (1t) T_5:Vermicompost (0.5t)+NPK (40:20:20), T_6:FYM(5t), T_7 :FYM (2.5t) + NPK (40:20:20) r-correlation with yield, DAS-days after sowing, N/S-Non-Significant.

Yield

The results (Table-5) show that the yield of rape seed has been found to be increased with treatments. The highest yield (10.16 q/ha) was recorded in the treatment where 50% vermicompost and 50% NPK fertilizers was applied combinedly which might be due to greater amount available nutrients including iron available in the soil solution contributing towards the yield of rape. The results of the present study also finds supports from the reports of Chand *et.al*, (2006).

Table 5: Combined application of difference source of organic manures and N. P.K Yield of rapeseed (*Brassica campestries.* L)

Treatment	Seed Yield (q/ha)	Stover Yield (q/ha)
T_1-NPK 100%	7.53	20.30
T_2- Compost100%	5.72	16.14
T_{3-} Compost50%+NPK50%	7.89	21.98
T_4-Vermi.100%	8.98	24.53
T_5-Vermi.50%+NPK50%	10.16	29.24
T_6-FYM 100%	8.20	22.18
T_7-FYM 50%+NPK50%	9.48	25.88
C.D at 5%	0.652	1.337
SE(m)$\pm$	0.209	0.429
C.V	3.908	3.210
R		0.990

T_1: NPK (80:40:40), T_2:Compost (2.5t),T_3:Compost (1.25)+NPK (40:20:20), T_4:Vermicompost (1t)T_5:Vermicompost (0.5t)+NPK (40:20:20), T_6: FYM(5t), T_7:FYM (2.5t)+NPK(40:20:20) r-correlation with yield, DAS-days after sowing, N/S-Non-Significant.

However, the following trend with respect to increase in yield was in the order of; T5 > T7 > T4 > T6 > T3 > T1 > T2

The stover yield was also followed a similar trend of changes to that of yield of rape. Only the absolute amount was higher than that of seed yield.

References

Chand, S., Anwar, M., Patra, D.D. 2006. Influence of long-term application of organic and inorganic fertilizer to build up soil fertility and nutrient uptake in mint-mustard cropping sequence. *Communications in soil science and plant analysis*, 37 (2) pp. 63-76.

Das, D.K. 2016. In: Introductory Soil Science. Chapter, Nutrient transformation in relation to soil-plant systems. Kalyani publishers. *4th edn.* pp-504-600.

Jckson, M.L. 1973. Soil Chemical Analysis. Prentice Hall of India (P) Ltd., New Delhi.

Pathak, D.V. and Yadav, S.S.2014. .Soil properties and seed yield as influenced by organic amendments in mustard-clusterbean cropping sequence .*Forage Res.,* 40 (3): pp. 186-191

Ramesh, P., Singh, M. and Subbaro, A. 2005. Organic farming: Its relevance to the Indian context. *Current Science*, 88: 561-569.

28

Performance of *kharif* potato cv. Kufri Pukhraj Grown Through Organic Farming in Chhota Nagpur Plateau

***Subhasis Mondal*[1] *and Avijit Kr. Dutta*[2]**

[1]*Azim Premji University, School of Development, Bangalore, India*
[2]*Assistant Professor (Horticulture), School of Agriculture and Rural Development, Ramakrishna Mission Vivekananda University Morabadi, Ranchi-834008, Jharkhand, India*
E-mail: mondalsubhasis1996@gmail.com

Abstract

From very ancient time the farming was practiced organically without using any chemical pesticides or fertilizers. In the mid-19th century, the chemical farming was introduced and the adverse impact of chemical fertilizers and pesticides in agriculture is in sight with deterioration of soil quality, quality of food and so on.The indiscriminate use of agro-chemicals in different crops causes numerous health hazards as well. In this context, organic farming came into existence in 20th century due mainly to produce quality foods through the rapid change in agriculture sector. Therefore, an attempt was taken through present investigation to cultivate potato without using any synthetic agro-chemicals by employing Kufri Pukhraj cultivar during *kharif* season. Potato is basically *rabi* season crop but due to climatic advantages of the south Chhotanagpurplateau of Jharkhand it can successfully grow during *kharif* season too. The experiment was designed after CRBD by employing seventreatments*viz.*,T_1: *Azolla* mixed *Shasyagavya* (10%); T_2: *Azolla* mixed *Shasyagavya* (15%); T_3: Vegetable waste mixed *Shasyagavya* (10%); T_4: Vegetable waste mixed *Shasyagavya* (15%); T_5: Potato peels mixed *Shasyagavya* (10%); T_6: Potato peels mixed *Shasyagavya* (15%) and T_7: Absolute Control with their four replications. *Shasyagavya* (an organic liquid formulation) serves as the source of plant nutrients andit was prepared by mixing cow dung, cow urine, agricultural wastes and water @ 1:1:1:2 proportions. Results showed that T_4as the best treatment with maximum yield potential (13.94t.ha^{-1})but T_1appeared as the best in terms of expression of quality attributes namely higher TSS (6.00^{0}Brix) and starch content (10.17%) in tubers.When benefit cost ratio was taken into account then again T_4materialized as best treatment with higher B: C ratio (4.49). Thus, T_4 treatment [Vegetable waste mixed *Shasyagavya* (15%)]may be recommended for commercial cultivation of *kharif* potato in plateau regions of Jharkhand.

***Keywords*:** Potato; (*Solanum tuberosum* L.); kufri Pukhraj; *Shasyagavya*; organic farming,; growth; yield; quality

Introduction

The word 'Potato' comes from Spanish word 'patata'. Potato is a starchy, tuberous crop from the perennial nightshade *Solanum tuberosum* L. The word "potato" may refer either to the plant itself or to the edible tuber. It has emerged as the fourth most important food crop in India after rice, wheat and maize. It is usually called as the king of vegetables of the family Solanaceae and extensively grown all over the world (Mustafa, 1997). The green leaves and green skins of tubers exposed to the light are toxic. Potato is a balanced staple food (Mehdi *et al.*, 2008), as it contains about 75 to 80% water, 16 to 20% carbohydrates, 2.5 to 3.2% crude protein, 1.2 to 2.2% true protein, 0.8 to 1.2 % mineral matter, 0.1 to 0.2% crude fats, 0.6 % crude fibre and some vitamins (Schoenemann, 1977). Potato is grown in the *Kharif* season in the hills and plateau. In the hills it is grown in J&K, Uttarakhand, HP and NE and Southern hills. In the plateau it is grown in Madhya Pradesh, Bihar (now Jharkhand), Chhattisgarh, Orissa, Maharashtra and Karnataka. In Jharkhand, some areas especially Ranchi (most specifically Bero and Nagri blocks) and surrounding areas have enormous potentiality to grow the crop twice in a year (during *kharif* and *rabi* seasons). The climatic condition of the South Chhotanagpur plateau permits potato cultivation during *kharif* season. Ranchi district of Jharkhand particularly Bero and Nagri blocks have remarkable share in *kharif* potato production. The local consumers prefer *kharif* potato over the *rabi* grown one due mainly to its higher level of palatable quality. Vegetable cultivation especially potato is considered one of the major sources of food security and income generation among the rural community of Jharkhand. Potato cultivation is suitable for small and marginal farmers as it is labor-intensive and provides high yields, with regular cash income. Based on above valuable information, the present study was conducted based on the following objectives.

- To study the performance of *Kharif* Potato in terms of different yield and quality attributes under organic management conditions.
- To estimate the economics of such potato grown through organic intervention.
- To study the present scenario of *Kharif* Potato grown by the traditional farmers of Ranchi district in Jharkhand

Materials and Methods

The experiment was conducted at experimental farm of the faculty center Integrated Rural and Tribal Development and Management under the school of Agriculture Rural and Tribal Development of Ramakrishna Mission Vivekananda University, Ranchi, Jharkhand during September-2015 to December-2015. According to the geography Ranchi (23.23^{0}N latitude and 85.23^{0}E longitude) is situated on the southern part of the Chhotanagpur Plateauunder the typical humid subtropical climatic region. The experiment was designed with Complete Randomized Block Design by employing seven treatments [Table 1] with their four replications in 28 beds through providing 20 cm intra row and 45 cm inter row spacing with bed size 3.0 m (length) x 0.9 m (breadth) [Table 2]. Potato cv. Kufri Pukhraj was used as experimental planting material for the study. 50g *Azotobacter* + 50g PSB mixture was used in experimental beds before sowing seed tubers. Treatment materials were applied four times at 15 days interval starting after 30 days of sowing. After forming the shoots and leaves the whey water along with turmeric powder @ 10g per litre was applied thrice at 15 days interval as prophylactic measures against pathogens especially blight and wherever required the *kerosene* oil (@ 50 ml/kg) soaked wood-ash was also applied for checking *Bihari Hairy Caterpillar*. Plant height (cm), No. of tuber(s).plant^{-1}, tuber yield (kg.plot^{-1}) and tuber yield (t.ha^{-1}) were taken as growth and yield attributes and TSS (^{o}Brix), total sugar (%), starch (%) and ascorbic acid (mg.100g^{-1}) content of tubers were considered as quality parameters of the crop. Data thus obtained were subjected to statistical analyses by following standard statistical tools.

Table 1: Details of treatments used for the experiment

Treatments	Solution	Ingredients	Fermentation (days)
T_1	*Azolla* mixed *Shasyagavya* (10%)	Cow dung: Cow urine: *Azolla*: Water @1:1:1:2	11
T_2	*Azolla* mixed *Shasyagavya* (15%)	Cow dung: Cow urine: *Azolla*: Water @1:1:1:2	11
T_3	Vegetable waste mixed *Shasyagavya* (10%)	Cow dung: Cow urine: Vegetable waste: Water @1:1:1:2	11
T_4	Vegetable waste mixed *Shasyagavya* (15%)	Cow dung: Cow urine: Vegetable waste: Water @ 1:1:1:2	11
T_5	Potato peels mixed *Shasyagavya* (10%)	Cow dung: Cow urine: Potato peels: Water @ 1:1:1:2	11
T_6	Potato peels mixed *Shasyagavya* (15%)	Cow dung: Cow urine: Potato peels: Water @ 1:1:1:2	11
T_7	Control	Only water	

Note: Normal ratio of *Shasyagavya* preparation ingredients = 1:1:1:2 (cow dung: cow urine: vegetable waste: water)

Results and Discussion

Significant results were recorded in growth, yield and quality parameters under the influences of different organically designed treatment conditions. Table 2 showed different growth and yield attributes of potato cv. Kufri Pukhraj under the influences of different organic treatments. Results showed significant differences (@ 0.05 probability level) under different treatment conditions. In this context, maximum plant height [61.33 cm], No. of tubers.plant^{-1} [4.95], tuber yield [3.77 kg.plot^{-1}] and tuber yield [13.94 t.ha^{-1}] were recorded in T_3, T_3, T_4 and T_4 treatments, respectively [Table 2]. Findings also highlighted that different organic inputs had significant role in increasing yield of tuber in *kharif* grown potato in the south Chhota Nagpur plateau of Jharkhand.

Table 2: *Pre se* performance of growth and yield attributes of *kharif* potato as influenced by the different treatments

Parameters	T_1	T_2	T_3	T_4	T_5	T_6	T_7	$SE_{m(\pm)}$	CD@ 5%
Plant height (cm)	57.43	57.25	61.33	60.50	56.58	56.08	56.90	3.56	8.74
No. of tuber(s).plant^{-1}	3.65	4.93	4.95	4.85	4.10	4.03	4.55	0.36	0.88
Tuber yield (kg.plot^{-1})	2.73	3.38	2.55	3.77	3.00	2.82	2.13	0.36	0.87
Tuber yield (t.ha^{-1})	10.12	12.51	9.44	13.94	11.12	10.44	7.88	1.32	3.23

Different quality attributes of *kharif* grown potato also influenced by different organic treatments [Table 3]. In this context, total soluble solid (TSS) noted maximum [6.00^{0}Brix] in T_1 treatment and the finding is closely matched with the earlier findings of Abbasi *et al.* (2012). T_7 treatment showed the best result in terms of total sugar content [11.04%] whereas the starch content recorded higher [10.71%] in T_1 treatment. However, ascorbic acid content [22.81 mg.100g^{-1}] documented higher in T_4 treatment [Table 3]. Notable findings regarding expression of different quality traits of the crop highlighted that different treatments performed independently. As the experiment was conducted in already organically converted experimental condition, thereby, T_7 with any additional organic input also showed higher total sugar content in the harvested tubers of the treatment.

Table 3: *Pre se* performance of quality attributes of *kharif* potato as influenced by the different treatments

Parameters	T_1	T_2	T_3	T_4	T_5	T_6	T_7	$SE_{m(\pm)}$	CD@ 5%
TSS (^{0}Brix)	6.00	5.75	5.75	5.25	5.75	4.75	5.93	0.41	0.99
Total sugar (%)	6.11	7.18	8.51	7.11	8.48	6.59	11.04	1.13	2.77
Starch (%)	10.71	9.81	9.74	9.77	8.24	8.06	9.70	0.73	1.78
Ascorbic acid (mg.100g^{-1})	21.78	22.50	20.63	22.81	21.88	16.88	22.50	1.31	3.22

Economics is an essential as well as integral part of agricultural production which depicted the viability or the acceptance of the experiment from farmer's point of view. The economics of growing potato under organically managed condition by utilizing different concentrations of *Shasyagavya* and their different supplemented input sources as different treatments illustrated in Table 4. The total cost of cultivation of *kharif* potato per hectare varies in between Rs. 31100.00 [T_7] and Rs. 35100.00 [T_3]. The findings accentuate the feasibility of organic intervention for growing potato varieties organically even in commercial scale under the Chhotanagpur plateau of Jharkhand. From the results, it was found that T_4 [15% *Shasyagavya* (vegetable wastes mixed)] is the most effective organic formulations for growing potato cultivars Kufri Pukhraj under sub-topical climatic conditions of Ranchi and its surrounding areas. T_4 [15% *Shasyagavya* (vegetable wastes mixed)] appeared with the highest benefit: cost ratio (4.49) [Table 4].

Table 4: Economics of growing potato in one hectare of land through different organically designed treatment conditions

Cost Components	T_1	T_2	T_3	T_4	T_5	T_6	T_7
Land preparation (Rs.)	3300.00	3300.00	3300.00	3300.00	3300.00	3300.00	3300.00
Inputs (Rs.)	8000.00	7500.00	10000.00	9000.00	8500.00	9500.00	6000.00
Planting material (Rs.)	1400.00	1400.00	1400.00	1400.00	1400.00	1400.00	1400.00
Weeding and earthing up (Rs.)	1300.00	1300.00	1300.00	1300.00	1300.00	1300.00	1300.00
Irrigation (Rs.)	2500.00	2500.00	2500.00	2500.00	2500.00	2500.00	2500.00
Plant protection (Rs.)	2100.00	2100.00	2100.00	2100.00	2100.00	2100.00	2100.00
Harvesting and post-harvest operations (Rs.)	13000	13000	13000	13000	13000	13000	13000
Miscellaneous Rs.)	1500	1500	1500	1500	1500	1500	1500
Total cost of cultivation (A)	33100.00	32600.00	35100.00	34100.00	33600.00	34600.00	31100.00
Production ($t.ha^{-1}$)	10.12	12.51	9.44	13.94	11.12	10.44	7.88
Selling price ($Rs.kg^{-1}$)*	11.00	11.00	11.00	11.00	11.00	11.00	11.00
Total income($Rs.ha^{-1}$) (B)	111320.00	137610.00	103840.00	153340.00	122320.00	114840	86680
Net profit ($Rs.ha^{-1}$) (B-A)	78220.00	105010.00	68740.00	119240.00	88720.00	80240.00	55580.00
B:C Ratio (B/A)	3.36	4.22	2.95	4.49	3.64	3.31	2.78

*Considering prevalent market price of potato @ Rs. 11.00kg^{-1} [premium price of organically grown produce is not considered here due to absenteeism of organic certification].

Conclusion

From the above study, it may be concluded that in terms of yield T_4 (vegetable wastes mixed *Shasyagavya* 15%) is the best treatment but when quality attributes were taken into consideration then T_1 (*Azolla* mixed *Shasyagavya* 10%) emerged as the best although T_4 treatment materialized as average to above average performer in terms of almost all studied quality attributes. Similarly, when benefit: cost ratio was taken into account then T_4 again emerged as the best treatment. Hence, T_4 treatment may be recommended for even commercial scale organic *kharif* potato cultivation in the South Chhotanagpur plateau regions of Jharkhand.

References

Abbasi, N. A., Zahoor, M.; Khan, H. A., and Qureshi, A. A. 2012. Effect of Encapsulated Calcium Carbide Application at Different Growth Stages On potato (*Solanumtuberosum* L.) Growth, Yield and Tuber Quality. *Pak. J. Bot.*, 44(5): 1543-1550.

Mustafa, G. 1997. Input management of potato crop in Northern Areas of Pakistan. Unpublished M.Sc. thesis, Department of Agri. Farm Management Unit Uni. of Reading. UK. P. 14.

Mehdi, M., Saleem, T.; Rai, K. H., Mir, S. M. and Rai, G. 2008. Effect of nitrogen and FYM interaction on yield and yield traits of potato genotypes under Ladakh condition. *Potato J.*, 35 (3&4): 126-129.

Schoenemann, J. A. 1977. Grading, packaging and Marketing potatoes, In: *Potatoes: Production, Storing, Processing*, Smith, O. (Ed.), 2nd Edition. The AVI Publishing Company Inc., West Port, pp. 470-505.

29

System of Rice Intensification Reducing Cost of Cultivation and Increasing Productivity of Rice

Manik Lal Takudar[1], *A. Zaman*[2] *and Partha Roychudhry*[3]

[1,2&3]*WBADMI (World Bank Project), Department of Water Resources Investigation and Development (DWRID) Government of West Bengal Krishnagar-711100, West Bengal, India*
Email: maniklaltalukder1975@gmail.com

Abstract

The on-farm programme on SRI was carried out in the village Narayanpur under RLI command area in Karimpur II Block of Nadia district in the year 2016 during summer season. The soil of farmland is clay loam and situated having latitude of 23°54'26.5"Nlongitude 88°30'24.1"Eat an average elevation of 9.75 m above the mean sea level. Monsoon ceases during October and cool season sets in by November. The average rainfall of this region is about 1500 mm and normal monsoon breaks in the first week of June. Though water resources in this area are generally satisfactory and 90% of these resources are tapped form the ground water source through deep and shallow tube wells. However, due to faulty crop planning and lack of scientific water management approach, a significant portion of agricultural area is identified as gray zone from hydrological point of view. Even many tube wells are becoming dry due to over exploitation of valuable water from the lower depths of confined aquifers. Rice is predominant crop during *kharif* season becoming less profitable because of high input costs involved. Summer rice gave satisfactory yield but also profitable proposition because of more input cost and less farm gate prices. In this stated background as above, the traditional rice growing technology has been shifted to System of Rice Intensification (SRI), a proven methodology to save water in growing rice were tried as intervention and adoption for the purpose of creating more man days and more income per drop of water and increasing agricultural productivity. Rice is pre-dominantly main crop in the area both during rainy as well as summer season. Rainy season rice with HYV gave an average yield of 2.5-3.5 tons / ha. Farmers usually grow summer rice (*Boro* season) to get higher yield at the higher inputs cost. The average yield of summer obtained was about 4-5 ton/ha. Presently, the cost of cultivation of rice did not match with the sale price due to exorbitant input cost hike, as such, rice growing became non-profitable.

Keywords: Water and nutrient use efficiency; grain yield; B: C ratio; System of Rice Intensification (SRI)

Introduction

SRI is a methodology for increasing the productivity of irrigated rice by changing the management of plants, soil, water and nutrients. These practices contribute to both healthier soil and plants supported by greater root growth and the nurturing of soil microbial abundance and diversity. It is based on a number of agro-ecological principles with good scientific foundations. SRI concepts and practices have also been successfully adapted to upland rice. System of RiceIntensification (SRI) involves the use of certain management practices, which together provides better growing conditions for rice plants, particularly in the root zone, than those plants grown under traditional practices. Four components of SRI include early planting (12 days' old single seedlings, wider spacing), limited irrigation (2-3 cm depth after the appearance of hairline cracks), weeding and application of more compost and building soil organic matter content. Increased weeding is required, because rice fields are not kept continuously flooded. But farmers report that with SRI methods, their rice plants are better able to resist damage from pests and diseases, making agrochemicals usually unnecessary. Compost gives even better results than does fertilizer with SRI methods. With SRI there can be water savings of around 50%. System of Rice Intensification (SRI) is a system of rice production in which synergistic interactions lead to much higher grain yield. SRI technology is a "Less Water" method of production which is suitable to poor farmers who have relatively more labour than land and capital. It involves the use of certain management practices of plants, soils, water and nutrients, which together provides better growing conditions for rice plants, particularly in the root zone, than those plants grown under traditional practices. Four components of SRI include early planting (12 days old single seedling, wider spacing), limited irrigation (2-3 cm depth after the appearance of hairline cracks), weeding and application of more compost rather than chemical fertilizers and pesticides.

The principles of the SRI have been well taken care of and the management practices are appropriately adjusted to suit the farmers' need and local situations. Age of seedling, spacing, fertilizer application, weeding or inter cultivation for soil aeration and water management practices are suitably modified for farmer friendly adoption and yield improvement. SRI is being practiced in states like Tamil Nadu, Andhrapradesh, Karnataka in South India and sporadically followed in few Eastern states like Tripura and Assam. Under present situation, our country cannot afford to grow rice with intensive use of scarce natural resources like seed, chemicals, water and labour. Simplification of SRI methodology and up scaling this innovative approach throughout the country alone could sustain the irrigated rice cultivation in future.

Benefits of SRI technology

Due to its wide applicability and feasibility in varied soil and climatic condition and the readily acceptance of the resource poor farmers for the benefits accrued from SRI technology for longer duration, it is thus sustainable and remunerative for rice production. In addition, this system of cultivation has the following advantages:

- **Less seed rate**: A seed rate of 5-8 kg depending on 1000 grain weight is adequate enough to plant one hectare of land under SRI. In contrast, in conventional method of planting depending upon the duration group, 60 kg ha^{-1} for short duration, 40 kg ha^{-1} for medium duration and 30 kg ha^{-1} for long duration varieties and 20 kg ha^{-1} for hybrid is recommended.
- **Less nursery area**: A mat nursery area of 2.5 cents (100 m^2) is sufficient to raise seedlings to cover one hectare of land in SRI while in conventional methods, 20 cents per hectare is essentially required.
- **Labour saving**: The labour required for nursery period is less (12 labourers) for SRI nursery as compared to conventional nursery that requires a minimum of 30 labourers.
- **Water saving**: Water requirement under SRI method is only 600 to 700 mm applied through intermittent irrigation for maintaining soil moist condition, while in conventional method, about 1200–1500 mm of irrigation water is required for continuous flooding. Thus it could save at least 50% of irrigation water for raising irrigated paddy.
- **Aeration**: Heavier weeding encourages profuse aeration to the root zone leading to healthier plant growth and thereby crop yields.
- **Enhanced yield**: The additional yield advantage in SRI ranges from 500 to 1500 kg/ha over conventional method of planting. The reason is mainly

attributed to large root volume, profuse and lengthy productive tillers, non-lodging, big panicle with increased number of filled grains per panicle and higher grain weight.

- **Control of diseases and pests**: It resists insects and diseases because it allows rice to grow naturally.
- **Labour intensive**: It creates 50% more man day's employments for transplanting and weeding. It offers an alternative to resource poor that puts in their family labour.
- **Less chemicals**: As emphasis has been given to use compost adequately for crop nourishment, less amount of chemical fertilizers and pesticides is required. It is applied instead of or in addition to chemical fertilizer. It gives good results without chemical fertilizers because of better soil health and structure and more balanced nutrients supply.
- **Control of water pollution**: By avoiding flood irrigation and adopting limited irrigation, the breeding of malarial mosquito in rice fields is checked.

In the recent past, no single technology had revealed such a tremendous impact in increasing crop productivity without adopting improved variety or hybrid. The success is mainly due to the ideal crop establishment with effective soil, water, nutrients and crop management.

Approaches in successful SRI programme

A. Improving drainage in rice fields

The major limitations for the low adoption rate of this management-oriented technology is the adverse landscape and poor drainage facility existing in the rice growing regions especially in the deltaic situation. Crop establishment with young seedlings, nutrient use and soil aeration vis-a-via crop growth and yields are primarily depending on the extent of drainage facility that could be improved substantially by -

1. Mass desiltingprogramme to improve the storage of inland reservoirs
2. Cleaning of the canals for easy drainage
3. Construction of water harvesting structures to store the draining water at appropriate place
4. Restructuring field bunds to have proper inlet and outlet for easy drainage
5. Farmers' participatory research cum demo on improved water management techniques

B. Production of healthy rice seedlings

1. Training to the extension personnel and farmers on production of healthy and robust rice seedlings through modified rice mat nursery or improved conventional nursery method,
2. Development of community mat nursery for larger scale distribution and motivating farmers through awareness programme, and
3. Training to the farm labourers on raising mat nurseries

C. Crop establishment

1. Organizing demonstrations on proper land leveling for successful crop establishment
2. Enhancing the accessibility of farmers to modern leveling equipments on rental basis/at subsidiary rates
3. Training to the farm labourers on square transplanting rice seedlings
4. Intensifying research on evolving simple mechanized transplanter to plant single seedling per hill
5. Developing direct sowing technology so as to establish rice crop with single seedling per hole to by-pass nursery and transplanting stages.

D. Soil and water management

1. Supply of cono rotary weeders at economic price.
2. Demonstrations to illustrate the significance of the conoweeder operation
3. Fabrication of motorized conoweeder for easy operation
4. Training to the farmers on the adoption of intermittent irrigation
5. Large scale demonstrations to illustrate the benefits of intermittent irrigation in comparison with continuous flooding.

The measures suggested are only extension approaches for promoting large-scale adoption of the SRI method of rice cultivation, except the fabrication of motorized conoweeder. Research efforts are already initiated to simplify the inter cultivation tool for easy operation in the field. Breaking the mind-set of the farmers through motivation and 'seeing is believing' approach is the pre-requisite for up-scaling this innovative crop, soil and water management approach.

Success Stories

Tamil Nadu

As success in Tamil Nadu, the SRI was demonstrated in large-scale at the Edamelaiyur village of New Cauvery Delta Zone during 2002-2003. On seeing the initial response and record yields from the Cauvery Delta farmers, the Government of Tamil Nadu had sanctioned a scheme during 2003-2004 for popularizing this innovative rice cultivation method in the Cauvery and Tamirabarani river basin area. The adoption of SRI technology during 2005-06 was only 10-15% of the rice area in Tamil Nadu. 2500 numbers of one-acre demonstrations trails in farmer's field have been organizedduring the year (2006-07). It is proposed to cover 25% of the irrigated area in the state with SRI method of cultivation. A study conducted by the state agricultural university of Andhra Pradesh documented at the tune of 21.3% increase in rice yield accompanied by reduced irrigation application due to application of SRI. The demonstration of SRI was confined to 15,000 plots last year, was increased to 32,000 plots in 2006-07. A similar study in Tamil Nadu calculated almost a four-fold increase in the productivity of irrigation water (kg/ha of rice produced per cubic meter of water).

West Bengal

Purulia: In the Purulia district, the results of the SRI in the area have been very encouraging in 2005. The average productivity was as high as 7.7 t/ha in spite of high incidence of pests and diseases coupled with dry spells, both during transplanting and in the grain filling stages of the crop. About 90% of the farmers had yields above 5 t/ha, around 2.5 times the district average, with 60% of the farmers recording yields in the range of 5-7 t/ha. This achievements leaves scope the participating farmers to bring more land under SRI.All the State Governments should initiate action in popularizing the SRI wherever possible. In particular, Agricultural universities and state agricultural departments should develop special programs on SRI including training and capacity building components. The irrigation departments should ensure proper water control through efficient water allocation methods. Similarly, The SRI method should be explored for other crops where potential exists. The aerobic and upland rice technologies should also be expanded wherever suitable.

Nadia: The on-farm programme on SRI was carried out in the village Narayanpur under RLI command area in Karimpur II Block of Nadia district in the year 2016 during summer season. The soil of farmland is clay loam and situated having latitude of 23°54'26.5"Nlongitude 88°30'24.1"Eat an average elevation of 9.75 m above the mean sea level. Monsoon ceases during October and cool season sets in by November. The average rainfall of this region is about 1500 mm and normal monsoon breaks in the first week of June. Though water resources in this area are generally satisfactory and 90% of these resources are tapped form the ground water source through deep and shallow tube wells. However, due to faulty crop planning and lack of scientific water management approach, a significant portion of agricultural area is identified as gray zone from hydrological point of view. Even many tube wells are becoming dry due to over exploitation of valuable water from the lower depths of confined aquifers. Rice is predominant crop during *kharif* season becoming less profitable because of high input costs involved. Summer rice gave satisfactory yield but also profitable proposition because of more input cost and less farm gate prices. In this stated background as above, the traditional rice growing technology has been shifted to Systematic Rice Intensification) SRI, a proven methodology to save water in growing rice were tried as intervention and adoption for the purpose of creating more man days and more income per drop of water and increasing agricultural productivity. Rice is pre-dominantly main crop in the area both during rainy as well as summer season. Rainy season rice with HYV gave an average yield of 2.5-3.5 tons / ha. Farmers usually grow summer rice (*Boro* season) to get higher yield at the higher inputs cost. The average yield of summer obtained was about 4-5 ton/ha. Presently, the cost of cultivation of rice did not match with the sale price due to exorbitant input cost hike. As such, rice growing became non-profitable.

Name of the WUA: Narayanpur Barikpara

Total Beneficiaries: Male: 107, Female: 15, Total: 122, SC/ST: 0. The area was primarily rainfed before implementation of the scheme and single crop of kharif rice was usually grown. The summer season, there was no possibility of growing crop due to scarcity of water. Summer season rice (*Boro*) was grown adopting traditional method in some pockets become non-profitable proposition. So the SRI methodology of rice cultivation during summer season was provided to the beneficiaries members of this WUA with objectives:

1. To reduce the cost of cultivation by reducing input requirement in the method;
2. To increase yield of crop and water productivity
3. To increase total income of the farmer by the process of adoption.

Storage Capacity achieved: Farmers used to develop local storage system as Dhan Gola, Storage for onion, aurat system (a temporary storage room) for banana and vegetables crops;

The Climates

The climate of this region is fall in humid tropic. Top soil texture of this area is clay loam .The summer is very hot & winter is moderate and May is generally the hottest month of the year. The average temperature ranges from 39.6^{0} to 25.4^{0}C during summer and 23.7^{0}C to 8.5^{0}C during winter. Monsoon ceases during the month of October and cool period started during the month of November.

ASS activities

The critical inputs were supplied in terms of seed and fertilizers along the methodology for growing rice cultivation (SRI); The Area covered with 40 ha RLI irrigation command with 9 sprouts approximately 4.40 ha coverage by each sprout. The farmers received 2 exposure visits and 6 lead farmers/farmers meeting. After implementation of project, Jute is the prevalent pre-*kharif* crop with sesame, rice during *kharif*and during winter/rabi season the crops like wheat, lentil, vegetables, onion, are growing as a part of crop diversification. 15-20% increase in average yield observed replacing 100% cropping intensity to 190% cropping intensity. The market interventions of the produce were done with arrangement of buyer-sellers meet and a fixed market available with 5 km radius distant.

Technology adoption

The farmer contacted with WBADMI-personnel being member beneficiary of the WUA in RLI command. He was given 1 kg of hybrid rice (cv*Arize*6444) during summer season to grow under SRI method with water as well as fertilizer saving technologies. The seed bed was prepared in December 19 and single plant transplanting was done at 13 days old seedling in puddled soils without allowing standing water. The water and fertilizer along with organic manures were applied as per recommendation less than half as compared to traditional transplanting. The control plots with traditional method was kept separately with recommended doses of fertilizer as well standing water allowed up to flowering at the depth of 5±2 cm depth. 13-14 days old single seedling was transplanted on single plant basis allowing spacing of 25 cm x 25 cm in puddled soil where no standing water allowed with recommended fertilizer doses of 40:20:20 kg N, P_2O_5 and K_2O, per acre during final land preparation, respectively along with FYM @2 tons/Acre. Because of higher spacing and single seedling, the cost of cultivation of rice including seed and fertilizer was severely cut down from the beginning.

Table 1: Cost and income difference for SRI and Conventional paddy cultivation at Narayanpur by Sri SagarHalsana for 1 Bigha land. Variety of Seed Used ARIZE-6444

SL	Package & practices	Under SRI Method(1 Bigha)		Under Conventional Method(1 Bigha)	
		Quantity/Rate	Cost (In Rs.)	Quantity/Rate	Cost(In Rs.)
1.	Cost of Seed	1 Kg@ Rs. 350/-	350	6 Kg.	2100
2.	Date of seed bed	19/12/15	n.a	19/12/15	n.a
3.	Labour Charge for planting	6 persons @ Rs. 300/-	1800	4 persons @ Rs. 300/-	1200
4.	Dt of Plantation	02/01/2016	n.a	15/01/2016	n.a
5.	Survival rate of Plants	98%		98%	
	Application of Chemical fertilizers				
1.	Urea @ Rs.8.16/Kg	30Kg.	245	45Kg	370
2.	SSP @ Rs.6.00/Kg	42Kg	250	52Kg	312
3.	MOP @ Rs.19.00/Kg	12Kg.	220	14Kg	260
4.	Seed Treatment	DM-45 (5gm)	3	DM-45(5gm)	3
	Organic fertilizers &pestisides				
1.	Vermi compost	3- Cart @ Rs. 900/-	2700	2- Cart @ Rs. 900/-	1800
2.	Pestisides	n.a		n.a	
	Weeding & Irrigation				
1.	Weeding number of times and cost	6 Labour @ 150/-	900	4 Labour @ 150/-	600
2.	Number of times Irrigated & Cost	20 times(60cm)	1200	12 times(120cm)	2000
	Harvesting				
1.	Harvesting Cost	5 labour @Rs. 200/-	1000	4 labour @Rs. 200/-	800
2.	Thresing & packing Cost	5 labour @Rs. 200/-	1000	4 labour @Rs. 200/-	800
	Total Cost of Production		9668		10245
	Sale Price of Produce				
1.	Quantity of Paddy produced	1200 Kg. @ 1400/Qtl.	16800	900 Kg. . @ 1400/Qtl.	12600
2.	Quantity of straw sold		2200		2000
	Totla Sale Price		19000		14600
	Total Income		8600		4355

Unlike traditional rice, the crop under SRI was irrigated as and when required basis, just up to soil saturation (Field Capacity), no standing water was allowed. Thus the number of irrigation required up to 30 with total water requirement only of 60 cm wherein as compared to number of irrigation required for conventional planting up to 15 with total water requirement of 120 cm even under alternate wetting and drying basis. Thus 50% of water cost along with 50% cost of fertilizer application was saved under SRI method along cost cut in seed requirement.

The rice gave 60-75 tillers/plant having 90% of effective tillers under SRI whereas traditionally rice gave maximum of 25 numbers of tillers. The length of panicle, number of grain per panicle, straw yield and test weight were observed more under SRI method than that of conventional method. The crop harvested at 130 days maturity little earlier than traditionally grown rice. The crop yielded 9.05 tons/ha under SRI as compared to 6.03 ton/ha in traditional method, thereby 50% yield increase over conventional method. Total cost of inputs and labours required for 1 bigha of land was (7.5 bighas = 1 ha) for rice cultivation was I Rs.10245 in conventional as compared to IRs. 9668 under SRI wherein the total sale price of rice obtained was IRs. 16800 under SRI as compared that of IRs. 12600 produced in conventional method (considering price of rice at IRs. 14/kg).

Therefore, incremental income from one bigha enjoyed by using SRI came to Rs. 4245/-. In *Kharif* season Sri Sagar Halsana cultivated 5 (five) bigha of paddy under SRI and his incremental income was nearly Rs.21225/- on adoption of SRI system of paddy cultivation.

Conclusion

The rice gave 60-75 tillers/plant having 90% of effective tillers under SRI whereas traditionally rice gave maximum of 20 numbers of tillers. The length of panicle, number of grain per panicle, straw yield and test weight were observed more under SRI method than that of conventional method. The crop harvested at 130 days maturity little earlier than traditionally grown rice. The crop yielded 9.05 tons/ha under SRI as compared to 6.03 ton/ha in traditional method, thereby 50% yield increase over conventional method. Total cost of inputs and labours required for 1 bigha of land was (7.5 bighas = 1 ha) rice cultivation was IRs.7500 in conventional as compared to IRs. 4500 under SRI wherein the total sale price of rice obtained was IRs. 14400 under SRI as compared that of IRs. 9600 produced in conventional method (considering price of rice at IRs. 12/kg). Thus obtained BCR of SRI method was 3.20 as compared 1.28 under conventional rice cultivation.SRI cultivation is summarily eco-friendly paddy cultivation system that requires 50% less water, 80% cut in paddy seed requirement, 40% less fertilizer having less infestation with pests and diseases which cut down crop duration by 8-10 days than that of conventional practices with 20-40% more grain and straw yield at less cost of cultivation with highest B: C ratio of 3.20.

More information

Norman Uphoff, Director. Cornell International Institute for Food, Agriculture and Development (CIIFAD); Box 14 Kennedy Hall, Cornell University, Ithaca NY 14853, USA Tel: 01-607-255-0831; Fax: 01-607-225-1005
Email: NTU1@cornell.edu.

References

Adhya, Tapan K., Bruce A. Linquist, Tim Searchinger, Reiner Wassmann, and Xiaoyuan Yan. 2014. "Wetting and Drying: Reducing Greenhouse Gas Emissions and Saving Water from Rice Production." Working Paper, Installment 8 of Creating a Sustainable Food Future. Washington, DC: World Resources Institute. http://www.wri.org/sites/default/files/wetting-drying-reducing-greenhouse-gas-emissionssaving-water-rice-production.pdf.

Africare, Oxfam America, WWF-ICRISAT Project. 2010. "More Rice for People, More Water for the Planet." Hyderabad, India: WWF-ICRISAT Project. https://www.africare.org/wpcontent/uploads/2014/04/more-rice-for-people-more-water-for-the-planet-sri.pdf.

Ahmad, Shahrear, Chengfang Li, Guangzhao Dai, Ming Zhan, Jinping Wang, Shenggang Pan, and Cougui Cao. 2009. "Greenhouse Gas Emission from Direct Seeding Paddy Field under Different Rice Tillage Systems in Central China." *Soil and Tillage Research,* 106 (1): 54–61. doi:10.1016/j.still.2009.09.005.

Ali, RI, TH Awan, M Ahmad, U Saleem, M Akhtar, and others. 2012. "Diversification of Rice-Based Cropping Systems to Improve Soil Fertility, Sustainable Productivity and Economics." *Journal of Animal and Plant Sciences,* 22 (1): 108–112.

Aulakh, Milkha S., Reiner Wassmann, C. Bueno, and Heinz Rennenberg. 2001. "Impact of Root Exudates of Different Cultivars and Plant Development Stages of Rice (Oryza Sativa L.) on Methane Production in a Paddy Soil." *Plant and Soil,*230 (1): 77–86.

Barrett, Christopher B., Christine M. Moser, Oloro V. McHugh, and JoeliBarison. 2004. "Better Technology, Better Plots, or Better Farmers? Identifying Changes in Productivity and Risk among Malagasy Rice Farmers."*American Journal of Agricultural Economics,* 86 (4): 869–88.

Bouman, B. A. M., Xiaoguang Yang, Huaqi Wang, Zhimin Wang, Junfang Zhao, and Bin Chen. 2006. "Performance of Aerobic Rice Varieties under Irrigated Conditions in North China." Field Crops Research, Preparing Rice for a Water-Limited Future: from Molecular to Regional Scale. *International Rice Research Congress,*97 (1): 53–65.

Choi, Joongdae, Gunyeob Kim, Woonji Park, Minhwan Shin, Yonghun Choi, Suin Lee, Sunjoo Kim, and Dongkoun Yun. 2014. "Effect of SRI Water Management on Water Quality and Greenhouse Gas Emissions in Korea." *Irrigation and Drainage,* 63 (2): 263–270.

Galloway, James N., John D. Aber, Jan Willem Erisman, Sybil P. Seitzinger, Robert W. Howarth, Ellis B. Cowling, and B. Jack Cosby. 2003. "The Nitrogen Cascade." *BioScience,* 53 (4): 341–56.

Gathorne-Hardy, DR, Narasimha Reddy, and others. 2013. "A Life Cycle Assessment (LCA) of Greenhouse Gas Emissions from SRI and Flooded Rice Production in SE India." *Taiwan Water Conservancy,* 61 (4): 110–125.

Ghosh, Sabyasachi, DeepanjanMajumdar, and M. C. Jain. 2003. "Methane and Nitrous Oxide Emissions from Irrigated Rice of North India." *Chemosphere,* 51 (3): 181–95.

Gutierrez, Jessie, Sang Yoon Kim, and PilJoo Kim. 2013. "Effect of Rice Cultivar on CH4 Emissions and Productivity in Korean Paddy Soil." *Field Crops Research,* 146 (May): 16–24. doi:10.1016/j.fcr.2013.03.003.

Itoh, Masayuki, ShigetoSudo, Shizuka Mori, Hiroshi Saito, Takahiro Yoshida, Yutaka Shiratori, ShinobuSuga, et al. 2011. "Mitigation of Methane Emissions from Paddy Fields by Prolonging Midseason Drainage."*Agriculture, Ecosystems & Environment,* 141 (3–4): 359–72.

Jacinthe, P. -A., and R. Lal. 2005. "Labile Carbon and Methane Uptake as Affected by Tillage Intensity in a Mollisol." *Soil and Tillage Research,* 80 (1–2): 35–45.

Jain, Niveta, RachanaDubey, D. S. Dubey, Jagpal Singh, M. Khanna, H. Pathak, and Arti Bhatia. 2013. "Mitigation of Greenhouse Gas Emission with System of Rice Intensification in the Indo-Gangetic Plains." *Paddy and Water Environment,* 12 (3): 355–63.

Jarecki, Marek K., and Rattan Lal. 2003. "Crop Management for Soil Carbon Sequestration." *Critical Reviews in Plant Sciences,* 22 (5): 471–502.

Jia, Zhongjun, ZucongCai, and HaruoTsuruta. 2006. "Effect of Rice Cultivar on CH4 Production Potential of Rice Soil and CH4 Emission in a Pot Experiment." *Soil Science and Plant Nutrition,* 52 (3): 341–48.

Kumar, Virender, and Jagdish K Ladha. 2011. "Direct Seeding of Rice: Recent Developments and Future Research Needs." *Advances in Agronomy*, No. 111: 297–413.

Li, Changsheng, William Salas, Benjamin DeAngelo, and Steven Rose. 2006. "Assessing Alternatives for Mitigating Net Greenhouse Gas Emissions and Increasing Yields from Rice Production in China Over the Next Twenty Years." *Journal of Environment Quality,* 35 (4): 1554. doi:10.2134/jeq2005.0208.

Liang, Wei, Yi Shi, Hua Zhang, Jin Yue, AndGuo-Hong Huang. 2007. "Greenhouse Gas Emissions from Northeast China Rice Fields in Fallow Season.*" Pedosphere,* 17 (5): 630–38..

McDonald, A. J., P. R. Hobbs, and S. J. Riha. 2006. "Does the System of Rice Intensification Outperform Conventional Best Management?: A Synopsis of the Empirical Record." *Field Crops Research,* 96 (1): 31– 36.

Minamikawa, Kazunori, and Naoki Sakai. 2006. "The Practical Use of Water Management Based on Soil Redox Potential for Decreasing Methane Emission from a Paddy Field in Japan." *Agriculture, Ecosystems & Environment,* 116 (3–4): 181–88.

Mishra, S., A. K. Rath, T. K. Adhya, V. R. Rao, and N. Sethunathan. 1997. "Effect of Continuous and Alternate Water Regimes on Methane Efflux from Rice under Greenhouse Conditions." *Biology and Fertility of Soils,* 24 (4): 399–405.

Sinha, Shekhar Kumar, and JayeshTalati. 2007. "Productivity Impacts of the System of Rice Intensification (SRI): A Case Study in West Bengal, India." *Agricultural Water Management,* 87 (1): 55–60.

Thakur, Amod K. 2010. "Critiquing SRI Criticism: Beyond Scepticism with Empiricism."*Current Science,* 98 (10): 1294–1299.

Wang, B., Y. Xu, Z. Wang, Z. Li, Y. Guo, K. Shao, and Z. Chen. 1999. "Methane Emissions from Ricefields as Affected by Organic Amendment, Water Regime, Crop Establishment, and Rice Cultivar." *Environmental Monitoring and Assessment,* 57 (2): 213–28.

Win, KhinThawda, RyokoNonaka, Aye Thida Win, Yu Sasada, Koki Toyota, and Takashi Motobayashi. 2013. "Effects of Water Saving Irrigation and Rice Variety on Greenhouse Gas Emissions and Water Use Efficiency in a Paddy Field Fertilized with Anaerobically Digested Pig Slurry." *Paddy and Water Environment,* 13 (1): 51–60.

Xiaoguang, Yang, B. A. M. Bouman, Wang Huaqi, Wang Zhimin, Zhao Junfang, and Chen Bin. 2005. "Performance of Temperate Aerobic Rice under Different Water Regimes in North China." *Agricultural Water Management,* 74 (2): 107–22.

Zhang, Wenju, MinggangXu, Xiujun Wang, Qinhai Huang, Jun Nie, Zuzhang Li, Shuanglai Li, SeonWoong Hwang, and Kyeong Bo Lee. 2012. "Effects of Organic Amendments on Soil Carbon Sequestration in Paddy Fields of Subtropical China." *Journal of Soils and Sediments,* 12 (4): 457–70.

Zou, Jianwen, Yao Huang, Yanyu Lu, Xunhua Zheng, and Yuesi Wang. 2005. "Direct Emission Factor for N_2O from Rice–winter Wheat Rotation Systems in Southeast China." *Atmospheric Environment,* 39 (26): 4755–65.

30

Pusa Hydrogel: A Superabsorbent for Soil Moisture Conservation to Enhance Crop Productivity

Jnana Bharati Palai, Sagar Maitra and A. Zaman

M.S. Swaminathan School of Agriculture, Centurion University of Technology and Management, Paralakhemundi, Odisha, India
Email: jnana@cutm.ac.in

Abstract

In view of the fast depletion of ground water reserves, uncertainty of rains and growing food demands due to human population explosion, efficient use of water available for crops has become highly relevant. In recent years, the use of superabsorbent polymers (saps) has been viewed as an innovative tool to improve water use efficiency in agricultural operations. Pusa Hydrogel, an indigenous superabsorbent hydrogel technology, has been developed for improving water use efficiency of agricultural and horticultural crops. They can absorb a minimum of 400 times of their dry weight of pure water and gradually release it according to the needs of the crop plant. So, it is used as soil moisture conservator and reduces number of irrigation. Because of their neutral ph, they do not affect nutrient availability, soil chemical composition, action of other agro chemicals, viz. fertilizers, herbicides, fungicides, insecticides, etc. Pusa hydrogels are found to improve the physical properties of soils. Increase in porosity results in improvement in seed germination and rate of seedling emergence, root growth and density, and reduced soil erosion due to reduction in soil compaction. It also increases biological/microbial activities in the soil, which increase oxygen/air and nutrient availability in root zone of the plant. So, ultimately it is helpful for increasing crop productivity.

***Keywords*:** Crop productivity; Pusa Hydrogel; soil moisture conservator and nutrient availability

Introduction

In view of the fast depletion of ground water reserves, uncertainty of rains in arid and semiarid regions of the world, along with the growing food demands of the burgeoning human population, efficient use of water available for crops has become highly relevant. Nowadays, the use of superabsorbent polymers (SAPs) has been viewed as an innovative tool to improve water use efficiency in agricultural operations. Hydrogels potentially influence infiltration rates, density, soil structure,compaction, soil texture, aggregate stability and crust hardness (El-Hady and Abo-Sedera 2006), and evaporation rates (Choudhary, Shalaby, and Al-Ornran 1995; Taban and Naeini 2006). It has been reported that addition of polymers to sandy soils improved water availability to plants, not only due to polymer swelling but also by decreasing the hydraulic conductivity of the sand and it decreases drainage of water below the root zone and increases retention pores (Kazanskii and Dubrovskii 1992; El-Shafei *et al*., 1992; Al-Darby 1996; Al-Omran and Al-Harbi 1998; El-Hady and Sedera 2006). It is also reported that applying organic materials mixed with theproper hydrogel to the soil will be more effective and economic than using each of them alone (Abd El-Hameed *et al*., 1995; El-Hady *et al.,* 2000b; El- Hady and Camilia, 2006). However, because of high cost and greater application rates, hydrogel application in agriculture did not get wider attention. Use of hydrogels for raising agriculture crops could be an economically viable option only if they are applied in smaller quantities at shallow depths in rows just below the seeds. For sandy soils, plant productivity is limited due to low water-holding capacity, low fertility and excessive deep percolation losses, which result in reductions in the efficiencyof water and fertilizer use by plants (Sivapalan, 2006; Sen *et al*., 1995; Sabrah, 1994). Moreover the plants in these soils often suffer from water stress, as a result of drought conditions, which leads to a loss in crop production (Rigas *et al*., 1999; Sen *et al*., 1995). Pusa Hydrogel is an indigenous product designed and developed by IARI, New Delhi to enhance the crop productivity per unit available water and nutrients, particularly in moisture stress agriculture. Patil *et al.* (2014) has been reported that Pusa hydrogel, which absorbs about 80g of water/gram of xerogel helped in conserving soil moisture in root zone, plants to withstand extended moisture stress, delays onset of permanent wilting point, reduction in irrigation and fertigation requirements of the crops.

Role of Pusa Hydrogel on soil moisture conservation

Water retention by the polymers increases and it provides extra available water to crops means less frequent irrigation and facilitates better crop growth. More available water in the soil also Hydrogel reduces irrigation frequency of field crops. It also reduces irrigation amount from 100 to 85% of the crop water requirements and increase crop yield (El-Hady *et al.*, 2006). Johnson (1984) also reported that use of hydrogels increase the amount of available moisture in root zone and reduces intervals between irrigations. Ekabafe *et al.* (2011) reported an increase of 171 to 402% in water retention capacity when polymers were incorporated in coarse sand. Addition of polymer to peat soil decreases water stress and increases the time to wilt (Karimi *et al.*, 2009). Soil physical properties can be improved by the incorporation of superabsorbent polymer with soil (El- Amir *et al.*, 1993). Application of superabsorbent polymer, increases water retention capacity and protection against drought was observed by Nazrali *et al.* (2010). The farmer's field studies shown that Pusa hydrogel application @ 2.5 kg/ ha save two irrigations in wheat with comparable yield (17.7q/acre) against the recommended normal six irrigations (ICAR 2014).

The best result was observed in loamy soilsregarding the available moisture, application of PR3005A polymer in levels of 4 and 8 g kg^{-1}. The moisture amount in this situation was increased by 2 to 4 times, respectively (Ghaiour, 2000). Sivapalan (2006) stated that with application of polymer at 0.03 and 0.07% of its weight, the retained water in sandy soil was equal to 23 and 95% respectively. Water availability of loamy and clay soil with the application of highest 0.67% PAM dose showed highly significant increase (108 and 105%, respectively) due to increase in water content at field capacity and decrease in water content at wilting point. Meanwhile, plant available water content of sandy soil increased by 55% since water content at wilting point increased (Uz *et al.*, 2008).

Role of Pusa Hydrogel to enhance productivity

The results of field experiment in wheat in different wheat-growing zones of India show that with all the levels of irrigation (viz. no irrigation, two and four irrigations) application of 5 kg/ha of hydrogel produced significantly higher grain yield. Also, the equivalent yield of four irrigations with no hydrogel was obtained with only two irrigations when 5 kg/ha of hydrogel was applied (DWR, 2013). Application of 2.5 kg/ha of hydrogel in aerobic rice produced significantly higher growth attributing characters and yield compared to control in all the types of lands (viz. flat bed sowing, ridge sowing and raised bed sowing) (Rehman *et al.*, 2011). Singh *et al,* (2012) reported that compared to control and water soaking treatment coating of pearl millet seed with 10 and 20 g of hydrogel/kg of seed resulted in the production of significantly higher effective tillers, ear length, test weight, grain and stover yield. Results of application of 200 kg/ha of hydrogel in peanut were found to be significantly superior in respect of all the growth and yield characters (viz. seed yield, biomass yield, pod yield, number of branches per plant and 100 seed weight) in sandy soil of Iran with hot and arid climate (Langaroodi *et al.,* 2013). Borivoj *et al.,* (2006) also reported that with the application of 5 kg/ha of hydrogel yield of wheat was found to increase by 8.48% over control in clay loam soil with 100% recommended dose of fertilizers. The results obtained from farmers field demonstration conducted by ICAR at different locations in Uttar Pradesh evidenced that soil application of hydrogel @ 5 kg/ha along with three irrigations in different wheat varieties is able to produce grain yield equivalent to irrigating wheat crop with five times without hydrogel application. It indicates that soil application of hydrogel can save two irrigations in wheat without reducing the grain yield.

Conclusion

Pusa Hydrogel application increases productivity in almost all the test crops (cereals, vegetables, oilseeds, flowers, spices etc.) in terms of crop yield. It also helps improve the quality of agricultural produce in terms of plant biomass, fruit and flower size and colour with improvement in hydro-physical and biological environment of the soil. Hence, Pusa hydrogels may become a practically convenient and economically feasible option in water-stressed areas for increasing agricultural productivity with environmental sustainability.

References

Abd El-Hameed, M.W., El-Hady, O.A., Hammad, S.A. and Kotb, M.T. 1995. Effect of incorporating organic manure or/and hydrogels in sandy soil on soil structurization and stabilization. Paper presented at the Egypt Soil Science Society Fifth National Congress (Bioagriculture in Relation to Environment), Nov. 20–21, Cairo, Egypt.

Al-Darby, A. M. 1996. Unsaturated hydraulic conductivity for sandy soil treated with gel-forming soil conditioner. *Soil Technology* 9:15–28.

Al-Omran, A. M., and Al-Harbi, A. R. 1998. Improvement of sandy soils with soil conditioners. In *Handbook of soil conditioners: Substances that enhance the physical properties of soil*, A. Wallace and R. E. Terry, 363–384. New York: Marcel Dekker.

Borivoj, S., Rak, L. and Bubenikova, I. 2006. The effect of hydroabsorbent on selected soil biological and biochemical characteristics and its possible use in revitalization *Ecologia*25(4):422– 429.

Choudhary, M.I., Shalaby, A.A. and Al-Ornran, A.M. 1995. Water-holding capacity and evaporation of calcareous soils as affected by four synthetic polymers.*Communications in Soil Scienceand Plant Analysis* 26 (13–14): 2205–2215.

Ekabafe, I.O.; Ogbeifun, D.E. and Okieimen, F.E. 2011.Polymer applications in agriculture.*Biokemistri* 23(2): 81-89.

El-Amir, S., Helalia, A.M. and Shawky, M. E.1993.Effects of acryhope and aquastore polymers on water regime and porosity in sandy soils.*Egyptian Journal of Soil Science*4: 395–404.

El- Hady, O. A., and Camilia, Y. E.. 2006. The conditioning effect of composts (natural) or/and Acrylamide hydrogels (synthesized) on a sandy calcareous soil 1: Growth response, nutrients Uptake, and water and fertilizers use efficiency by tomato plants. *Journal of Applied Sciences Research* 2:890–898

El-Hady, O.A., El-Sherif, A.F., Saleh, A.L. and. El-Kader, A.A. 2000. Effect of incorporating Organic manure and /or acrylamide hydrogels in sandy soil on growth response, nutrients uptake and water and fertilizers use efficiency by rye grass plants. Paper presented at the Tenth International Colloquium on Plant Nutrition for the Next Millennium, April 8–13, Cairo, Egypt.

El-Hady, O.A., and Abo-Sedera, S. A. 2006. Conditioning effect of composts and acrylamide Hydrogels on a sandy calcareous soil, II: Physico-bio-chemical properties of the soil. *International Journal of Agriculture and Biology* 8:876–884.

El-Shafei, Y.Z., Al-Ornran, A.M., Al-Darby, A.M., Shalaby, A.A., 1992. Influence of upper layer treatment of gel-forming conditioner on water movement in sandy soils under sprinkler infiltration. *Arid Soil Research and Rehabilitation*6: 217–231.

Ghaiour, F.A. 2000. Effect of moisture absorbent materials on soil water holding potential.Ministry of agriculture. Isfahan research center of animal and natural resources. Isfahan, Iran: 515-517.

ICAR. 2014. *Proc. 23rd Meeting ICAR Regional Committee*. No-V, Ludhiana, India, 139p.

Johnson, M.S. 1984. The effect of gel-forming polyacrylamides on moisture storage in sandy soils.*Journal of the Science of Food and Agriculture*35: 1196–1200.

Kalhapure, A., Kumar, R., Singh, V.P. and Pandey, D.S. 2016. Hydrogels: a boon for increasing agricultural productivity in water-stressed environment. *Current Science*111(11):15

Karimi, A.; Noshadi, M. and Ahmadzadeh, M. 2009.Effects of superabsorbent polymer (igeta) on crop, soil waterand irrigation interval. *Journal of Science and Technology of Agriculture and Natural Resources*12: 415–420.

Kazanskii, K. 1992. Influence of upper layer treatment of gel-forming conditioner on water movement in sandy soils under sprinkler infiltration. *Arid Soil Research and Rehabilitation* 6:217–231.

Kazanskii, K.S., Dubrovskii, S.A. 1992. Chemistry and physics of agricultural hydrogels. *Advances in Polymer Science*104: 97–133.

Langaroodi, N.B.S., Ashouri, M., Dorodian, H.R. and Azarpour, 2013. Study effects of super absorbent application, saline water and irrigation management on yield and yield components of peanut (*Arachishypogaea* L.). *Annals of Biological Research* 4(1): 160–169.

Nazarli, H., Zardashti, M.R., Darvishzadeh, R. and Najafi, S. 2010. The effect of water stress and polymer on water use efficiency, yield and several morphological traits of sunflower under greenhouse condition. *Not SciBiol*2: 53-58.

Patil M.D., Dhindwal A.S. and Rajanna G.A. (2014), Integrated moisture stress management in wheat (Triticumaestivum). *Indian Journal of Agronomy* 59(4): 35-39.

Rehman, A., Ahmad, R. And Safdar, M., Effect of hydrogel on the performance of aerobic rice sown under different techniques.*Plant Soil Environ.*, 2011, 57(7): 321–325.

Rigas, F., E. Sachini, G. Chatzoudis, and N. Kanellopoulos. 1999. Effects of a polymeric soil conditioner on the early growth of sunflowers. *Canadian Journal of Soil Science* 79: 225-231.

Sabrah, R. E. A. 1994. Water movement in a conditioner-treated sandy soil in Saudi Arabia.*Journal of Arid Environments*27: 363–373.

Sen, K.K., Bhadoriaand, P.B.S. and Datta. B. 1995. Influence of soil conditioners on soil physical properties and maize growth. *Tropical Agriculture (Trinidad)* 72(1): 23–27.

Singh, H. 2012. Effect of hydrogel on growth, yield and water use efficiency in pearl millet (*Pennisetumglaucum*) production. *ForageResearch*38(1): 27–28.

Sivapalan, S. 2006. Some benefits of treating a sandy soil with a cross-linked type polyacrylamide. *Australian journal of experimental agriculture*46: 579-584.

Taban, M., and Naeini, S.A.R.M. 2006. Effect of aquasorb and organic compost amendments on soil water retention and evaporation with different evaporation potentials and soil textures. *Communications in Soil Science and Plant Analysis* 37:2031–2055.

Uz, B.Y.; Sabit, E.; Demiray, E. And Erta, E. 2008. Analyzing the soil texture effect on promoting water holding capacity by Polyacrylamide. International meeting on soil fertility land management andagroclimatology, Turkey : 209-215.

31

Changes in the Availability of Some Selected Nutrients *vis-a-vis* Some Soil Properties in Relation to Yield of Rape (*Brassica campestris* L.)

Soumya Sundar Tewary, Prasenjit Adhikary, Masud Alam, Priyanka Naskar and D.K. Das

Department of Agricultural Chemistry and Soil Science
Institute of Agricultural Science, University of Calcutta,
51/2,Hazra Road, Kolkata-700 019, India
Email: soumyasundartewary@gmail.com

Abstract

A field experiment was conducted in an aericendoaquept (pH, 6.67; organic carbon, CEC,) to study the relative efficiency of different organic manures on the changes in some selected soil properties growing rape (cv. B-9) in a randomised block design replicated thrice using different combinations of organic and inorganic chemical fertilizers (T1- NPK 100%, T2- Compost 100%, T3- Compost 50% + NPK 50%, T4- Vermicompost 100%, T5- Vermicompost 50% + NPK 50%, T6- FYM 100% and T7- FYM 50% + NPK 50%). However, the objectives of the study was to assess the relative efficiency of organic manures with respect to availability of nutrients in relation to yield of rape affected by combined application of organic manures and chemical fertilizers. The results show that the amount of available N, organic carbon content, CEC and K contents were found highest in the treatment T5 where Vermicompost 50% + NPK 50% was applied combinedly. The results show that the periodic changes soil pH has been found to be slightly changed due to combined application of organic manures and chemical fertilizers. The magnitude of increase in P content of soil varied with treatments, being maintained highest (38.30 kg/ha) in the T1 treatment where 100% NPK fertilizer was applied during 60 days of crop growth. The results show that the yield of rape seed has been found to be increased with treatments. The highest yield (10.16 q/ha) was recorded in the treatment where 50% vermicompost and 50% NPK fertilizers was applied combined.The overall results suggested that the amount of available N,P,K and S content, CEC etc. in soil was recorded an increase in most treatments where combined application of organic manures and chemical fertilizers were applied, being maintained higher amount with the application of vermicompost along with chemical fertilizers with the simultaneous increase in yield of rape.

Keywords: Availability of nutrients; organic manures; soil properties; yield of rape

Introduction

One of the most important challenges facing humanity today is to conserve/ sustain natural resources, including soil and water, for increasing food production while protecting the environment. As the world population grows, stress on natural resources increases, making it difficult to maintain food security. Long term food security requires a balance between increasing crop production, maintaining soil health and environmental sustainability.

Rapeseed-Mustard are the third important oilseed crop in the world after soybean (*Glysine max*) and palm (*Elaeis guineensis*) oil. In India, it is the second most important edible oilseed after groundnut among nine annual oilseed crops grown. Indiscriminate use of chemical fertilizers for the supply of major nutrients and declining use of secondary nutrients and organic sources of inputs over time led to the deficiency of secondary and micronutrients (De *et al.,* 2014). Through the approach of integrated nutrient management, it is important to exploit the potential of organic manures, composts, crop residues, agricultural wastes, biofertilizers and synergistic effect with chemical fertilizers for increasing balanced nutrient supply and their use efficiency for increasing productivity, sustainability of agriculture, and improving soil health and environment safety. Judicious combination of FYM, bio-fertilizers and chemical fertilizers facilitate profitable and sustainable production (De *et al.,* 2009). The bio-fertilizers have an important role to play in improving the nutrient's supplies (by way of nitrogen fixation or solubilization of phosphorus), their availability in crop production; production of growth hormones and antifungal substances favorably effect crop growth (De and Sinha. 2011). Therefore, the present investigation was carried out to study the effect of integrated nutrient management on the changes in some selected physico-chemical properties of soil growing rape (*Brassica campestris* L.).

Materials and Methods

On station experiment on rape (*brassica campestris* L.) Cv. B – 9 was conducted in *rabi* season during the year 2016-2017 in the Calcutta University Research Farm, Baruipur, 24-parganas (south), West Bengal. The soil of the experimental field was sandy clay loam whose relevant physico-chemical properties are pH,6.46; organic carbon, 0.65%; CEC, 15.60 cmol (p^+) / kg). The experiment was laid out in a Randomised Block Design (RBD) and the respective treatments were applied in each plot. There were altogether seven (7) treatment combination was replicated thrice (3) consisting of 21 plots. Rape (Cv. B-9) was sown in lines at a spacing of 15 cm and subsequently seedlings were thinned with spacing 5-6 cm plant to plant. Fertilizer application was done in two split dose. NPK (80:40:40 kg/ha) 1/2 N out of total N and full dose of P & K was applied @40-40-40 kg/ha at sowing time and 40 kg/ha N was applied 30 days after sowing by use of Urea, SSP and MOP respectively. Manures were applied once to their respective plots two weeks before sowing.

Treatments Details

T1 - NPK 100%	(80:40:40 kg/ha)
T2 - Compost 100%	(2.5 t/ha)
T3 - Compost 50% + NPK 50%	(1.25 t/ha + 40:20:20 kg/ha)
T4 - Vermi Compost 100%	(1 t/ha)
T5 - Vermi Compost 50% + NPK 50%	(0.5 t/ha + 40:20:20 kg/ha)
T6 - (FYM 100%)	(5 t/ha)
T7 - (FYM 50% + NPK 50%)	(2.5 t/ha + 40:20:20 kg/ha)

Periodic collected soil samples were analysed for different nutrient contents, pH, organic carbon, CEC etc. by following the method described by Jackson (1973).

Results and Discussion

Changes in pH of soil

The results (Table no.4.1) show that the periodic changes soil pH has been found to be slightly changed due to application of organic manures and chemical fertilizers in combinations. The pH value almost remained unchanged in most treatment, only 0.09-0.14 unit decrease in pH over T1 treatment at 10 days (6.58) was recorded in the treatments when 100 % FYM (6.44) and 100% vermicompost (6.49) was applied at 60 days of crop growth.

Table 1: Periodic changes in pH content of soil affected by integrated nutrient management on growing rape (*Brassica campestris* L.) (mean of three replications).

Treatment	10 DAS	20 DAS	30 DAS	40 DAS	60 DAS	MEAN
T1 (NPK 100%)	6.58	6.63	6.62	6.62	6.53	6.59
T2 (Compost 100%)	6.56	6.57	6.57	6.60	6.56	6.57
T3 (Compost 50%+NPK 50%)	6.57	6.56	6.52	6.55	6.53	6.54
T4 (Vermicompost 100%)	6.49	6.50	6.49	6.51	6.49	6.49
T5 (V.C 50% + NPK 50%)	6.52	6.51	6.49	6.51	6.50	6.50
T6 (FYM 100%)	6.47	6.47	6.45	6.49	6.44	6.46
T7 (FYM 50% + NPK 50%)	6.54	6.53	6.51	6.54	6.54	6.53
MEAN	6.53	6.53	6.52	6.54	6.51	
C.D at 5%	0.038	0.039	0.026	0.025	0.30	

Changes in Cation Exchange Capacity (CEC) in soil

The results (Table 2) show that the CEC of soil has been found to be changed or rather increased over the initial value (15.60 cmol (p+)/kg) due to combined application of organic manures and NPK fertilizers. However, such changes varied with treatments. The increase in CEC was continued upto 20 days of crop growth and thereafter, the amount of the same decreased with the progress of crop growth. The CEC was maintained highest (17.20 cmol (p+)/kg) with the combined application of 50% VC and 50% NPK fertilizers followed by the treatment T7 (16.88 cmol (p+)/kg) where 50% FYM and 50% NPK fertilizers were applied combinedly during 60 days of crop growth. Such increased effect of CEC in organic manure amended treatments might be explained by the increased negative charge density onto soil colloids.

Table 2: Periodic Changes in (CEC) cation exchange capacity (cmol (p+) /kg) affected by integrated nutrient management on growing rape (*Brassica campestris* L.) (mean of three replications)

Treatment	10 DAS	20 DAS	30 DAS	40 DAS	60 DAS	MEAN
T1	15.72	16.50	14.52	14.34	14.16	15.04
T2	16.50	16.08	16.01	15.53	15.06	15.83
T3	16.66	17.25	16.58	15.74	15.50	16.34
T4	17.87	17.95	17.10	16.69	16.26	17.17
T5	18.29	18.15	17.95	17.38	17.20	17.79
T6	17.39	17.42	17.20	16.58	16.24	16.96
T7	18.05	17.89	17.58	17.26	16.88	17.53
MEAN	17.21	17.32	16.70	16.21	15.90	
C.D at 5%	0.80	1.38	0.71	0.56	0.68	

Changes in Organic Carbon content in soil

The results (Table 3) reveal that the amount of organic carbon has been found to be increased over the initial value (0.65%) due to application of various kinds of organic manures with the progress of crop growth upto 40 days and thereafter, the amount of the same decreased irrespective of treatments. The magnitude of such changes varied with treatments, being maintained highest amount (0.98%) of organic carbon throughout the growth period of the crop upto 60 days in the treatment when 50% vermicompost along with 50% NPK fertilizers were applied. However, during the 60 days period of crop growth, the following trend of changes with respect to organic carbonic content was found in the order of V.C 50% + NPK 50% > 100% V.C > 50% FYM + 50% NPK > 100% NPK > 100% FYM > 50% Compost+50% NPK > 100% compost

Table 3: Periodic changes in Organic Carbon content (%) in soil affected by integrated nutrient management on growing rape (*Brassica campestris* L.) (mean of three replications)

Treatment	10 DAS	20 DAS	30 DAS	40 DAS	60 DAS	MEAN
T1	0.55	1.15	1.20	1.26	1.10	1.05
T2	0.56	1.14	1.15	1.09	1.12	1.01
T3	0.55	1.15	1.20	1.23	1.09	1.04
T4	0.61	1.25	1.21	1.20	1.15	1.08
T5	0.77	1.23	1.21	1.18	1.18	1.12
T6	0.58	1.25	1.18	1.18	1.13	1.06
T7	0.69	1.17	1.18	1.17	1.18	1.08
MEAN	0.62	1.19	1.19	1.19	1.14	
C.D at 5%	0.033	0.059	NS	0.085	0.025	

Changes in Nitrogen (N) content in soil

The results (Table 4) show that the amount of available N content has been found to be increased substantially over the initial amount (250.88 kg/ha) with the progress of crop growth irrespective of treatments. However, the magnitude of such increase varied with treatments, being maintained highest with the T5 treatment where combined applications of 50% vermicompost along with 50% NPK fertilizers were made. Such increased effect might be partly due to greater rate of mineralization of organic nitrogen as well as possibly higher amount of N present in vermicompost compared to other composts.

Table 4: Periodic Changes in N content (kg/ha) in soil affected by integrated nutrient management on growing rape (*Brassica campestris* L.) (mean of three replications).

Treatment	10 DAS	20 DAS	30 DAS	40 DAS	60 DAS	MEAN
T1	407.68	501.76	470.4	344.96	282.24	401.408
T2	250.88	313.6	250.88	219.52	282.24	263.424
T3	313.6	470.4	376.32	282.24	250.88	338.688
T4	282.24	407.68	376.32	282.24	344.96	338.688
T5	376.32	470.4	501.76	407.68	439.04	439.04

T6	250.88	407.68	376.32	313.6	313.6	332.416
T7	282.24	376.32	439.04	376.32	344.96	363.776
MEAN	309.12	421.12	398.72	318.08	322.56	
C.D at 5%	4.95	5.42	4.06	4.30	4.29	

Changes in phosphorus (P) content in soil

The results (Table - 5) reveal that the amount of available P has been found to be increased over the initial amount (15.53 kg/ha) initially up to 20 days of crop growth and thereafter, the amount of P decreased up to 60 days period of crop growth irrespective of treatments. The magnitude of such changes of P content in soil varied with treatments, being maintained highest (38.30 kg/ha) in the T1 treatment where 100% NPK fertilizer was applied during 60 days of crop growth. However, the following trend was maintained during 60 days of crop growth: T1 > T5 > T4 > T3 > T2 > T7 > T6

Table 5: Periodic changes in available P content (kg/ha) in soil affected by integrated nutrient management on growing rape (*Brassica campestris* L.) (mean of three replications).

Treatment	10 DAS	20 DAS	30 DAS	40 DAS	60 DAS	MEAN
T1	23.97	48.39	36.51	32.26	38.30	35.88
T2	15.90	43.23	41.89	31.13	23.08	31.04
T3	23.08	46.60	45.92	31.59	24.64	34.36
T4	16.58	51.29	43.00	34.04	27.10	34.40
T5	21.28	53.76	46.81	33.38	27.78	36.602
T6	15.46	45.48	29.34	26.62	13.00	25.98
T7	19.49	47.27	36.51	21.96	18.37	28.72
MEAN	19.39	48.00	39.99	30.14	24.61	
C.D at 5%	3.74	5.61	4.42	5.18	4.63	

Changes in potassium (K) content in soil

The results (Table -6) show that the amount of exchangeable K has been found to be increased initially upto 20 days of crop growth over that of the initial amount (290.08 kg/ha) and then decreased throughout the growth period of the crop irrespective of treatments. However, such pattern of changes varied with treatments, being maintained highest (310.24 kg/ha) in the treatment T5 where 50% VC and 50% NPK fertilizers was applied combinedly followed by the treatment T7 (276.64 kg/ha) receiving 50% FYM and 50% NPK fertilizers in combination at 60 days of crop growth. Such highest amount of maintenance in exchangeable K in the T5 treatment might be attributed to greater amount of K present in vermicompost.

Table 6: Periodic changes in exchangeable K content (kg/ha) in soil affected by integrated nutrient management on growing rape (*Brassica campestris* L.) (mean of three replications).

Treatment	10 DAS	20 DAS	30 DAS	40 DAS	60 DAS	MEAN
T1	407.68	420	374.08	293.44	245.28	348.09
T2	282.24	360.64	320.32	271.04	231.84	293.21
T3	303.52	399.84	344.96	295.68	247.52	318.30
T4	299.04	378.57	338.24	292.32	241.92	310.01
T5	389.76	416.64	388.64	332.64	310.24	367.58
T6	290.08	372.96	327.04	284.48	257.6	306.43
T7	299.04	376.32	349.44	305.76	276.64	321.44
MEAN	324.48	389.2814	348.96	296.48	258.72	
C.D at 5%	9.92	7.85	5.38	8.98	8.62	

Changes in sulphur (S) content in soil

The results (Table 7) show that the amount of sulphate-sulphur content was also followed a similar trend of changes to that of available P content in soil. However, the amount of absolute value was different among different treatments. The amount was recorded highest (13.48 mg/kg) in the treatment T5 followed by the treatment T6 (12.50 mg/kg) at 30 days of crop growth.

Table 7: Periodic changes in sulphate–sulphur content (mg/kg) in soil affected by integrated nutrient management on growing rape (*Brassica campestris* L.) (mean of three replications)

Treatment	10 DAS	20 DAS	30 DAS	40 DAS	60 DAS	MEAN
T1	9.68	9.88	10.14	8.86	8.99	9.51
T2	9.10	9.44	9.69	8.98	9.00	9.24
T3	10.08	10.26	10.98	9.88	9.47	10.13
T4	11.49	11.98	12.42	11.34	11.02	11.65
T5	11.95	12.68	13.48	12.5	11.59	12.44
T6	11.25	11.65	12.5	11.28	11.04	11.54
T7	11.89	12.55	13.39	12.38	11.48	12.33
MEAN	10.77	11.20	11.80	10.74	10.37	
C.D at 5%	0.47	2.20	2.36	2.49	0.31	

Yield

The results (Table -8) show that the yield of rape seed has been found to be increased with treatments.

Table 8: Combined Application of difference source of Organic Manures and N. P.K ,Yield of rapeseed (*Brassica campestries* L.)

Treatment	Seed Yield (q/ha)	Stover Yield (q/ha)
T1- NPK 100%	7.53	20.30
T2- Compost 100%	5.72	16.14
T3- Compost 50% + NPK50%	7.89	21.98
T4- Vermicompost 100%	8.98	24.53
T5- Vermicompost 50% + NPK50%	10.16	29.24
T6- FYM 100%	8.20	22.18
T7- FYM 50% + NPK 50%	9.48	25.88
C.D at 5%	0.65	1.34

The highest yield (10.16 q/ha) was recorded in the treatment (T_5) where 50% vermicompost and 50% NPK fertilizers was applied combinedly which might be due to greater amount available nutrients including iron available in the soil solution contributing towards the yield of rape.

References

De, B., and Sinha, A.C. 2011. International Nutritional Management on Rapeseed (Yellow sarson) *Lambert Academic Publisher, Germany.*

De, B., Das, B., Das, B., and Sinha, A.C.2014. Effect of integrated nutrient management on yield, nutrient uptake and economics of rapeseed (*Brassica campestris* var. yellow sarson) in terai region of West Bengal. *Journal of Crop and Weed* 10(1): 69-72.

De, B., Sinha, A.C. and Patra, P.S. 2009. Effect of organic and inorganic sources of nutrient on rapeseed (*Brassica campestris* L.) under terai region. *Journal of Crop and Weed* 5: 281-284

Jackson, M.L.1973. Soil Chemical Analysis. Prentice Hall of India (P) Ltd., New Delhi.

32

Aerobic Rice Cultivation on Adoption of Water Saving Technologies during Summer Season under Conservation Agriculture

A. Zaman[1] and Gangarani Th.[2]

[1]*Emeritus Professor, CUTM, Paralakhemundi-761211, Odisha, India*
Email: aftab@cutm.ac.in
[2]*Ph D Scholar, Bidhan Chandra Krishi Viswavidyalaya*
Mohanpur-741252, India
Email:thgangarani@gmail.com

Abstract

A field investigation entitled, "Studies on growth and yield of aerobic rice in summer season under various moisture regimes and planting techniques in upland condition" was carried out at Regional Research Station, Gayeshpur, Nadia, West Bengal of Bidhan Chandra Krishi Viswavidyalaya. The field experiment was laid out in a split plot design with three replications. The treatment consisted of three irrigation regimes in main plots *viz.,* IW/CPE =1.0, IW/CPE =1.5 and IW/CPE = 2.0 and four treatments on planting techniques in sub plots *viz.,* P_1: Sprouted seeds, P_2: Non-sprouted seed, P_3: Soaking seeds overnight (12 hrs), $P_{4:}$ Soaking seeds overnight (12 hrs) followed by shade drying. Pooled data revealed that scheduling of irrigation at IW/CPE = 2.0 registered significantly maximum growth attributes in plant height (86.05 cm), tiller number (369.79 m^{-2}), leaf area index (3.26), dry matter accumulation (522.28 g m^{-2}), crop growth rate (8.22 g $m^{-2}d^{-1}$) whereas minimum was recorded when irrigation scheduling was done at IW/CPE=1.0. Root characters *viz.,* root length (24.00 cm), root volume (12.72 cc $hill^{-1}$) and root weight (121.95 g m^{-2}) also registered maximum when irrigation was scheduled at IW/CPE = 2.0 and recorded minimum in the treatment where irrigation was provided at IW/CPE = 1.0. The yield attributes panicle length (21.29 cm), filled grains per panicle (84.46 panicle^{-1}), test weight (16.72 g) grain yield (4.13 t ha^{-1}), straw yield (5.84 t ha^{-1}), and harvest Index (41.16 %) were significantly higher with scheduling of irrigation at IW/CPE = 2.0.Amongst the interaction effects, irrigation scheduled at IW/CPE of 2.0 coupled with sprouted seeds showed highest plant height (94.33 cm), tiller number (412.50 m^{-2}), leaf area index (5.35), dry matter accumulation (654.89 g m^{-2}), crop growth rate (9.92 g m^{-2} d^{-1}) whereas minimum was recorded when irrigation scheduling was done at IW/CPE = 1.0 combined with dry seeds. Similarly, highest root length (25.74 cm), root volume (13.80 cc hill^{-1}) and root weight (131.67 g m^{-2}) was observed at IW/CPE of 2.0 coupled with sprouted seeds. The yield attributes panicle length (24.00 cm), filled grains per panicle (94.50), test weight (18.03 g), grain yield (5.03 t ha^{-1}), straw yield (6.28 t ha^{-1}) and harvest Index (44.43 %) were significantly higher with scheduling of irrigation at IW/CPE = 2.0 coupled with sprouted seeds.On the basis of results it could be concluded that, scheduling of irrigation at IW/CPE = 2.0 irrigation regime coupled with sprouted rice seeds was found suitable as upland dry seeded aerobic rice for maximum growth, yield and monetary benefit.

***Keywords*:** Aerobic rice; water saving; productivity; rice yield

Introduction

In particular, increasing physical water scarcity is a main constraint for irrigated rice (*Oryza sativa)* production (Peng *et al.,* 2006). By 2025, the per capita available water resources in Asia are expected to decline by 15–54 percent compared with 1990 (Moya *et al.*, 2001). The supply of water for irrigation is endangered by declining water quality, declining resource availability, increased competition from other users, and increasing costs.

At the farm level, water inputs can be reduced by decreasing the relatively large and unproductive losses from seepage, percolation, and evaporation. Water-saving irrigation technologies such as saturated soil culture and alternate wetting and drying can drastically diminish these losses. In Asia, upland rice is aerobically grown with minimal inputs and it is usually planted as a low yielding subsistence crop in the adverse upland conditions (Lafitte *et al.,* 2002). With predictions suggesting that many Asian countries will have severe water problems by 2025, aerobic rice under conservation agriculture gives hope to farmers who do not have access to enough water to grow flooded lowland rice. A new concept of growing rice using less water is aerobic rice: high-yielding rice grown in non-puddled aerobic soil using supplementary irrigation just like upland crops. Aerobic rice is crop grown in well-drained, non-puddled & non-saturated soils without ponded water (Bouman *et al.*, 2007). Growing rice in conservation agriculture, with the use of external inputs such as supplementary irrigation, fertilizers and aiming at high yields (Bouman *et al*, 2007) has been established. Main driving force behind aerobic rice is economic water use. Farmers in Brazil, China, and India are pioneering this system where water is scarce or costly. Preliminary studies in Italian climatic environments have shown promising results when rice was grown under dry land conditions, using sprinkler or flushing irrigation rather than flooding, indicating that rice does not necessary require flooded conditions for high yield and good grain quality (Losavio *et al.*, 1997; Russo and Nardi, 1996). However, new aerobic rice varieties and specially designed management strategies are needed if this system is going to be successful. Through the adoption of water-saving irrigation technologies, rice land will shift away from being continuously anaerobic to being partly or even completely aerobic.

Materials and Methods

A field experiment on growth and yield of aerobic rice in summer season under various moisture regimes and planting techniques in upland condition was carried out at the Regional Research Station of Bidhan Chandra Krishi Viswavidyalaya, Mohanpur (India) during summer seasons of 2011 and 2012. The station located in a sub-tropical region at 23°N latitude, 89°E longitude and at an elevation of 9.75 m above sea level. The soil of the experimental field is sandy clay loam in texture and the depth of the soil is shallow to medium with total nitrogen (0.072 %), moderate in phosphorus (15.70 kg ha^{-1}) and potassium (193.58 kg ha^{-1}) content. The soil was moderately alkaline in reaction (pH 6.8). Organic carbon content of soil was 0.67 per cent, respectively. The bulk density of soil was 1.47 g cc^{-1}, respectively. The experiment was laid out in a split plot design and replicated three times. The treatment consist of three irrigation regimes *viz.*, IW/CPE =1, IW/CPE =1.5 and IW/CPE = 2 and four treatments on planting techniques *viz.*, P_1: Sprouted seeds, P_2: Non-sprouted seed, P_3: Soaking seeds overnight (12 hrs), $P_{4:}$ Soaking seeds overnight (12 hrs) followed by shade drying.

Result and Discussion

The results revealed that scheduling of irrigation at IW/CPE = 2 registered significantly maximum growth attributes in *viz.*, plant height (85.98 cm), tiller number (374.17 m^{-2}), leaf area index (3.26), dry matter accumulation (974.97 g m^{-2}), crop growth rate (8.22 g $m^{-2}d^{-1}$). Root characters *viz.*, root length (24 cm), root volume (12.72 cc $hill^{-1}$) and root weight (121.95 g $hill^{-1}$) also registered maximum when irrigation was scheduled at IW/CPE = 2. The yield attributes panicle length (21.29 cm), filled grains per panicle (84.46) test weight (16.80 g), and also grain yield (4.13 t ha^{-1}) and straw yield (5.84 t ha^{-1}) were significantly higher with scheduling of irrigation at IW/CPE = 2 irrigation regime. Amongst planting technique, sprouted seeds showed maximum growth attributes *viz.,* plant height (85.98 cm), tiller number (376.89 m^{-2}), leaf area index (3.26), dry matter accumulation (1018 g m^{-2}) crop growth rate (8.18 g $m^{-2}d^{-1}$). Root characters *viz.,* root length (24.24 cm), root volume (12.59 g $hill^{-1}$) and root weight (124.05g $hill^{-1}$) was significantly higher when sprouted seeds were used. The same trend was followed in yield attributes *viz.,* panicle length (21.38 cm), filled grains per panicle (88.11) test weight (17.12 g), grain yield (4.16 t ha^{-2}) (Table 1) and straw yield (5.84 t ha^{-1}. Application of irrigation at IW/CPE ratio of 2.0 along with sprouted seeds recorded significantly maximum water use efficiency (3.84 kg ha^{-1}-mm) than rest of the irrigation regimes. With regard to the production economics in pooled data, highest gross return, net return and consequently highest B:C ratio was obtained in the treatment combination IW/CPE of 2.0 coupled with sprouted seeds (IRs 61000, Rs 2574 and 1.73 respectively). On the basis of results it could be concluded that, scheduling of irrigation at IW/CPE = 2 irrigation regime coupled with sprouted rice seeds soaked for 48 hrs was found suitable as upland aerobic summer rice for maximum growth, yield and monetary benefit. And that saved up to 40% of irrigation requirement for rice cultivation during summer season under aerobic situation.

Table 1: Effect of moisture regimes and planting techniques on grain yield of rice under aerobic condition in conservation agriculture Grain Yield (t ha^{-1})

Treatment	2011	2012	Pooled
Moisture regimes			
I_1	2.93	3.08	3.00
I_2	3.46	3.61	3.54
I_3	4.00	4.26	4.13
SEm(+)	0.06	0.04	0.04
CD at 5%	0.25	0.14	0.12
Planting techniques			
P_1	4.07	4.25	4.16
P_2	3.03	3.24	3.14
P_3	3.30	3.34	3.32
P_4	3.45	3.76	3.60
SEm(+)	0.08	0.08	0.06
CD at 5%	0.23	0.24	0.16

I_1: Irrigation when IW/CPE = 1; I_2: Irrigation when IW/CPE = 1.5; I_3: Irrigation when IW/CPE = 2; P_1: Sprouted seeds; P_2: Non-sprouted dry seeds (control); P_3: Soaking seeds (12hrs); P_4: Soaking seeds overnight (12hrs) followed by shade drying.

References

Bouman, B.A. M, Humphreys, E., Tuong, T.P., Barker R. 2007. Rice and water.*Adv Argon.*, 92:187–237.

Lafitte R. H; Courtois, B., Arraudeau M., 2002. Genetic improvement of rice in aerobic systems: progress from yield to genes. *Field Crops Res,* 75:171-190.

Losavio N.; Ventrella D.; Vonella A. V.; Russo S.; Arcuti P.,1997. Coltivazione del riso "in asciutta". In : *L'InformatoreAgrario*, 16: 7-8

Moya, P; Hong, L; Dawe, D., Chen, C.D., 2001.Comparative assessment of on-farm water saving irrigation techniques in the Zhanghe Irrigation System. Water saving irrigation for rice: *Proceedings of an International Workshop* Wuhan, China. 81–96.

Peng Shao Bing; Bouman, B., Visperas, R., Castaneda, A, M; Nie Li Xiao; Park Hong

Kyu, 2006.Comparison between aerobic and flooded rice in the tropics: agronomic performance in an eight-season experiment. *Field Crops Research,*96 (2/3): 252-259.

Russo, S., Nardi, S., 1996. Effett idellacolturaasciuttasullaqualità del granello. In: *L'Informatore Agrario* LII, n. 12: 75-78.

33

Influence of *Sasyagavya* on Growth, Yield and Quality Attributes of Cabbage and Bio-Chemical Properties of Soils of the Experimental Site

***Souvik Chatterjee*[1]*, Avijit Kumar Dutta*[2]*, Swami Bhaveshananda*[3] *and Pritom Das*[4]**

[1&4]*Azim Premji University, School of Development, Bangalore-560100 Karnataka, India*
[2&3]*Ramakrishna Mission Vivekananda Educational and Research Institute School of Agriculture and Rural Development, F/C: IRTDM, Ramakrishna Mission Ashrama, Morabadi, Ranchi-834008, Jharkhand, India*
Email: souvik.meet@gmail.com

Abstract

Chhotanagpur plateau is well-known for its quality vegetable production, however, indiscriminate uses of chemical fertilizers and pesticides in different vegetables lead to more yields but of their inferior quality of produces. The over uses of agro-chemicals also drastically affect the soil bio-physico-chemical properties. In this context, an alternative approach was taken into account through present study in order to address such problems. For the purpose, cabbage cv. Golden Acre was grown in organic experimental plots of the university by using '*Sasyagavya* liquid organic formulation' to observe its effects over the crop and soil bio-chemical properties of the experimental site.Theexperiment was designed after CRBDby adoptingseven treatments *i.e.*T_1: 1% Sasyagavya; T_2: 5% Sasyagavya; T_3: 10% Sasyagavya; T_4: 15% Sasyagavya; T_5: 20% Sasyagavya; T_6: 25% Sasyagavya and T_7: Absolute Control withtheir three replications. Different growth, yield and quality attributes of the crop and soil bio-chemical properties of the experimental soils were studied. The findings revealed that T_6 (25% *Sasyagavya*) is the best treatment in terms of different studied growth, yield and quality attributing characters of the crop.

It was also observed that application of organic liquid formulation (*Sasyagavya*) improved the soil health by enhancing beneficial microbial population, NPK, organic carbon content as well as maintaining the pH status of the experimental soil towards the neutral range. *Sasyagavya* with its 25% concentration positively influenced all growth and yield parameters of the crop resulting higher yield (73.39 ± 3.76 $t.ha^{-1}$) in T_6 treatment. As a consequence, higher benefit: cost ratio (5.06) was also documented in the case of T_6 treatment. From the findings of the present investigation, it may be concluded that organic package of practices along with the incorporation of *Sasyagavya* (25%) is well suited for growing cabbage even in commercial scale under the Chhotanagpur plateau of Jharkhand.

***Keywords*:** Cabbage; Growth; Yield; Quality; *Sasyagavya*; Soil Microbes; NPK; Organic Carbon.

Introduction

Cabbage (*Brassica oleracea* L.) is an important cole crops of the family Brassicaceae with somatic chromosome number of 2n=2x=18 (Karpetchenko, 1924). It is a biennial crop for seed production (Ryder, 1979) but grown as annual crop for vegetable purpose. Cabbage is generally believed to have originated from the wild, leafy, non-heading types which are found to be distributed in Europe (Grubben and Denton, 2004) and *Brassica oleracea* var. *sylvestris* L. is the wild progenitor of all *cole* crops including cabbage (Stan and Munteanu, 2001). It is an excellent source of vitamin C. In addition to containing some B vitamins, cabbage supplies some potassium and calcium to the diet. Cabbage as a vegetable mainly cultivated in *Rabi* season in our country. Growth and yield of this vegetable crop are influenced by organic and inorganic fertilizer management. It is an well-known fact that use of inorganic fertilizer for the crops is not so good for health as well as environment because of their residual toxic effect but in the case of organic fertilizer such problem does not arise and on the other hand it increase the productivity of soil as well as crop quality and yield.

After green revolution due to intervention of modern farming techniques, agriculture become resource intensive, resulting resources are over exploited. Almost 80 percent population of Jharkhand is directly or indirectly depends on agriculture. But the chemical fertilizer affects human health as well as soil textural properties. Due to high use of chemical fertilizer the soil texture is broken down. The beneficial bacteria which transform the nutrient to the plant uptake able form is no more due to the effect of chemical fertilizer. As a result when there is no chemical fertilizer plant cannot get any nutrient to uptake. The chemical fertilizer stored up into the fruit and plant body are taken by the human at the time of consume the fruit or plant.

The present investigation has conducted with the objective of helping the farming communities through low cost involving organic farming practices. In India, since more than 82% farmers belong to small and marginal category and about 37% people live below poverty line. Thereby, organic farming is the suitable alternative for such under privilege section of the farming community of our country. The organic liquid manure like *Sasyagavya* can easily be produced by utilizing locally available resources. The requirement of such nutrient solution (*Sasyagavya*) is comparatively lower (about 250 litre/ha when applied at 10% concentration) as compared to bulky organic manures (20-25 t/ha). The ingredients for making *Sasyagavva* are: cow dung, cow urine, vegetable waste/ crop residues and water. It is estimated that in India nearly 700 million tonnes of organic waste is generated annually which is either burned or land filled (Bhiday, 1994). These huge quantities of organic wastes can easily be utilized for the preparation of *Sasyagavya*. Keeping view the above facts, the present study was conducted with emphasizing on the following objectives:

- To study the growth and yield attributes of cabbage under different doses of liquid organic formulation (*Sasyagavva*).
- To study about the quality attributes of cabbage under such organic growing condition.
- To check the soil bio-chemical properties under the influence of liquid organic formulation.
- To find out the benefit: cost ratio of cabbage grown through non-chemical intervention.

Materials and Methodology

The present investigation was carried out at the Organic Experimental Farm of Integrated Rural & Tribal Development and Management Faculty Centre under the School of Agriculture and Rural Development of Ramakrishna Mission Vivekananda University, Morabadi, Ranchi during *Rabi* season of 2016-2017 by employing cabbage cv. Golden Acre. The experiment was designed after CRBD by adopting seven treatments with their three replications. The treatments were T_1: 1% *Sasyagavya*; T_2: 5% *Sasyagavya*; T_3: 10% *Sasyagavya*; T_4: 15% *Sasyagavya*; T_5: 20% *Sasyagavya*; T_6: 25% *Sasyagavya* and T_7: Absolute Control. The above treatments were applied as spot application near root zone of the crop 6 times at 15 days interval starting from 21 days after transplanting

(DAT).Four weeks old seedlings of the crop variety were transplanted in experimental plots at 60 cm x 45 cm spacing in a plot of 2.40 m x 1.80 m. For insect pest management neem leaf extract and tobacco stalk decoction were alternately applied at fortnight interval for four times. Cheese water mixed with turmeric powder @ 10 g per litre of cheese water at 15 days interval for three timeswere applied as prophylactic measure against pathogenic infection. Different growth, yield and quality attribute of the crop and soil bio-chemical properties were studied time to time. Data thus obtained were subjected to statistical analysis for their interpretation.

Results and Discussion

Different growth attributes of cabbage showed significant differences under the influences of *Sasyagavya* with its different concentrations but in few cases non-significant differences also observed as the the cases of number of leaves both wrapper and non-wrapper and width of the leaves (Table-1). Organic liquid formulation (*Sasyagavya*) with its higher concentration greatly influenced almost all growth parameters resulting higher plant height and leaves sizes in the cases of the treatment (T_6). Higher photosynthetic areas as recorded in T_6may probably due to the influences of possibly higher quantity of plant nutrients present in *Sasyagavya* (25%) solution.

Table 1: *Per se* performance of different studied growth attributing characters of cabbage as influenced under different treatments

Treatments	Plant height (cm)	Number of Leaves		Leaf length with petiole (cm)	Leaf breadth (maximum) (cm)	Leaf area (cm^2)
		Wrapper head^{-1}	Non wrapper plant^{-1}			
T_1	23.31	58.22	13.78	20.61	21.92	443.39
T_2	23.69	56.00	13.22	21.52	20.25	427.81
T_3	28.18	61.56	14.11	22.60	20.94	464.92
T_4	24.53	60.00	12.11	22.44	20.88	459.86
T_5	24.29	58.11	14.11	20.59	21.06	430.36
T_6	29.03	59.00	14.56	23.66	22.59	525.11
T_7	24.16	58.89	14.11	19.60	19.85	384.26
SEm (±)	1.01	2.08	1.28	0.81	0.64	24.92
CD $_{Pd"0.05}$	2.53	NS	NS	2.03	NS	62.79

Table 2: *Per se* performance of different studied yield attributing characters of cabbage as influenced under different treatments

Treatments	Head Weight		Head Diameter		Harvest index (%)	Yield t.ha^{-1}
	Fresh weight (kg)	Dry weight (%)	Polar (cm)	Equatorial (cm)		
T_1	0.75	8.39	12.43	12.10	66.07	43.70
T_2	0.93	8.41	13.84	13.72	64.03	60.29
T_3	0.91	8.41	14.94	14.57	66.57	64.91
T_4	0.96	8.66	13.19	12.61	64.37	68.45
T_5	0.78	8.36	11.94	11.32	68.12	57.84
T_6	1.08	8.50	14.20	13.33	68.66	73.39
T_7	0.78	8.13	11.70	11.63	60.77	57.85
SEm (±)	0.02	0.19	0.69	0.90	1.33	3.76
CD Pd"0.05	0.05	NS	1.74	NS	3.35	9.48

Table 3: *Per se* performance of different studied quality attributes of cabbage as influenced by different treatment.

Treatments	TSS (^{0}Brix)	Reducing Sugar (%)	Total Sugar (%)	Ascorbic Acid (mg.100g^{-1})
T_1	4.27	5.58	6.60	190.00
T_2	4.62	4.67	6.38	200.00
T_3	4.71	5.02	5.53	228.52
T_4	5.60	5.12	5.36	195.56
T_5	4.89	4.49	6.60	215.56
T_6	4.57	5.41	6.30	242.22
T_7	4.44	4.44	6.93	195.18
SEm (±)	0.17	0.33	0.33	8.09
CD Pd"0.05	0.43	NS	0.82	20.39

Table 2 highlighted the yield attributes of cabbage and results also revealed the feasibility of this liquid organic formulation (*Sasyagavya*) in alternative approach of cabbage cultivation. However non-significant differences were observed in dry weight and equatorial diameter of cabbage head. The most important trait (yield) is highly influenced by the higher concentration of organic liquid formulation culminated with the higher yield in T_6 treatment. The findings also showed at par effect in terms of production per unit area in the cases of T_3 and T_4. But an interesting findings as documented here in case of yield expression the control treatment (T_7) as higher than T_1 treatment or as same as in other treatment like (T_5). Such observation may probably due to presence of higher levels of organic carbon and other sources of plant nutrients in the organic experimental plots of the university.

Different qualitative attributes performed independently under the influences of different concentrations of *Sasyagavya* (Table 3). However, T_6 once again emerged as the best treatment in terms of expression of different quality attributes of the crop. Higher level of ascorbic acid content recorded in almost all cases of organically designed treatments (Table 3) and it has already been established that synthesis of antioxidant compounds like ascorbic acid is more in plants grown under stress condition or limited nutrient supply conditions *i.e.* organic growing conditions.

Sasyagavya being alkaline in nature may cause negative impact over soil reaction when it is applied continuously with its excess quantity but our finding revealed that limited quantity (250 lit/ha) with 15 days interval application is beneficial for maintaining soil pH towards neutral or slightly acidic range (Table-4) in the acidic soils of the south Chhotanagpur plateau of Jharkhand. Almost all cases bio-chemical properties of the experimental soils showed positive influences after application of *Sasyagavya* (Table 4) with moderate to higher level of beneficial soil microbial build up. These beneficial soil microbes help in the mineralization process of soil applied plant nutrient sources and thereby plant nutrients become available for the plants.

Economics of Growing Cabbage through Organic Management Condition

Economics is an important as well as integral part of agricultural production which depicted the viability or the acceptance of the experiment from farmer's point of view. The economics of growing cabbage under organically managed condition by utilizing different concentrations of Sasyagavya solution as different treatments illustrated in Table-5. The total cost of cultivation (A) documented highest in T_6 treatment (Rs. 50800.00) followed by T_5 (Rs. 49620.00) and T_4 (Rs. 49400.00) treatments as against the lowest as documented in T_7 (Rs. 43000.00) treatment due to comparatively less cost required for different inputs used in this treatment. Sasyagavyawith its higher concentration (25%) significantly influenced growth and yield attributes of cabbage. As a consequence, the higher total income (B) documented as in T_6 (Rs. 256865.00) treatment followed by T_4 (Rs. 239575.00) and T_3 (Rs. 227185.00) as against the lowest (Rs. 152950.00) in T_1 treatment (Table-5) culminated with the higher benefit cost ratio (5.06) in T_6 treatment, while the lowest (3.21) as documented in T_1 treatment. It was also established that the conventional farming input cost (especially variable costs) is higher than organic farming. Though the findings on benefit cost ratio of the present study did not match with the earlier findings of several scientists (Nandi and Nayak, 2008; Amoabeng et al., 2014).

Table 4: Different studied bio-chemical properties of soils as influenced by different treatment conditions

Parameters	Initia After Stage	Completion of the Experiment							SEm (±)	CD (Pd"0.05)
		T_1	T_2	T_3	T_4	T_5	T_6	T_7		
pH	6.32	6.56	6.39	6.48	6.44	6.51	6.52	6.19	0.09	0.13
Organic Carbon (OC) [%]	4.58	4.67	3.30	3.91	4.35	3.63	3.43	3.88	0.45	0.66
Nitrogen (N) [kg.ha^{-1}]	263.42	278.42	283.37	292.67	293.42	302.52	304.92	270.15	3.22	8.11
Phosphorus (P) [%] (*10^{-3})	19.38	16.13	16.67	17.61	18.50	19.29	19.79	15.93	0.62	1.56
Potassium (K) [%] (*10^{-3})	52.89	33.86	38.07	39.59	43.81	46.63	51.06	29.19	3.01	7.59
Total bacterial population (cfu.ml^{-1}) (*10^{-5})	1116.00	98.00	120.33	126.00	153.67	162.00	173.33	90.33	8.92	22.46
Total fungal population (cfu.ml^{-1}) (*10^{-5})	89.00	82.67	121.00	119.00	115.67	128.00	137.67	82.00	9.68	24.37

Table 5: Economics of growing cabbage organically in one hectare of land under different treatment conditions

Components	T_1	T_2	T_3	T_4	T_5	T_6	T_7
Cost of land preparation (Rs.)	5000.00	5000.00	5000.00	5000.00	5000.00	5000.00	5000.00
Cost of inputs (Rs.)	12000.00	12400.00	12650.00	13000.00	13400.00	13700.00	7000.00
Cost of seedlings (Rs.)	15800.00	15800.00	15800.00	15800.00	15800.00	15800.00	15800.00
Cost incurred for weeding and hoeing (Rs.)	1200.00	1200.00	1200.00	1200.00	1200.00	1200.00	1200.00
Watering/irrigation cost (Rs.)	4000.00	4000.00	4000.00	4000.00	4000.00	4000.00	4000.00
Cost for plant protection (Rs.)	900.00	900.00	900.00	900.00	900.00	900.00	900.00
Harvesting and post-harvest operations cost (Rs.)	6200.00	6450.00	7250.00	7000.00	6820.00	7700.00	6600.00
Miscellaneous costs (Rs.)	2500.00	2500.00	2500.00	2500.00	2500.00	2500.00	2500.00
Total cost of cultivation (A)	47600.00	48250.00	49300.00	49400.00	49620.00	50800.00	43000.00
Production ($t.ha^{-1}$)	43.70	60.29	64.91	68.45	57.84	73.39	57.85
Selling price ($Rs.kg^{-1}$)*	3.50	3.50	3.50	3.50	3.50	3.50	3.50
Total income ($Rs.ha^{-1}$) (B)	152950.00	211015.00	227185.00	239575.00	202440.00	256865.00	202475.00
Net profit ($Rs.ha^{-1}$)	105350.00	162765.00	177885.00	190175.00	152820.00	206065.00	159475.00
B:C Ratio (B/A)	3.21	4.37	4.61	4.85	4.08	5.06	4.71

*Prevailing and average market price of cabbage during January-2017 was taken into consideration and the premium price of organically grown produce was not considered here due to absenteeism of organic certification of the crop.

Conclusion

From the above findings, it may be concluded that T_6 is the best treatment in terms of growth and yield of cabbage. However in the case of expression of quality attributes it was observed that different treatments performed independently. Among different quality attributes considered for the study in the present investigation, ascorbic acid and TSS are the most important. In this context, T_6 treatment emerged as the best in terms of synthesis of ascorbic acid in cabbage headbut for TSS, T_4 treatment emerged as the best. When soil health, one of the most important factors for organic farming, is taken into account then it was also observed that treatment T_6 has greater influenced over bio-chemical properties of the experimental soils after conducting the experiment and in this perspective higher microbial population also detected in soils treated with *Sasyagavya* (25%) solution *i.e.* T_6 treatment. The results revealed that *Sasyagavya* with its 25% concentration is highly effective to grow cabbage even in commercial scale and it may be recommended as an alternative and safe approach of cabbage cultivation under the conditions of the south Chhotanagpur plateau regions of Jharkhand.

References

Amoabeng, B.W., Gurr, G.M., Gitau, C.W. and Stevenson, P. C. 2014. Cost: benefit analysis of botanical insecticide use in cabbage: Implications for small holder farmers in developing countries. *Crop Protection*, 57: 71-76.

Bhiday, M.R. 1994. Earthworms in Agriculture. *Indian Farming*, 43(12): 31-34.

Grubben, G.J.H. and Denton, O.A. (2004). Plant resource of Tropical Africa and vegetables. PROTA foundation, Wageningen, Netherlands/Backlmys Publishers, Leiden, Netherlands / CTA Wageningen, Netherlands. pp.668.

Karpetchenko, G.D. 1924. The number of chromosomes and the genetic correlation of cultivated Cruciferae. *Trudy Prikl. Bot. Selekts.*, 13: 3-14 (In Russian).

Nandi, A. and Nayak, S.C. 2008. Performance of hybrid cabbage (*Brassica oleracea* var. *capitata*) as influenced by foliar micronutrient sprays. *Vegetable Science*, 35(1): 45-48.

Ryder, E. J. 1979. Leafy salad vegetables. AVI Publishing Company. Inc., Westport, Connecticut.

Stan, N. and Munteanu, N. 2001. Legumiculturăspecială, Vol.II. Editura "*Ion Ionescu de la Brad*" Ia°i.